Fish Cognition and Behavior

Fish Cognition and Behavior

Edited by

Culum Brown
Macquarie University

Kevin Laland
University of St Andrews

Jens Krause
University of Leeds

Blackwell
Publishing

© 2006 by Blackwell Publishing Ltd

Editorial Offices:
Blackwell Publishing Ltd, 9600 Garsington Road, Oxford OX4 2DQ, UK
  Tel: +44 (0)1865 776868
Blackwell Publishing Professional, 2121 State Avenue, Ames, Iowa 50014-8300, USA
  Tel: +1 515 292 0140
Blackwell Publishing Asia Pty Ltd, 550 Swanston Street, Carlton, Victoria 3053, Australia
  Tel: +61 (0)3 8359 1011

First published 2006 by Blackwell Publishing Ltd
2   2007

ISBN: 978-1-4051-3429-3

Library of Congress Cataloging-in-Publication Data
  Fish cognition and behavior / edited by Culum Brown, Kevin Laland, Jens
  Krause.— 1st ed.
      p. cm.
  Includes bibliographical references and index.
  ISBN 978-1-4051-3429-3 (hardback : alk. paper)
  1. Fishes—Behavior. 2. Fishes—Psychology. 3. Cognition in animals. I. Brown, Culum.
  II. Laland, Kevin N. III. Krause, Jens, Dr.
    QL639.3.F575 2006
    597.1513—dc22
2006004742

A catalogue record for this title is available from the British Library

Set in 10/12pt Times Ten by Graphicraft Limited, Hong Kong
Printed and bound in India by Replika Press, Pvt. Ltd

The publisher's policy is to use permanent paper from mills that operate a sustainable forestry policy,
and which has been manufactured from pulp processed using acid-free and elementary chlorine-free
practices. Furthermore, the publisher ensures that the text paper and cover board used have met acceptable
environmental accreditation standards.

For further information on Blackwell Publishing, visit our website:
www.blackwellpublishing.com

Cover acknowledgements:
Artist: Robert Gibbings (1889–1958)
Title: St Brendan and the sea monsters
Medium: Woodcut
Date created: 1934
Credit line: Collection Christchurch Art Gallery Te Puna o Waiwhetu.
Presented by Mrs Rosalie Archer, 1975.
Copyright permission kindly granted by Reading University Library.

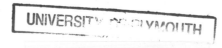

Contents

Preface and Acknowledgements

This book is a follow-up from our successful volume of *Fish and Fisheries* dedicated to learning in fishes. All of the contributors to that volume have updated their work here, and we have added several more contributions covering a broad range of fish behaviour. It is encouraging to see a range of contributions from both established and emerging experts in fish behaviour. The editors would like to thank all of the contributors for their hard work and enthusiasm while producing this volume. Such an undertaking would be far too big a task for one person alone, given the increasing volume of behavioural research conducted on fishes. There is also a long list of reviewers whose comments have made valuable contributions to each of the chapters. We would like to thank Nigel Balmforth and Laura Price of Blackwell Publishing for their valuable support and Tony Pitcher for writing the series foreword.

Financial support for our research on learning in fishes was provided by the Leverhulme Trust, BBSRC, NERC and the Royal Society.

CB: To Felicity
JK: To Hasso and Barbara

List of Contributors

Michael S. Alfieri
Department of Biology, Viterbo University, 900 Viterbo Drive, La Crosse, Wisconsin, USA

Victoria Braithwaite
Institute of Evolutionary Biology, School of Biological Sciences, University of Edinburgh, Edinburgh, Scotland, UK

Cristina Broglio
Laboratorio de Psicobiología, Universidad de Sevilla, Campus Santiago Ramón y Cajal, s/n 41005-Sevilla, Spain

Culum Brown
Department of Biological Sciences, Macquarie University, Sydney, Australia

Grant E. Brown
Department of Biology, Concordia University, 7141 Sherbrooke St. W., Montreal, Quebec, Canada

Redouan Bshary
Université de Neuchâtel, Rue Emile-Argand 11 CH-2007, Neuchâtel, Switzerland

Douglas P. Chivers
Department of Biology, University of Saskatchewan, Saskatoon, Saskatchewan, Canada

Iain D. Couzin
Department of Zoology, University of Oxford, Oxford, UK

Darren Croft
School of Biological Sciences, University of Wales, Bangor, UK

Lee Alan Dugatkin
Department of Biology, University of Louisville, Louisville, KY, USA

Emilio Durán
Laboratory of Psychobiology, Department of Psychology, University of Huelva,
Campus de El Carmen, 21071 Huelva, Spain

Ryan L. Earley
Department of Biology, Georgia State University, Atlanta, Georgia, USA

Anders Fernö
Department of Biology, University of Bergen, Box 7800, N-5020 Bergen,
Norway

Antonia Gómez
Laboratorio de Psicobiología, Universidad de Sevilla, Campus Santiago
Ramón y Cajal, s/n 41005-Sevilla, Spain

Siân Griffiths
Cardiff School of Biosciences, Main University Building, PO Box 915, Cardiff,
Wales, UK

Yuying Hsu
Department of Life Science, National Taiwan Normal University, No. 88,
Section 4, Tingchou Rd, Taipei 116, Taiwan

Geir Huse
Institute of Marine Research, Box 1870, Nordnes, N-5817 Bergen, Norway

Per Johan Jakobsen
Department of Biology, University of Bergen, Box 7800, N-5020 Bergen,
Norway

Richard James
Department of Physics, University of Bath, Bath, UK

Jennifer Kelley
School of Biological, Earth and Environmental Sciences, University of
New South Wales, Sydney, NSW, Australia

Jens Krause
Institute of Integrative and Comparative Biology, University of Leeds,
Leeds, UK

Tore S. Kristiansen
Institute of Marine Research, N-5292 Storebø, Norway

Kevin Laland
Centre for Social Learning and Cognitive Evolution, School of Biology,
University of St Andrews, St Andrews, Fife, Scotland, UK

Anne E. Magurran
Gatty Marine Laboratory, University of St Andrews, St Andrews, Fife,
Scotland, UK

David Mawdsley
Department of Physics, University of Bath, Bath, UK

Lucy Odling-Smee
Nature Publishing Group, 4 Crinan Street, London, UK

Fernando Rodríguez
Laboratorio de Psicobiología, Universidad de Sevilla, Campus Santiago
Ramón y Cajal, s/n 41005-Sevilla, Spain

Cosme Salas
Laboratorio de Psicobiología, Universidad de Sevilla, Campus Santiago
Ramón y Cajal, s/n 41005-Sevilla, Spain

Stephen D. Simpson
Institute of Evolutionary Biology, School of Biological Sciences, University
of Edinburgh, Edinburgh, Scotland, UK

Kevin Warburton
School of Integrative Biology, University of Queensland, Brisbane, Queensland,
Australia

Ashley Ward
Department of Biology, University of Leicester, Leicester, UK

Klaudia Witte
University of Bielefeld, Postfach 100131, D-33501 Bielefeld, Germany

Larry L. Wolf
Department of Biology, Syracuse University, Syracuse, New York, USA

Foreword

Fish researchers like to explain, to the bemused bystander, how fish have evolved an astonishing array of adaptations; so much so that it can be difficult for these fish freaks to comprehend why anyone would study anything else. Yet, at the same time, fish are among the last wild creatures on our planet that are hunted by humans for sport or food. As a consequence, today we recognize that the reconciliation of exploitation with the conservation of biodiversity provides a major challenge to our current scientific knowledge and expertise. Even evaluating the trade-offs that are needed is a difficult task. Moreover, solving this pivotal issue calls for a multi-disciplinary consilience of fish physiology, biology and ecology with social sciences such as economics and anthropology in order to probe the frontiers of applied science. In addition to food and recreation (and inspiration for us fish freaks), it has, moreover, recently been realized that fish are essential components of aquatic eco-systems that provide vital services to human communities. Sadly, virtually all sectors of the stunning biodiversity of fishes are at risk from human activities. In freshwater, for example, the largest mass extinction event since the end of the dinosaurs has occurred as the introduced Nile perch in Lake Victoria eliminated over 100 species of endemic cichlid fish. But, at the same time, precious food and income from the Nile perch fishery was created in a miserably poor region. In the oceans, we have barely begun to understand the profound changes that have accompanied a vast expansion of human fishing over the past 100 years. The *Blackwell Publishing Fish and Aquatic Resources Series* is an initiative aimed at providing key, peer-reviewed texts in this fast-moving field.

Many years ago, teaching a series of practical animal behaviour classes to under-graduates, I asked my students to try to train goldfish to feed on aquarium pellets delivered from coloured tubes. It all worked well enough, the goldfish learned to feed from tubes painted with, to our eyes, subtly different colours, and after running the usual controls for colour intensity, the students discovered that the fish had very effective colour vision. After the classes, the goldfish were returned to a stock aquarium and were left alone for a year, although some of them they may have taken part in other experiments. The following year, it was evident right at the start of the student practical that each goldfish remembered the exact colour and location of its feeding tube from one year before, a remarkable cognitive feat from an animal that is supposed to have only a three-second memory, as satirized in the Pixar *Finding Nemo* film.

Fishermen, anglers, and most of the general public, encounter live fish only when they are flopping helplessly, and apparently dumbly, on the boat deck or seashore.

In such circumstances it is hard to believe that fish are intelligent sentient beings: even in the least speciesist[1] science fiction, in Douglas Adams' (1919) otherwise splendid *Hitchhikers Guide to the Galaxy* for example, fish are merely food for whales, except for one brilliant automaton, the universal translator known as Babelfish. On the other hand, watching fish hunting for food, engaging with mates, or raising young, aquarists and divers gain a very different view of the behavioural complexities, elegant adaptations and cognitive abilities that lie behind the actions of fish. It is this latter view, of fish with a complex evolved neural capacity, that is the scientific reality.

This volume, the 11th in the *Blackwell Publishing Fish and Aquatic Resources Series*, grew out of a series of eight articles that comprised a special issue of the Blackwell journal *Fish and Fisheries* in 2003 (Brown *et al.* 2003), which itself built upon a pioneering review paper published in the early 1990s (Kieffer & Colgan 1992). In today's book, we have a set of 14 expanded and updated chapters by 27 internationally renowned authors that review this important area. The book has been put together and edited by three of the world leaders in this field: Culum Brown from Sydney, Australia, Kevin Laland from St Andrews in Scotland, and Jens Krause from Leeds, England.

The editors point out that cognition includes perception, attention, memory formation and executive functions related to information processing such as learning and problem solving. Fish, it turns out, are not primitive in these respects. Bony fish have, after all, had over 60 million years for their genes to evolve the capacity to build and run fish brains that can deal flexibily with the diverse but volatile underwater environment – this is a ten times longer time span than our human line. Indeed, the editors of this book show that, far from being primitive automatons as had once been thought, fish 'have evolved complex cultural traditions, pursue Machiavellian strategies of manipulation, deception and reconciliation, can monitor the social prestige of others, and can cooperate during foraging, navigation, reproduction and predator avoidance'.

This volume presents fascinating, timely and comprehensive 'state of the art' reviews of the cognitive abilities of fish, and readers will find the elements of a fresh synthesis in this field. It therefore should find a home on the bookshelves and in the libraries of a broad set of practitioners and students concerned with fish evolution, behaviour and ecology, including those, like myself, who might still wish to call themselves ichthyologists.

Professor Tony J. Pitcher
Editor, Blackwell Publishing Fish and Aquatic Resources Series
Fisheries Centre, University of British Columbia, Vancouver, Canada
May 2006

---

[1] 'Speciesism' involves assigning different values or rights to beings on the basis of their species. The term was coined by Richard D. Ryder in 1970 and is used to denote prejudice similar in kind to sexism and racism.

References

Adams, D. (1979) *The Hitchhiker's Guide to the Galaxy*. Heinemann, London, UK.

Brown, C., Laland, K. and Krause, J. (2003) Special Issue on Learning in Fishes: Why they are smarter than you think. *Fish and Fisheries*, **4**(3): 197–288.

Kieffer, J.D. and Colgan, P.W. (1992) The role of learning in fish behaviour. *Reviews in Fish Biology and Fisheries*, **2**, 125–143.

Chapter 1
Fish Cognition and Behaviour

Culum Brown, Kevin Laland and Jens Krause

1.1 Introduction

The field of animal cognition is the modern approach to understanding the mental capabilities of animals. The theories are largely an extension of early comparative psychology, with a strong influence of behavioural ecology and ethology. Cognition has been variously defined in the literature, with some confining it to higher order mental functions including awareness, reasoning and consciousness. A more general definition, however, includes perception, attention, memory formation and executive functions related to information processing such as learning and problem solving. The study of animal cognition has been largely centred on birds and mammals, particularly non-human primates. This bias in the literature is, in part, a result of the approach taken in the 1950s when cognitive psychologists began to compare known human mental processes with other closely related species. This bias was reinforced by an underlying misconception that learning played little or no role in the development of behaviour in reptiles and fishes.

During the majority of scientific history fishes have been viewed as automatons. Their behaviour was thought to be largely controlled by unlearned predispositions. Ethologists characterised their behaviour as a series of fixed action patterns released on exposure to appropriate environmental cues (sign stimuli). While there is no doubt that fishes are the most ancient form of vertebrates, they are only 'primitive' in the sense that they have been on earth for in excess of 500 million years and that all other vertebrates evolved from some common fish-like ancestor (around 360 million years ago). It is important to note, however, that fishes have not been stuck in an evolutionary quagmire during this time. Their form and function has not remained stagnant over the ages. On the contrary, within this time frame they have diversified immensely, to the point where there are more species of fish than all other vertebrates combined (currently over 29 000 described species), occupying nearly every imaginable aquatic niche.

The erroneous view that both behavioural and neural sophistication is associated in a linear progression from fish through reptiles and birds to mammals, is largely attributable to a heady mix of outdated and unscientific thinking. Aristotle's concept of *Scala naturae* (the scale of nature) and a Christian fundamentalist view that man is the pinnacle of the natural world, have dominated conceptions of animal intelligence for millennia. However, Darwin's theory of evolution is fundamentally inconsistent with a gradual progression of behavioural flexibility and cognitive complexity from 'primitive' to 'advanced' life forms, leading inevitably to humans at the peak (i.e. the wrong-headed notion of an evolutionary ladder). There is nothing progressive about evolution, and any semblance of progression merely reflects our anthropocentric bias to tracking evolutionary lineages that culminate in our species, and to evaluating other species by their similarity to ourselves. The cognitive capabilities of a species will reflect the history of selection among its ancestors, rather than phylogenetic proximity to humanity.

Among the vertebrates, fishes have suffered the most from the common misconception of the evolutionary ladder. Over the past few decades, however, this fallacy has begun to be redressed. We now realise that, like the rest of the vertebrates, fishes exhibit a rich array of sophisticated behaviours and that learning plays a pivotal role in the behavioural development of fishes. Gone, or, at least, redundant, are the days where fishes were looked down upon as pea-brained machines whose only behavioural flexibility was severely curtailed by their infamous three-second memory (as portrayed by Dory in Disney's *Finding Nemo*). As this book will reveal, fishes do in fact have impressive long-term memories comparable to most other vertebrates (Brown 2001; Warburton 2003). Their neural architecture has both analogous and homologous components with mammals, and is capable of much the same processing power (Broglio *et al.* 2003). Their cognitive capacity in many domains is comparable with that of non-human primates (Bshary *et al.* 2002; Laland & Hopitt 2003; Odling-Smee & Braithwaite 2003). Fish have evolved complex cultural traditions and pursue Machiavellian strategies of manipulation, deception and reconciliation (Bshary *et al.* 2002; Brown & Laland 2003). They not only recognize one another, but can monitor the social prestige of others (Griffiths 2003; McGregor 1993) and cooperate in a variety of ways during foraging, navigation, reproduction and predator avoidance (Huntingford *et al.* 1994; Johnstone & Bshary 2004; Fitzpatrick *et al.* 2006). It is clear that the recent developments in our understanding of fish behaviour and their behavioural flexibility warrant further investigation.

Since the 1960s there has been a rapid increase in the number of papers published on learning in fishes, and the number of papers published since 1991 has undergone a further dramatic rise (Fig. 1.1). In the early 1990s James Kieffer and Patrick Colgan published the first comprehensive review of the role of learning in the development of fish behaviour (Kieffer & Colgan 1992). In their review, they were able to draw on some 70 published papers on learning in fishes, a vast improvement over previous works (Thorpe 1963; Gleitman & Rozin 1971). In 2003, we published a collection of reviews on the topic in a special edition of the journal *Fish and Fisheries* (Brown *et al.* 2003). The special issue contained eight reviews on various aspects of learning in fishes referring to over 500 papers. Many of these reviews have been revised and extended in this book, which contains 14 chapters focusing on the role of cognition in every major aspect of fish biology, from foraging and predator avoidance to fighting and social relationships.

1.2   Contents of this book

In addition to this opening introduction, Chapter 2, by Kevin Warburton, investigates the role of learning in foraging behaviour, drawing on both psychological and behavioural ecology literature. He suggests that learning and memory play

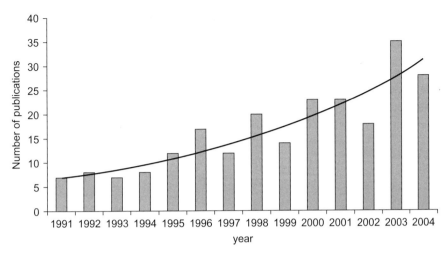

Fig. 1.1 The number of publications produced on fish learning and cognition since 1991 has increased dramatically. Data based on keyword search (fish, fishes, learn, learning and cognition) in *Web of Science* (http://scientific.thomson.com/products/wos).

significant roles in the foraging activities in fish and that memory, like many traits, seems to be highly adapted to the specific requirements of each species. Interestingly, Warburton suggests that in some circumstances forgetting might be just as important as remembering. The chapter highlights the fact that the similarities between vertebrate learning systems are far more striking than the differences, and that fishes rely on a wide array of learning mechanisms in their daily lives. The literature shows that learning is vital in many aspects of fish foraging behaviour, from the formation of foraging search images, to prey capture and handling. Warburton also outlines various experiments that explore foraging theory and points out that fishes are ideal candidates for such research.

It is often assumed that anti-predator behaviour should have a significant unlearned component to it because fish need to be able escape predators from the moment they hatch. The penalty for failure in this instance is death, so there is an expectation that natural selection will exert significant evolutionary pressure in this respect. Jennifer Kelley and Anne Magurran (Chapter 3) point out that while this is the case to some degree, learning still plays a key role in the fine tuning of predator recognition and response systems. In environments that are unpredictable from moment to moment and, perhaps more importantly, from generation to generation, it is essential that prey species have some general template for predator recognition, but that this template be flexible enough to enable fine tuning to match the prevailing predatory threats. Kelley and Magurran discuss the various ways in which fishes learn about predators and the need for prey species to be able accurately to assess potential risks and act accordingly. They cover the evolutionary arms race between predators and prey, highlighting the role learning plays in this race from both perspectives.

There are many ways in which prey can learn about predators without high risk exposure, including the observation of conspecifics as they interact with, or detect, predators. One such method is the reliance on predator odours and prey alarm cues that may be detected from some distance and this is the focus of Chapter 4. Here, Grant Brown and Doug Chivers explore how fish use chemical cues both to assess risk and to learn about predators. There are obviously great fitness advantages to be had by the accurate assessment of risk, primarily because it frees the individual time budget from unnecessary anti-predator behaviour (Lima & Dill 1990). Fishes not only learn from conspecifics but may also respond to the alarm signals generated by heterospecifics that are part of the same prey guild, thus enabling the recognition of predators and dangerous habitats alike. It is interesting to note that fishes often undergo massive growth from larval to adult stages and in doing so pass through a series of predatory guilds, each with its own specific threats. In this scenario, Brown and Chivers point out, learning may play a larger role in the development of anti-predator behaviour than previously suspected.

In Chapter 5, Klaudia Witte explores the role of learning in mate choice decisions. In her review, Witte examines the evidence for the influence of imprinting during the critical period of early life-history stages on later mate choice decisions. She reveals that imprinting is most likely to occur in those species that show some kind of extended parental care, such as the cichlids. However, it is also evident that other social influences can affect mate choice decisions later in life; for example, naive male guppies can learn to discriminate between conspecifics and heterospecifics and alter their mating strategy to concentrate on courting conspecifics. Part of this alteration in behaviour may be mediated by their mating success and feedback from the females they are attempting to court. Species recognition may be reinforced by learning in those areas where multiple closely related species coexist. While mate choice often relies on some predetermined innate recognition and preference system, Witte reveals that these unlearned preferences can be overcome by learning and especially by copying the mate-choice decisions of others. As discussed in many of the chapters, fishes are capable of relying on a mixture of eavesdropping and social information to help them make important decisions, and mate choice is no exception. Reliance on public information may enable females to gauge the quality and aggression levels of a potential mate, without having to suffer any negative consequences associated with the early stages of courtship.

Yuying Hsu, Ryan Earley and Larry Wolf examine the modulation of aggression through prior experience in Chapter 6. Many factors combine to influence the outcome of aggressive encounters, including size, motivation and prior residency, and, as Hsu and his colleagues highlight, prior experience with fights can also play a large role. The outcome of fights can have considerable consequences including access to food, mates or territories so it is important to understand how experience can influence their outcome of fights. Recent literature suggests that fishes that have recently lost a fight are more likely to lose a second encounter compared with winners, all else being equal. Therefore, an individual's history must be considered when predicting the outcome of a current fight. All of us know that confidence can influence our behaviour considerably and this is likely to be mediated both through physiological as well as psychological mechanisms. Relying on both modelling and empirical data, Hsu *et al.* explore how previous experiences combine or interact to shape an individual's present fighting capability.

In Chapter 7, Lucy Odling-Smee, Steve Simpson and Victoria Braithwaite explore the role of cognition in spatial orientation, navigation and migration. The authors point out that, like most animals, the resources fishes utilize are often widely separated in space. Many of these biologically important locations are relatively temporally and spatially stable and as such can be reliably found by learning and memory retrieval. As Warburton previously pointed out, here it is also the case that natural selection has favoured learning strategies to closely match the needs of the species under consideration. As in all animals, cue reliance is constrained by the species' perception, and fishes display a huge array of perceptual capabilities, many of which are only just beginning to be understood, such as electroreception and ultraviolet vision. It is evident that fishes rely on a wide array of navigation cues and mechanisms, ranging from egocentric turns to the formation of cognitive maps, to move accurately around their environments. Natural selection would favour the ability to select the most efficient movement pathways possible so as to reduce any potential waste of time and energy. Thus accurate navigation is a key component to an individual's fitness landscape. In the final part of their chapter, Odling-Smee *et al.* concentrate on large-scale migration in salmon as a case study, highlighting both the recall of long-term memory and initial imprinting processes.

Siân Griffiths and Ashley Ward review the evidence for individual recognition in Chapter 8. When closely examining social interactions it is apparent that not all individuals are treated equally by a given fish; for example, as discussed by Hsu *et al.*, closely related fish often receive less aggression than non-relatives. Individual recognition has several implications on multiple levels, including predicting species dispersal patterns, which has conservation and fisheries management outcomes. But how do fish recognize one another? Siân and Ashley review the ever increasing body of publications that attest that fish not only recognize kin, but can distinguish between familiar and unfamiliar individuals. This process seems to build up over 10–14 days, although it may vary from species to species. Being able to recognize, and preferentially associate with, kin or familiar individuals potentially has substantial direct and indirect fitness benefits; for example, there is evidence that shoals comprised of familiar individuals show more efficient schooling behaviour than those comprised of strangers. Such benefits may accrue as a result of an increase in the ability of an individual fish to predict the response of familiar individuals across a variety of contexts. Individual recognition is germane to other aspects of fish behaviour, including cooperation (Chapter 11), the exploitation of social cues and signals (Machiavellian intelligence; Chapter 12) and social learning (Chapter 10).

In Chapter 9 Iain Couzin, Richard James, Darren Croft and Jens Krause develop mathematical approaches and review current literature that links the behaviour of individuals to the higher-order properties at the group and population levels. It is evident that the behaviour of individuals within a social group is largely influenced by their fellow group members. Through the rapid transfer of information between group members, shoals of fish often seem to behave as a single collective. However, a few individuals within a groups can assert undue influence on the behaviour of the majority, particularly if these 'leaders' are more motivated to perform some behaviour than the remainder of the shoal (i.e. they are more directed than the average). Such processes may have a significant impact on the three-dimensional structure and movement of shoals. Moreover, because information is shared between group members, a shoal as a whole may be able to solve problems more efficiently than singletons (e.g. navigation), for example by filtering environmental noise or collective detection and processing of external cues. In addition, Couzin *et al.*'s examination of association networks can be utilized to predict the path by which information is likely to be transferred within the group.

The transfer of information between individuals is reliant on social learning processes. Social learning refers to those situations where individuals acquire new information or behaviour by observation of, or interaction with, others. Social learning can occur across a wide variety of contexts and appears to be a ubiquitous form of learning among fishes. Social learning often enables individuals to acquire information more rapidly and efficiently than would be the case if they themselves had to explore their environment fully and learn by trial and error. Traditionally social learning was thought to be restricted to mammals and birds but in Chapter 10, Culum Brown and Kevin Laland explore the substantive body of evidence showing the widespread existence of social learning in fishes. Social learning that occurs across generations (vertical or oblique transmission) can lead to the establishment of localized, stable behavioural traditions and forms the very roots of animal culture. Such cultural evolution can operate in tandem with biological evolution and these processes interact in many interesting ways. Brown and Laland argue that social learning is likely to play a key role in the development of fish behaviour and point out that exploitation of such processes could be utilized in training regimes for fisheries and in conservation management programmes such as restocking.

Cooperation between individuals has long been considered something of an enigma within evolutionary biology. If Darwinian fitness is all about outcompeting others, then one might think all individuals ought to behave selfishly. This notion is central to many existing theories such as the selfish herd hypothesis, which is particularly pertinent to group-living animals such as fishes. However, it became clear that the evolution of cooperation could be explained through a number of alternative hypotheses, namely kin selection, reciprocity, byproduct mutualism and group selection, in which individuals gain long-term fitness benefits in spite of short-term costs. In Chapter 11, Michael Alfieri and Lee Dugatkin suggest that cooperation not only occurs in fishes but may be widespread in a number of contexts.

Redouan Bshary continues the social theme by examining the evidence for social or Machiavellian intelligence in fishes, largely stemming from his early observations of the behaviour of cleaning wrasse. In Chapter 12, Bshary extends his earlier review (Bshary *et al.* 2002) on the topic and presents an overview of the social strategic cognitive abilities of fishes. The primary thesis of the Machiavellian intelligence hypothesis (Whiten & Byrne 1997) is that one of the principal driving forces for the evolution of cognition was the challenge to cope with and exploit the complexity of an individual's social environment. For years the hypothesis was used almost exclusively to 'prop up' the apparent existence of the higher cognitive capacity of primates, including humanity. It soon became apparent, however, that the theory, if true, should apply equally to other vertebrate groups. Bshary examines the evidence for individual recognition, individualized group living, cooperation, manipulation, reconciliation and deception in fishes.

For decades the cognitive ability of fishes was highly underrated, largely because of a lack of direct experimentation. However, an additional factor here was a reliance on direct comparisons of the fish brain with that of mammals, in which the majority of studies on cognition had occurred (particularly primates and rodents) and for which a great deal was known about the function and connectivity of brain structures. Such comparison suggested that the brains of fish and mammals differ in many ways, with fish brains being typically smaller and less structured than those of mammals. Because of this it was often indirectly inferred that fishes must lack certain cognitive abilities observed in mammals because their brain structure was not the same as that of mammals. Not until very recently have scientists begun to study the brains of fishes and their function in any detail. It should be pointed out that virtually nothing is known about the vast majority of fish species, let alone anything about their brain structure and function. The results of these pioneering studies, as Rodriguez *et al.* (Cosme Salas and his laboratory members) highlight in Chapter 13, are startling. They reveal many similarities between the mammalian and fish brain in terms of its function. In the light of recent developmental, neuroanatomical and functional data it appears that many functions are highly conserved right across vertebrates, despite the fact that morphology can often differ substantially. Rodriguez *et al.* point out that these morphological differences stem from an entirely different developmental pathway; for example the fish telencephalon goes through a process of eversion (bending out) during embryonic development, whereas the brain of all other vertebrates develops by evagination (bending in). Through the results of their extensive research and their review of related literature, Rodriguez *et al.* challenge the prevailing notion that fishes lack most of the brain centres and neural circuits that support cognition capabilities in the other vertebrate groups.

The final chapter in this collection of reviews examines just one of the many practical applications that can stem from a greater understanding of fish cognition and behaviour. In Chapter 14, Anders Fernö and colleagues examine the role of fish learning in aquaculture and fisheries. For thousands of years humans have relied on a steady harvest of fish from the world's rivers and oceans as an important source of protein. Today fishes remain the only wild food source humans harvest on a large scale, and through a greater understanding of their behaviour we have begun to farm them and exploit natural populations at an ever growing rate. However, as Fernö and his colleagues point out, human fishing methods have evolved at a far greater rate than the fishes' response to this selection pressure and there is now a huge gap between fishes' natural responses to predators and our modern fishing techniques. However, fish can respond to the threat of fishing through learning. There is now some evidence that fish learn to respond to fishing gear, largely by avoidance of vessels, and such responses may interfere with our estimates of stock sizes. Fishing may also affect fish learning; for instance, removal of larger, more knowledgeable individuals from stocks may disrupt social transmission chains, thus breaking long-standing cultural traditions in some of the world's economically most important fish species (such as the location of feeding, migration routes or breeding grounds). Following the crash of the northern cod stocks, for example, an abrupt change was recognized in the stocks distribution. Fernö *et al.* also investigate the ways in which behavioural flexibility can be utilized in aquaculture scenarios. They highlight the fact that due consideration must be given to the large influence of early experience in the development of fish behaviour when managing hatchery stocks, particularly in those instances where the stocks are used for conservation reintroductions or to buffer existing natural stocks from the pressures of commercial and recreational fisheries.

Two themes emerge in this book. The first is that the learning abilities and complexity of behaviour of fishes are, in many respects, comparable to those of land vertebrates. The second is that fishes provide a flexible and pragmatic biological model system for studying many aspects of animal learning and cognition. These observations lead us to the view that interest in the topic of fish learning, cognition and behaviour is likely to continue to grow for the foreseeable future.

1.3   References

Broglio, C., Rodriguez, F. & Salas, C. (2003) Spatial cognition and its neural basis in teleost fishes. *Fish and Fisheries*, **4**, 247–255.

Brown, C. (2001) Familiarity with the test environment improves escape responses in the crimson spotted rainbowfish, *Melanotaenia duboulayi*. *Animal Cognition*, **4**, 109–113.

Brown, C. & Laland, K. (2003) Social learning in fishes: a review. *Fish and Fisheries*, **4**, 280–288.

Brown, C., Laland, K., and Krausse, J. (2003) Special Issues on Learning in Fishes: Why they are smarter than you think. *Fish and Fisheries*, **4**(3): 197–288.

Bshary, R., Wickler, W. & Fricke, H. (2002) Fish cognition: a primate's eye view. *Animal Cognition*, **5**, 1–13.

Fitzpatrick, J.L., Desjardins, J.K., Stiver, K.A., Montgomerie, R. & Balshine, S. (2006) Male reproductive suppression in the cooperatively breeding fish *Neolamprologus pulcher*. *Behavioral Ecology*, **17**, 25–33.

Gleitman, H. & Rozin, P. (1971) Learning and memory. In: W.S. Hoar & D.J. Randall (eds), *Fish Physiology*, volume 6, pp. 191–278. Academic Press, New York.

Griffiths, S.W. (2003) Learned recognition of conspecifics by fishes. *Fish and Fisheries*, **4**, 256–268.

Huntingford, F.A., Lazarus, J., Barrie, B.D. & Webb, S. (1994) A dynamic analysis of cooperative predator inspection in sticklebacks. *Animal Behaviour*, **47**, 413–423.

Johnstone, R.A. & Bshary, R. (2004) Evolution of spite through indirect reciprocity. *Proceedings of the Royal Society of London Series B – Biological Sciences*, **271**, 1917–1922.

Kieffer, J.D. & Colgan, P.W. (1992) The role of learning in fish behaviour. *Reviews in Fish Biology and Fisheries*, **2**, 125–143.

Laland, K.N. & Hoppitt, W. (2003) Do animals have culture? *Evolutionary Anthropology*, **12**, 150–159.

Lima, S.L. & Dill, L.M. (1990) Behavioral decisions made under the risk of predation: a review and prospectus. *Canadian Journal of Zoology*, **68**, 619–640.

McGregor, P.K. (1993). Signaling in territorial systems – a context for individual identification, ranging and eavesdropping. *Philosophical Transactions of the Royal Society of London Series B – Biological Sciences*, **340**, 237–244.

Odling-Smee, L. & Braithwaite, V.A. (2003) The role of learning in fish orientation. *Fish and Fisheries*, **4**, 235–246.

Thorpe, W.H. (1963) *Learning and Instinct in Animals*. Methuen, London.

Warburton, K. (2003) Learning of foraging skills by fish. *Fish and Fisheries*, **4**, 203–215.

Whiten, A. & Byrne, R.W. (1997) *Machiavellian Intelligence II: Extensions and Evaluations*. Cambridge University Press, Cambridge.

Chapter 2
Learning of Foraging Skills by Fishes

Kevin Warburton

2.1   Introduction

Investigations of the roles of learning and memory in the foraging behaviour of fishes are at an exciting stage. Although less thoroughly studied than traditional laboratory favourites such as rats, pigeons or bees, fish species are no longer to be consigned to the 'poorly understood' category. It is now possible to place the capacities of fishes in the context of learning and memory as a whole, as evidenced by previous reviews such as those by Hart (1986, 1993), Hughes *et al.* (1992) and Kieffer & Colgan (1992). This chapter attempts to integrate perspectives and findings from the fields of behavioural ecology and comparative psychology, an approach that has been taken for the following reasons.

Comparative psychology has revealed broad regularities in the general principles of learning across invertebrate and vertebrate taxa (Logue 1988; Domjan 1998) and across spatial and temporal domains (Cheng & Spetch 2001). The general principles that apply to learning in bees, pigeons and rats are likely to apply to fishes also. Psychology can clarify the mechanisms that underlie observed behaviour, while behavioural ecology can evaluate the adaptive significance of behavioural capacities demonstrated by psychology.

In several cases, static first-generation models based on optimal foraging theory (OFT) that do not represent temporal changes in internal state fail to predict observed behaviour (Hart 1993). More recent studies using more flexible (e.g. dynamic programming) models have been more successful because of their ability to represent changes in internal state (e.g. physiology, learning), interactions between intrinsic and extrinsic variables, and continuous behavioural adjustment in response to these factors (Ehlinger 1989; Kieffer & Colgan 1991; Hart 1993; Dall *et al.* 1999). Although standard OFT models predict that animals should exhibit all-or-nothing choice, experiments usually reveal partial preferences. It is likely that future models will draw increasingly on psychological effects (e.g. discrimination, memory, cue competition, interference, attention) to explain such divergences from the predictions of simple or idealized models and to test models based on risk and information (Shettleworth 1988).

A suggested conceptual framework for fish foraging is outlined in Figure 2.1. The main intention of this framework is to highlight how different contributory factors combine to affect foraging performance. Intrinsic and extrinsic (stimulus-related) factors affect motivation, which, in turn, influences the attention that is directed to relevant stimuli and the willingness to explore the general environment in which such

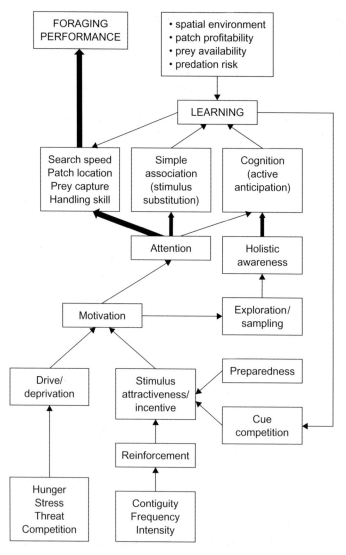

Fig. 2.1   Fish foraging: suggested relationships between foraging performance, learning and contributory factors. Bold arrows indicate main influences.

stimuli occur. Holistic environmental awareness is a key prerequisite for cognitive appraisal, while attention plays an important role in the formation of simple stimulus associations and in the development of foraging skills. Both association-formation and cognition contribute to learning. Such associative and cognitive information processing enhances the development of physical skills and thus improves foraging performance. This logic forms the basis for the organization of the first part of the present review. In later sections, the adaptive significance of learning and memory is considered in terms of the resulting ability of fishes to track environmental variation and improve competitive foraging success.

2.2   Some factors affecting the learning process

2.2.1   *Reinforcement*

Reinforcement is the increase in response probability following a stimulus event. Reinforcement appears to affect learning mainly by influencing what is learned (rather than how or how well it is learned), and appears to have a greater effect on motivation than on learning (Crespi 1942; Lieberman 1990). As in other animals, associative learning by fishes is strongly influenced by the frequency and intensity of reinforcement and the spatial and temporal contiguity of events. In goldfish (*Carassius auratus*, Cyprinidae), the formation of learned associations between new stimuli (e.g. visual cues) and rewards (e.g. food) occurs more efficiently when the delay between the stimulus and the reward is short (Breuning *et al.* 1981).

2.2.2   *Drive*

While the explanatory value of the term 'drive' is the subject of continuing debate, in the present context it seems useful to recognize drive (deprivation level) and stimulus attractiveness (incentive value) as two distinct components of motivation (Lieberman 1990). When fishes are hungry they are less distracted by other pressures (Milinski 1993), which enhances foraging learning. In social learning by guppies (*Poecilia reticulata*, Poeciliidae), food deprivation has a stronger effect on male than on female performance (Reader & Laland 2000). Isolation stress (as caused, for example, by predatory attacks) interferes with memory formation (goldfish; Laudien *et al.* 1986). Fishes may learn feeding skills more slowly when isolated than when in shoals (Jain & Sahai 1989). Similarly, foraging motivation is seriously reduced, or abolished entirely, by predator threat. There is evidence that fishes continually adjust their behaviour in accordance with a risk-balancing forage-refuge trade-off (Mittelbach 1981; Pitcher *et al.* 1988; Gotceitas & Colgan 1990). Thus, three-spined sticklebacks (*Gasterosteus aculeatus*, Gasterosteidae) learn to feed in profitable patches but abandon them when they are threatened (Huntingford & Wright 1989).

2.2.3   *Stimulus attractiveness*

Preparedness, which is the tendency to associate some stimulus combinations more readily than others, varies across taxa and populations. Preparedness is presumably adaptive but this can be hard to verify in anything more than correlative terms. Visual landmarks can vary in salience; for example, goldfish form feeding associations more readily with tall, coloured columns than with weed/plant combinations, and with vertically striped screens more readily than with horizontally striped screens (Warburton 1990, and K. Warburton unpublished observations). Patterns of retinal topography in fishes have been proposed to reflect habitat variation (Collin & Pettigrew 1988a,b), and in this connection it is interesting that goldfish have vertically oriented retinal cell patterns (Mednick & Springer 1988) and favour habitats with 'vertical' vegetation (McDowall 1996). Also noteworthy here is the fact that, in mammals, rearing in a striped environment affects the development of the visual cortex such that more surface area becomes devoted to the experienced orientation than to the orthogonal one (Sengpiel *et al.* 1999).

Generalization is the response to one stimulus as a result of training involving another stimulus. Learning in fishes is expected to conform to Shepard's law of generalization (Shepard 1987, 1988) in that fishes should respond to stimuli that are similar to those involved in pre-existing associations. Generalization gradients are typically exponential (i.e. the response falls off rapidly as the similarity to the original stimulus declines) if the relevant stimuli are clearly distinguishable, but are more likely to be Gaussian if the stimuli are not distinguishable (Shepard 1988). Area shift occurs when animals position themselves close to a rewarded location but on the side away from a nearby, unrewarded location. Cue competition is also a possibility; for example, the conditioning associated with a stimulus can be reduced or over-shadowed when it is reinforced in compound with another stimulus. Furthermore, prior conditioning to one element can block (i.e. prevent conditioning to) other elements of a compound stimulus. Thus, nearer landmarks may dominate over more distant landmarks (overshadowing) or landmarks encountered earlier may dominate (blocking). Blocking may also occur between different types of stimuli (e.g. colours and landmarks; Couvillon *et al.* 1997). Salient (e.g. large) cues, or those with intrinsic biological significance, are less susceptible to blocking (Denniston *et al.* 1996). Although relatively poorly documented in the case of fishes, these principles may be used to explain 'suboptimal' patterns of fish foraging, such as the consumption of non-preferred prey types and spatial biases caused by the visual environment. Cue competition may help explain why fifteen-spined sticklebacks (*Spinachia spinachia*, Gasterosteidae) and corkwing wrasse (*Crenilabrus melops*, Labridae), in a radial maze make associations between visual cues (coloured tiles) and food less efficiently as the diversity of cues increases (Hughes & Blight 2000).

2.2.4 *Exploration and sampling*

Fishes discover food through individual sampling and by observing other foragers (Pitcher & Magurran 1983; Pitcher & House 1987). There are good reasons to suppose that exploration and sampling are fundamental, integral aspects of foraging. This idea is captured in the information primacy hypothesis of Inglis *et al.* (2001), according to which a major determinant of behaviour is the need to obtain information continually in order to deal with environmental variability. The information primacy hypothesis helps to account for behavioural observations that cannot be explained by conventional reward theory, such as spontaneous alternation, patrolling, effects of hunger on the variability of learned results, latent learning, contrafreeloading, and behaviour following changes in food availability (see Inglis *et al.* 2001 for details). Such phenomena have not attracted much study in fish, but there is enough evidence to suggest that the information primacy hypothesis is valid for fish; for example, goldfish that spend more time in food patch sampling are able to switch patches faster when patch profitability changes (Pitcher & Magurran 1983; Warburton 1990). Such results indicate the existence of latent learning, where learning occurs on non-reinforced trials but remains unused until the introduction of a reinforcing factor provides an incentive for using it (Lieberman 1990). A contrasting effect, latent inhibition, occurs when exposure to an isolated stimulus reduces subsequent conditioning to that stimulus. When visual cues are consistently associated with food, cue fixation may occur and exploratory activity in other areas may decline markedly (Warburton 1990). While the value of sampling is in little doubt, it is important to separate purposeful sampling from inefficient patch use: at least some 'sampling' can be explained in terms of phenomena such as delay reduction and scalar expectancy rather than as a special type of information-gathering behaviour (Shettleworth 1988).

2.2.5 *Attention and simple association*

Only certain stimuli influence behaviour. Increases in foraging motivation tend to improve attention to relevant cues, such as features of food patches and prey. Fishes in larger groups are better able to sample their environment and overcome the confusion effect caused by random movements of swarming prey, presumably because increased group size allows individuals to benefit from shared vigilance and attend more to foraging (Pitcher & Magurran 1983; Smith & Warburton 1992). The mechanism underlying attention switching is a central concern in sensory physiology (Rice 1989). By restricting an animal's focus, attention helps to form associations when problems are simple: as a result, in conditioning experiments the conditioned stimulus effectively replaces the unconditioned stimulus in an animal's brain. Stimulus substitution is the formation of simple associations. However, if problems are more complex, increased focus may be counterproductive because subtle, although relevant, cues may be missed (Lieberman 1990).

Predators that take a wide range of prey may be expected to be relatively inefficient foragers and to suffer from divided attention (Bernays *et al.* 2004; see also Griffiths *et al.* 2004). Working with silver perch (*Bidyanus bidyanus*, Terapontidae) fingerlings foraging on brine shrimp and chironomid larvae, Warburton & Thomson (2006) found that when fishes that were familiar with one of the two prey types were offered both prey types simultaneously, the rate at which they captured both familiar and unfamiliar prey dropped progressively over succeeding trials. This result was not predicted by simple learning models, according to which a steady improvement in predation efficiency would be expected, but it could be explained in terms of an interaction between learning and attention. The authors postulated that when fishes were faced with mixed prey populations, cognitive constraints associated with divided attention impaired foraging efficiency. This effect increased over time because experience increased awareness of both prey types, which then competed for attention. The presence of two alternative prey types led to substantial fluctuations in reward rate over extended periods (20 days), even when prey densities at the start of each trial were kept constant. Warburton & Thomson (2006) interpreted these effects as examples of costs of learning.

2.2.6 *Cognition*

Fishes have extensive and diverse abilities for pattern discrimination and categorization (Douglas & Hawryshyn 1990; Chase 2001). Cognitive processing typically allows the subject to select from a wide range of preparatory responses and not just innate ones. This permits flexibility of response, and goal-directed action represents the most basic behavioural marker of cognition (Dickinson 1994). Classical conditioning involves two systems: stimulus substitution and a cognitive/awareness system. The latter system involves active anticipation, or expectation, of an unconditioned stimulus. Fishes can be trained to feed at a particular location and at a particular time of day; for example, golden shiners (*Notemigonus crysoleucas*, Cyprinidae) learned to expect food at midday in one of the brightly lit corners of their tank (Reebs 2000). They displayed daily food-anticipatory activity by leaving the shady area of the tank and spending more and more time in the food corner up to the normal feeding time. Mobile aggregations of stout chromis (*Chromis chrysurus*, Pomacentridae) feeding on zooplankton search specific foraging locations slowly and tortuously, but swim rapidly between these foraging locations, so that local search involves spatial memory and expectation of resource use (Noda *et al.* 1994). Dugatkin & Wilson (1992) found that individual bluegill sunfish (*Lepomis macrochirus*, Centrarchidae), could remember their feeding success with particular conspecifics and used that information to prefer or avoid those associates over a period of several weeks. The fishes therefore displayed cognitive abilities for strategic behaviour.

Psychological experiments indicate that retrieval of memories may be affected by interference from other memories. Interference may be proactive (caused by interference from experiences that preceded the event to be remembered) or retroactive (caused by interference from events that followed the event; Lieberman 1990). Fishes have difficulties in exploiting different feeding strategies (such as those required for different types of prey) simultaneously (Persson 1985). Fifteen-spined sticklebacks learn less efficiently and have shorter memories when fed alternating rather than pure diets: the mean handling time with amphipod (*Gammarus locusta*) prey was reduced to a greater extent (by 68%) in monospecific trials than in trials where amphipods and brine shrimp (*Artemia* spp.) were presented alternately (59%; Croy & Hughes 1991a). A short 'reorientation' lag to start of consumption occurred at the beginning of experimental sessions where silver perch were offered *Chironomus* sp. larvae and brine shrimp in alternating trials (Warburton & Thomson, 2006).

2.3   Patch use and probability matching

One of the central issues in foraging theory concerns the ability of animals to assess patch quality and adjust their behaviour accordingly. Do fishes have a memory for patch profitability that enables them to spend longer in patches with higher prey densities? If so, what decision rules do they use? To address these questions, Marschall *et al.* (1989) and DeVries *et al.* (1989) examined the behaviour of bluegill sunfish foraging for chironomid (*Chironomus riparius*) larvae among patches of artificial macrophytes. Both studies found that bluegills searched randomly within a patch, but each came to different conclusions with respect to the patch departure rules that the fishes used. Marschall *et al.* (1989) concluded that, of three different types of decision rules based on time, total prey caught, or capture rate in a patch, a constant residence time rule explained the observed data best. According to this rule, a forager is expected to stay in each patch for a constant optimal amount of time. Giving-up time (GUT) is the time spent in a patch after the final prey capture event. Although the optimal strategy in environments with patches of different profitability is to persist in high value patches, bluegill actually did the opposite, displaying shorter GUT in high value patches. This result may have occurred because fishes did not have perfect environmental knowledge and were unable to assess patches without sampling them.

In contrast, the study by DeVries *et al.* (1989) suggested that bluegills used a patch departure rule based on capture rate. Observed GUTs were longer than those predicted by optimality theory. However, this bias was in a direction that minimized the cost of poorly approximating the optimal solution. The different outcomes of the two studies were attributed to variations in experimental design: Marschall *et al.* (1989) varied patch quality within habitats but kept habitat quality (i.e. total prey density) constant across treatments, while DeVries *et al.* (1989) varied habitat quality but kept patch quality constant within a habitat. The main conclusion was that bluegills can assess environmental characteristics such as patch depletion and adjust their patch departure rules accordingly.

Further work by Wildhaber *et al.* (1994), where bluegills were offered food pellets in a partitioned, two-patch shuttle tank, indicated that patch GUTs were based on foraging experience. Bluegills had longer residence times and GUTs in relatively poor environments. However, within an environment, residence times and GUTs increased linearly with prey encounter and a variable GUT rule was the best predictor of departure. Individuals varied in terms of mean GUTs but there was little variation in overall patterns of patch use.

Taken together, these studies suggest that bluegills can adjust their foraging strategy as patterns of patch profitability change. Other generalist foragers may be the same. Flexible decision-making permits matching between patch exploitation and patch profitability. An example would be the use of a linear GUT rule coupled with a memory for patch quality.

Because tests of optimal foraging theory are essentially similar to studies of reinforcement schedules (Shettleworth 1988), it is relevant to consider whether fishes are capable of probability matching; that is, whether they are able to choose alternative food sources in proportion to their associated likelihood of reward. Mackintosh *et al.* (1971) trained goldfish either on visual probability discrimination with irrelevant spatial cues or on spatial problems with irrelevant visual cues. In the test tank fishes were presented with a screen containing two holes behind which were paddles onto which red, green or white lights could be projected. Food rewards were delivered via a magazine opening midway between the screen holes. When subjected to a 70:30 probability schedule (i.e. two patches that were rewarded on 70% and 30% of occasions respectively), fishes showed probability matching for both visual and spatial tests, in that they chose the majority stimulus on approximately 70% of trials. However, their responses were not random. Most subjects showed significant biases to one or other value of the irrelevant dimension, presumably by failing to maintain consistent attention to the inconsistently reinforced relevant cue. Birds and mammals that have been tested in the same way also show non-random responses but a greater tendency for maximizing (i.e. concentrating only on the most reinforced stimulus) than probability matching (Mackintosh *et al.* 1971). While these findings show that fishes are capable of probability matching, the reinforcement context appears to be critical, because probability matching was replaced by maximizing unless a 'guidance' procedure was used. In such a procedure, if on any trial the non-reinforced stimulus was chosen initially it was removed and the fish was allowed to earn a reward for response to the other stimulus (Behrend & Bitterman 1961).

Despite intensive research on a wide range of animal taxa, none of the major theories about the psychological processes that underlie matching satisfactorily explains all observed data, and there is still uncertainty as to whether matching or maximizing is the basic choice rule (Williams 1994). However, evidence from both psychology and behavioural ecology indicates that fishes are capable of flexible exploitation strategies in response to reinforcement variation. Detailed cross-interpretation of findings from the two fields is complicated by the use of different protocols and performance currencies, but valid analogies may be drawn in terms of the use of information by fishes. In a reinforcement study, the more frequently reinforced stimulus will be associated with both a higher mean reward rate and a higher degree of certainty, because it will be better sampled by the subject. Both these factors will encourage maximizing with respect to that stimulus. Guidance procedures increase return rates and certainty with respect to the less frequently reinforced stimulus, thus encouraging probability matching. In larger foraging areas and natural habitats, with more widely spaced food patches and higher travel costs, residence times per visit are likely to be higher and animals will acquire more accurate information on mean profitability – a process akin to guidance. Attention deficits (Mackintosh *et al.* 1971) will contribute to uncertainty. There is a need for more research on the impact of intrinsic and extrinsic factors on patch-use strategies by fishes.

2.4   Performance

Learning and motivation can both enhance performance. Although improvements in performance over a series of experimental trials are commonly described as 'learning', much of the observed improvement is the result of skill development or increased speed, and the period when associations are actually formed is likely to be much shorter (Lieberman 1990). Repeated experience can improve the efficiency of prey recognition, attack, manipulation and ingestion by fishes (e.g. Ware 1971; Colgan *et al.* 1986; Ranta & Nuutinen 1986; Mills *et al.* 1987; Croy & Hughes 1991a; Hughes & Croy 1993). Planktivores such as bluegill sunfish use past encounter rates to decide when to pursue prey (Werner *et al.* 1981). The speed with which discrimination is learned may vary between prey types (Croy & Hughes 1991b; Kaiser *et al.* 1992). Fifteen-spined sticklebacks find oligochaetes, whiteworms (*Enchytraeus albidus*), easier to catch and handle than amphipods, for which the capacity for learned improvements in attack efficiency, handling efficiency and handling time is greater (Mackney & Hughes 1995). Croy & Hughes (1991a) found that fifteen-spined sticklebacks required five to eight trials to learn to handle new prey. Similarly, bluegill sunfish required six to eight exposures to reach a peak handling efficiency on *Daphnia* (Werner *et al.* 1981). Silver perch fingerlings reached peak efficiency on chironomid larvae and brine shrimp after five trials (Warburton & Thomson 2006).

Reiriz *et al.* (1998) found that naive Atlantic salmon (*Salmo salar*, Salmonidae) increased their consumption of *Hydropsyche* drastically and decreased that of *Gammarus* after only three attempts, thus matching the selection pattern of wild fishes. The preference of pellet-reared juvenile Atlantic salmon for live prey increased within 16 attempts (Stradmeyer & Thorpe 1987). The ability of naive coho salmon (*Oncorhynchus kisutch*, Salmonidae) to capture shrimps (*Crangon* sp.) also increased with experience (Paszkowski & Olla 1985). Johnsen & Ugedal (1986, 1989, 1990) reported that hatchery-reared brown trout (*S. trutta*, Salmonidae) had broadened their diets to coincide with those of wild fishes after a few weeks at liberty.

Individual juvenile Atlantic salmon that were offered prey in a sequential encounter context increased their consumption of all prey types as they gained experience, but the improvement was higher for prey types that were initially less frequently consumed (Reiriz *et al.* 1998). The fast learning response towards novel prey was interpreted as a way of maintaining high foraging efficiency in the face of frequent changes in prey availability.

Day (1999) obtained evidence for context-dependent familiarity in rainbowfish (*Melanotaenia duboulayi*, Melanotaeniidae): individuals were able to become familiar with two shoals in two separate contexts (feeding and predation) at the same time. In later trials, when given a choice between both familiar shoals, fishes showed varying preferences that correlated with the context of the test situation. Fishes appeared to be motivated by an interaction between their external environment (e.g. threat level) and their internal motivational state (e.g. hunger state, curiosity).

The larvae of the cane toad (*Bufo marinus*), which has been introduced to Australia, are toxic to predatory native Australian fish species. On exposure to *B. marinus*, naive native fish species show different patterns of learned behaviour. In experiments most barramundi (*Lates calcarifer*, Centropomidae) rapidly learned to avoid toad tadpoles, while sooty grunter (*Hephaestus fuliginosus*, Terapontidae) showed wide intraspecific variation in behaviour; some individuals learned to avoid tadpoles after only a few attacks, while others continued to attack and reject tadpoles throughout the series of trials. Differences in fishes' learning ability, hunger levels and tadpole palatability may have contributed to the observed behavioural variation (Crossland 2001).

Learning can magnify the effect of small behavioural variations into pronounced individual foraging differences (Ehlinger 1989; Kieffer & Colgan 1991; Kohda 1994) and thus help to explain individual variation in the diets of fishes in the wild (Warburton *et al.* 1998). Kieffer & Colgan (1991) observed pumpkinseed sunfish (*Lepomis gibbosus*, Centrarchidae), feeding on whiteworms in structured and open habitats and noted significant individual variation in learning rates. Moreover, habitat structure had a significant effect on the efficiency of individuals with respect to the interval between captures and total time feeding. Such individual variation may represent adaptive flexibility in foraging behaviour (Magurran 1993). The finding that pumpkinseed sunfish exhibited a positive successive contrast effect, whereas fishes that were familiar with a low yield, vegetated habitat showed enhanced performance when exposed to an open habitat, indicates that differing background experience may affect the foraging efficiency and competitive success of individuals (Kieffer & Colgan 1991). By changing the efficiency of prey exploitation, and thus relative prey profitability, learning can also help to account for prey switching and frequency-dependent predation (Murdoch 1969; Ringler 1985). Experiments with fifteen-spined sticklebacks by Croy & Hughes (1991a) indicated that *Artemia* became more profitable than *Gammarus* as trials proceeded.

2.5   Tracking environmental variation

Food patch discrimination can be improved by remembering the spatial position of previously exploited patches, and predators must continually compare present information on patches and prey types with search images held in their memory (Hart 1986, 1993; Odling-Smee & Braithwaite 2003). There appear to be individual differences in food patch learning: some fishes seem to identify food patches by reference to local visual cues, while others rely more on global cues (Huntingford & Wright 1989). When tested in an eight-arm radial maze in the absence of spatial cues, fifteen-spined sticklebacks and corkwing wrasse adopted an algorithmic strategy of visiting every third arm, but in the presence of spatial cues (coloured tiles) algo-rithmic behaviour was largely subsumed by the use of spatial memory (Hughes & Blight 1999). The results suggested that fishes were able to memorize the spatial con-figuration of the cues and indicate that when spatial information is limited, reliance on search algorithms that reduce the likelihood of revisiting depleted patches tends to increase.

Fauchald (1999) modelled prey search in a stochastic hierarchical patch system where high-density patches at small scale were nested within low-density patches at larger scales. The scaling of the prey system was adapted for schooling marine fishes and krill. Simulations suggested that information flow and tracking efficiency were maximal at high levels of prey abundance but at intermediate levels of prey aggregation. These conditions describe patch systems that are well-structured and hierarchical, where clear structural cues are available to guide foraging behaviour.

The allocation of priorities to patches with changing profitability is made possible by patch sampling, which permits fishes to switch rapidly between win-shift and win-stay behaviour when revisiting preferred locations. This is especially valuable in habitats where food patch distributions are subject to frequent change, as found, for example, in soft-bottom intertidal environments (Hughes & Blight 2000).

Braithwaite *et al.* (1996) showed that juvenile Atlantic salmon have the ability to distinguish between two identical visual landmarks and to learn to track the move-ments of one of them to predict the location of food. They concluded that chemo-sensory cues, perhaps originating from substrate marking by the fishes themselves, can be used in conjunction with visual cues. Salmonids are known to mark substrates (Halvorsen & Stabell 1990) and switch to nocturnal foraging in winter (Fraser *et al.* 1993), possibly in response to changes in prey availability or predator threat.

Context-switching and the retention interval appear to have additive effects on memory (Bouton *et al.* 1999). A strong case can be made in support of the idea that forgetting is adaptive rather than a negative outcome of process failure (Kraemer & Golding 1997). Foraging covers a diverse set of activities in which the animal may draw on several learning and memory systems; for example, spatial memory to construct a cognitive map of its environment (Odling-Smee & Braithwaite 2003; Broglio *et al.* 2003; see also Chapters 7 and 13); a memory for cues associated with particular food patches and food types; an ability to recognize individual shoalmates (Griffiths 2003; and see Chapter 8) and to benefit from socially transmitted informa-tion (Brown & Laland 2003; and see Chapter 10); and a memory for predator-related cues that indicate when the risks of foraging are unacceptably high (Brown 2003; and see Chapter 4; Kelley & Magurran 2003; and see Chapter 3). The typical memory windows for these different systems are likely to reflect corresponding levels of environmental variation; for example, in a structured, relatively undisturbed environment spatial memory should be persistent, while associations involving varying patch profitability and ephemeral prey types should be relatively short.

Curves of forgetting against time are typically exponential in form (White 2001) and foraging theory has tended to assume that recently obtained information should be the most valued (e.g. Cuthill *et al.* 1990). However, little is known about rates of forgetting and the factors that lead to variations in such rates, including the effects of intervening experience. Fish species appear to vary widely in terms of their memory window, defined as the duration of learned foraging skills. In the absence of reinforcement, fifteen-spined sticklebacks start to forget foraging skills after 2 days and return to naive levels by 8 days (Croy & Hughes 1991a), while corresponding times for rainbow trout (*Oncorhynchus mykiss*, Salmonidae), are 14 days and 3 months (Ware 1971). Silver perch can retain learned foraging skills for at least 5 weeks (Warburton & Thomson in press). It seems likely that the skills to recognize and capture a number of prey types are retained simultaneously, but the decisions to draw on those skills are based on recent experience. These results are consistent with proposals that extinction is not equivalent to unlearning, and that in an appropriate context previous learning can be reinstated (Pearce & Bouton 2001).

Feeding habits, which are easily learned in early life, can be retained for a significant length of time, and early training on natural foods may develop preferences that persist through long periods of feeding on artificial foods (Suboski & Templeton 1989). Norris (2003) found that two groups of juvenile whiting (*Sillago maculata*, Sillaginidae), one raised on a diet of live food and the other on artificial pellets over a period of 4 months, diverged in terms of gross morphology, taste bud distribution and learned feeding responses. The observed differences in development were attributed to the different stimuli required to locate food on the two diets, pellets being approached primarily using chemical cues and live prey mainly via visual stimuli. The development of an individual's ability to use social cues in foraging also depends critically on experience (Huntingford 1993). More information is required on the influence of primacy effects on the ontogeny of feeding behaviour.

Evidence suggests that the memory window for prey is related to the predictability of the feeding environment, and there appear to be significant interspecific and interpopulation differences correlated with ecology. Although freshwater and anadromous three-spined sticklebacks and marine fifteen-spined sticklebacks exhibit similar rates of prey learning, their memory windows differ considerably, being >25 days, 10 days and 8 days, respectively. For residential or anadromous marine populations that move with the tide and where older information is rapidly devalued, a long memory window could be maladaptive by retarding behavioural adjustment to changing conditions (Mackney & Hughes 1995). In silver perch, a freshwater species, the relatively long memory window (>5 weeks) is consistent with evidence that individuals respond to slower cycles of food availability and unpredictable, intermittent peaks in particular prey types (Warburton *et al.* 1998).

In radial maze experiments, imposition of a delay within trials had no effect on foraging efficiency of wrasse and sticklebacks when memory for spatial cues could be used to guide foraging, but in the absence of cues the behavioural algorithm was reset, leading to reduced efficiency (Hughes & Blight 1999). Memory retention for previous choices lay within the range 0.5–5.0 min, consistent with the timescale of rapidly changing prey distributions during the tidal cycle.

Phylogenetic influences on the length of the memory window cannot be ruled out. However, systematic studies of a range of fish taxa tested with the same experimental protocol are needed before the importance of such influences can be properly assessed. Traditional assumptions that animal learning and memory are influenced mainly by phylogenetic relationships have been significantly revised as evidence for the importance of lifestyle and ecological context has emerged. Though very incomplete, our present knowledge of the role of learning and memory in the foraging behaviour of fishes is consistent with this revised interpretation; for example, the difference in memory window between two forms of the three-spined stickleback is as great as that between one form of three-spined stickleback and the fifteen-spined stickleback, a species from a different genus (Mackney & Hughes 1995).

2.6   Competition

Learning and memory can influence the distributional patterns of competing individuals. The relative pay-off sum learning rule predicts that good competitors will decide where to feed earlier and switch less between patches than poor competitors (Regelmann 1984). Within-shoal competition can cause subordinate fishes to abandon patch sampling (Croy & Hughes 1991b).

Hakayama & Iguchi (2001) recorded patterns in distribution, aggression, food intake and growth of the salmonid *Onchorhynchus masou ishika*, which had free access to two patches and was able to use long-term memory to assess patch quality. The within-group variation in body weight increased with time, and over the 4-week experimental period the pattern of resource use changed from a random distribution to an ideal free distribution and finally to an ideal despotic distribution. On average, the better patch was used by more individuals than predicted by a random distribution but by fewer than predicted by an ideal free distribution. Competition was therefore a contributory reason for the non-occurrence of an ideal free distribution.

A capacity for individual recognition allows foraging fishes to identify potential competitors (Griffiths 2003; and see Chapter 8). In rainbow trout, aggression, food intake and growth rate are positively correlated and aggressive dominants deny subordinates access to food (Johnsson 1997). Levels of aggression in contests between familiar individuals are lower than in contests between strangers, so that familiarity appears to reduce aggression and increase the foraging rate (Griffiths *et al.* 2004). This is consistent with the hypothesis that individual recognition is used to reduce the costs of contesting resources: after an initial contest, encounters between familiar individuals should be settled with less aggression and a lower probability of status reversal than encounters between unfamiliar fishes (Switzer *et al.* 2001). However, this effect decays rapidly with time since previously familiar individuals are not treated as such after 3 days of separation (Johnsson 1997).

2.7   Learning and fish feeding: some applications

Stimuli that elicit feeding responses such as approach and oral contact serve to initiate individual learning mechanisms based on the palatability of the novel food

(Suboski & Templeton 1989). In aquaculture, self-feeding devices rely on the fact that fishes can be trained to press a trigger to obtain food (Boujard & Leatherland 1992). The key stimuli (presence of potential food particles, the sight of other fishes feeding) are the same as those involved in the wild. Naive fishes appear to mistake the self-feeding triggers for food particles but later learn to distinguish between the trigger and food, and reduce the force they apply to the trigger (Alanara 1996). Trout reared in tanks in relatively small groups of 100–300 individuals reach a stable level of self-feeding in approximately 25 days, but under large-scale rearing conditions where 1000–2000 individuals are kept in cages, learned associations between trigger and food seem less prevalent, because the frequency of trigger actuations remains high even when food is not provided (Alanara 1996).

Also relevant to aquaculture is the ability of fishes to learn the time of day when food is available. This may be evidenced by food-anticipatory behaviour, in the form of a gradual increase in locomotory activity prior to feeding time (Sanchez-Vazquez *et al.* 1997). Amano *et al.* (2005) found that goldfish fed once a day showed approach behaviour to food odour only at the time when they were normally fed, but fishes fed three times a day showed no significant approach behaviour at any time of day.

There is growing interest in the possibility of life-skills training of hatchery-reared fishes before release into the wild. Compared to wild fishes, hatchery-reared individuals tend to suffer behavioural deficits that may significantly reduce post-release survivorship (Brown & Laland 2001; Brown & Day 2002). Such deficits include reduced food consumption, lower diversity of prey types, consumption of non-prey objects, longer delays in attacking and consuming prey, slower prey switching and atypical microhabitat choice (Sosiak *et al.* 1979; Ersbak & Haase 1983; Olla *et al.* 1998; Neveu 1999; Sundstrom & Johnsson 2001), all of which may be redressed if steps are taken to provide appropriate learning opportunities.

2.8   Conclusions

Learning and memory play a decisive role in the foraging activities of fishes. Learning and associated improvements in prey search, capture and handling efficiency can lead to significant enhancements in foraging performance after only a few exposures. Fishes are capable of adjusting their foraging strategy as patterns of patch profitability change. There is also evidence that forgetting seems to have adaptive significance, because the length of the memory window appears to be related to environmental variability in at least some species. Learning and memory thus permit rapid and flexible adaptation to changing prey densities, identity and location.

Fishes have been used to test the predictions of classic foraging theory. These studies show that fishes can match exploitation rates to patch profitability. To do this they draw on memories for general patch quality (based on previous sampling and exploitation) and on departure rules for leaving the current patch. However, it appears that such patch departure rules can be varied according to circumstances. In this respect fishes resemble higher vertebrates (e.g. jays, *Garrulus glandarius*, Corvidae; Shettleworth 1988).

Evidence from the literature on foraging in fishes is consistent with the proposal that, in terms of basic learning mechanisms, the similarities across animal taxa are more striking than the differences (Lieberman 1990). If cognition is viewed as a collection of adaptively specialized modules, then all extant species are equally intelligent in their own ways and it makes no sense to propose a linear evolutionary hierarchy of cognitive modules (Shettleworth 1998). Like other vertebrates, fishes draw on an array of learning and memory systems as part of a broad cognitive repertoire. There is much to be discovered about the ways in which individually and socially learned information about the feeding habitat, prey types, conspecifics and predator risk is combined to make foraging decisions.

Convergence between behavioural ecology and comparative psychology offers promise in terms of developing more mechanistically realistic foraging models and explaining apparently 'suboptimal' patterns of behaviour. Foraging decisions involve the interplay between several distinct systems of learning and memory, including those that relate to habitat, food patches, prey types, conspecifics and predators. Fish biologists therefore, face an interesting challenge in developing integrated accounts of fish foraging that explain how cognitive sophistication can help individual animals deal with the complexity of the ecological context.

Although progress has been made in identifying general psychological principles underlying animal behaviour, in most cases their impacts on the foraging efficiency of fishes remain to be explored; for example, how do visual stimuli bias movement patterns, independent of patch profitability? How is learning constrained by attention limitations? Which factors encourage latent learning and latent inhibition? How do order effects such as successive contrast affect the competitive foraging success of individuals in shoals? There is scope for imaginative experiments to address such questions. Almost all previous work has been carried out under laboratory conditions and there is a pressing need to assess the role of learning in natural, complex environments.

2.9   Acknowledgements

I would like to thank Paul Cunningham, Craig Hull, Ottmar Lipp and Culum Brown for their helpful comments on the manuscript, Andrew Norris for access to unpublished information on the sensory development of whiting, and Viviana Gamboa-Pickering for assistance with referencing.

2.10   References

Alanara, A. (1996) The use of self-feeders in rainbow trout (*Oncorhynchus mykiss*) production. *Aquaculture*, **145**, 1–20.

Amano, M., Iigo, M. & Yamamori, K. (2005) Effects of feeding time on approaching behaviour to food odor in goldfish. *Fisheries Science*, **71**, 183–186.

Behrend, E.R. & Bitterman, M.E. (1961) Probability-matching in the fish. *American Journal of Psychology*, **74**, 542–551.

Bernays, E.A., Singer, M.S. & Rodrigues, D. (2004) Foraging in nature: foraging efficiency and attentiveness in caterpillars with different diet breadths. *Ecological Entomology*, **29**, 389–397.

Boujard, T. & Leatherland, J.F. (1992) Demand-feeding behaviour and diel pattern of feeding activity in *Oncorhynchus mykiss* held under different photoperiod regimes. *Journal of Fish Biology*, **40**, 535–544.

Bouton, M.E., Nelson, J.B. & Rojas, J.M. (1999) Stimulus generalization, context change, and forgetting. *Psychological Bulletin*, **125**, 171–186.

Braithwaite, V.A., Armstrong, J.D., McAdam, H.M. & Huntingford, F.A. (1996) Can juvenile Atlantic salmon use multiple cue systems in spatial learning? *Animal Behaviour*, **51**, 1409–1415.

Breuning, S.E., Ferguson, D.G. & Poling, A.D. (1981) Second-order schedule effects with goldfish: a comparison of brief-stimulus, chained, and tandem schedules. *Psychological Record*, **31**, 437–445.

Broglio, C., Rodríguez, F. & Salas, C. (2003) Spatial cognition and its neural basis in teleost fish. *Fish and Fisheries*, **4**, 247–255.

Brown, C. & Day, R. (2002) The future of stock enhancements: bridging the gap between hatchery practice and conservation biology. *Fish and Fisheries*, **3**, 79–94.

Brown, C. & Laland, K.N. (2001) Social learning and life skills training for hatchery reared fish. *Journal of Fish Biology*, **59**, 471–493.

Brown, C. & Laland, K.N. (2003) Social learning in fishes: a review. *Fish and Fisheries*, **4**, 280–288.

Brown, G.E. (2003) Learning about danger: chemical alarm cues and local risk assessment in prey fishes. *Fish and Fisheries*, **4**, 227–234.

Chase, A.R. (2001) Music discriminations by carp (*Cyprinus carpio*). *Animal Learning and Behavior*, **29**, 336–353.

Cheng, K. & Spetch, M.L. (2001) Blocking in landmark-based search in honeybees. *Animal Learning and Behavior*, **29**, 1–9.

Colgan, P.W., Brown, J.A. & Orsatti, S.D. (1986) Role of diet and experience in the development of feeding behaviour in large-mouth bass (*Micropterus salmoides*). *Journal of Fish Biology*, **28**, 161–170.

Collin, S.P. & Pettigrew, J.D. (1988a) Retinal topography in reef teleosts. I. Some species with well-developed areae but poorly developed streaks. *Brain Behavior and Evolution*, **31**, 269–282.

Collin, S.P. & Pettigrew, J.D. (1988b) Retinal topography in reef teleosts. II. Some species with prominent horizontal streaks and high-density areae. *Brain Behavior and Evolution*, **31**, 283–295.

Couvillon, P.A., Arakaki, L. & Bitterman, M.E. (1997) Intramodal blocking in honeybees. *Animal Learning and Behavior*, **25**, 277–282.

Crespi, L.P. (1942) Quantitative variation in incentive and performance in the white rat. *American Journal of Psychology*, **55**, 467–517.

Crossland, M.R. (2001) Ability of predatory native Australian fishes to learn to avoid toxic larvae of the introduced toad *Bufo marinus*. *Journal of Fish Biology*, **59**, 319–329.

Croy, M.I. & Hughes, R.N. (1991a) The role of learning and memory in the feeding behaviour of the fifteen-spined stickleback, *Spinachia spinachia* L. *Animal Behaviour*, **41**, 149–159.

Croy, M.I. & Hughes, R.N. (1991b) The influence of hunger on feeding behaviour and on the acquisition of learned foraging skills by the fifteen-spined stickleback, *Spinachia spinachia* L. *Animal Behaviour*, **41**, 161–170.

Cuthill, I.C., Kacelnik, A., Krebs, J.R., Haccou, P. & Iwasa, Y. (1990) Starlings exploiting patches: the effect of recent experience on foraging decisions. *Animal Behaviour*, **40**, 625–640.

Dall, S.R.X., McNamara, J.M. & Cuthill, I.C. (1999) Interruptions to foraging and learning in a changing environment. *Animal Behaviour*, **57**, 233–241.

Day, J. (1999) Context-dependent familiarity in rainbowfish. Honours thesis, Department of Zoology and Entomology, University of Queensland.

Denniston, J.C., Miller, R.R. & Matute, H. (1996) Biological significance as a determinant of cue competition. *Psychological Science*, **7**, 325–331.

DeVries, D.R., Stein, R.A. & Chesson, P.L. (1989) Sunfish foraging among patches: the patch-departure decision. *Animal Behaviour*, **37**, 455–464.

Dickinson, A. (1994) Instrumental conditioning. In: N.J. Mackintosh (ed.), *Animal Learning and Cognition*, pp. 45–79. Academic Press, San Diego.

Domjan, M. (1998) The Principles of Learning and Behavior. Brooks/Cole, Pacific Grove, California.

Douglas, R.H. & Hawryshyn, C.W. (1990) Behavioural studies of fish vision: an analysis of visual capabilities. In: R.H. Douglas & M.B.A. Djamgoz (eds) *The Visual System of Fish*, pp. 373–418. Chapman and Hall, London.

Dugatkin, L.A. & Wilson, D.S. (1992) The prerequisites for strategic behaviour in bluegill sunfish, *Lepomis macrochirus*. *Animal Behaviour*, **44**, 223–230.

Ehlinger, T.J. (1989) Learning and individual variation in bluegill foraging: habitat-specific techniques. *Animal Behaviour*, **38**, 643–658.

Ersbak, K. & Haase, B.L. (1983) Nutritional deprivation after stocking as a possible mechanism leading to mortality in stream-stocked brook trout. *North American Journal of Fisheries Management*, **3**, 142–151.

Fauchald, P. (1999) Foraging in a hierarchical patch system. *American Naturalist*, **153**, 603–613.

Fraser, N.H.C., Metcalfe, N.B. & Thorpe, J.E. (1993) Temperature-dependent switch between diurnal and nocturnal foraging in salmon. *Proceedings of the Royal Society of London, Series B – Biological Sciences*, **252**, 135–139.

Gotceitas, V. & Colgan, P.W. (1990) The effects of prey availability and predation risk on habitat selection by juvenile bluegill sunfish. *Copeia*, **1990**, 409–417.

Griffiths, S.W. (2003) Learned recognition of conspecifics by fishes. *Fish and Fisheries*, **4**, 256–268.

Griffiths, S.W., Brockmark, S., Hoejesjoe, J. & Johnsson, J.I. (2004) Coping with divided attention: the advantage of familiarity. *Proceedings of the Royal Society of London, Series B – Biological Sciences*, **1540**, 695–699.

Hakayama, H. & Iguchi, K. (2001) Transition from a random to an ideal free to an ideal despotic distribution: the effect of individual difference in growth. *Journal of Ethology*, **19**, 129–137.

Halvorsen, M. & Stabell, O.B. (1990) Homing behaviour of displaced stream-dwelling trout. *Animal Behaviour*, **39**, 1089–1097.

Hart, P.J.B. (1986) Foraging in teleost fishes. In: T.J. Pitcher (ed.) *The Behaviour of Teleost Fishes*, pp. 211–235. Croom Helm, London.

Hart, P.J.B. (1993) Teleost foraging: facts and theories. In: T.J. Pitcher (ed.) *The Behaviour of Teleost Fishes*, 2nd edition, pp. 253–284. Chapman and Hall, London.

Hughes, R.N. & Blight, C.M. (1999) Algorithmic behaviour and spatial memory are used by two intertidal fish species to solve the radial maze. *Animal Behaviour*, **58**, 601–613.

Hughes, R.N. & Blight, C.M. (2000) Two intertidal fish species use visual association learning to track the status of food patches in a radial maze. *Animal Behaviour*, **59**, 613–621.

Hughes, R.N. & Croy, M.I. (1993) An experimental analysis of frequency-dependent predation (switching) in the 15-spined stickleback, *Spinachia spinachia*. *Journal of Animal Ecology*, **62**, 341–352.

Hughes, R.N., Kaiser, M.J., Mackney, P.A. & Warburton, K. (1992) Optimizing foraging behaviour through learning. *Journal of Fish Biology*, **41 (Suppl. B)**, 77–91.

Huntingford, F.A. (1993) Development of behaviour in fish. In: T.J. Pitcher (ed.), *Behaviour of Teleost Fishes*, pp. 57–83. Chapman and Hall, London.

Huntingford, F.A. & Wright, P.J. (1989) How sticklebacks learn to avoid dangerous feeding patches. *Behavioural Processes*, **19**, 181–189.

Inglis, I.R., Langton, S., Forkman, B. & Lazarus, J. (2001) An information primacy model of exploratory and foraging behaviour. *Animal Behaviour*, **62**, 543–557.

Jain, V.K. & Sahai, S. (1989) Learning behaviour of the black molly, *Molliensia sphenops*. *Environmental Ecology*, **7**, 337–344.

Johnsen, B.O. & Ugedal, O. (1986) Feeding by hatchery-reared and wild brown trout, *Salmo trutta* L., in a Norwegian stream. *Aquaculture and Fisheries Management*, **17**, 281–287.

Johnsen, B.O. & Ugedal, O. (1989) Feeding by hatchery-reared brown trout, *Salmo trutta* L. released in lakes. *Aquaculture and Fisheries Management*, **20**, 97–104.

Johnsen, B.O. & Ugedal, O. (1990) Feeding by hatchery- and pond-reared brown trout, *Salmo trutta* L., fingerlings released in a lake and in a small stream. *Aquaculture and Fisheries Management*, **21**, 253–258.

Johnsson, J.I. (1997) Individual recognition affects aggression and dominance relations in rainbow trout, *Oncorhynchus mykiss*. *Ethology*, **103**, 267–282.

Kaiser, M.J., Gibson, R.N. & Hughes, R.N. (1992) The effects of prey type on the predatory behaviour of the fifteen-spined stickleback *Spinachia spinachia* L. *Animal Behaviour*, **43**, 147–156.

Kelley, J.L. & Magurran, A.E. (2003) Learning of predator recognition and anti-predator responses in fishes. *Fish and Fisheries*, **4**, 216–226.

Kieffer, J.D. & Colgan, P.W. (1991) Individual variation in learning by foraging pumpkinseed sunfish, *Lepomis gibbosus*: the influence of habitat. *Animal Behaviour*, **41**, 603–611.

Kieffer, J.D. & Colgan, P.W. (1992) The role of learning in fish behaviour. *Reviews in Fish Biology and Fisheries*, **2**, 125–143.

Kohda, M. (1994) Individual specialized foraging repertoires in the piscivorous cichlid fish *Lepidolamprologus profundicola*. *Animal Behaviour*, **48**, 1123–1131.

Kraemer, P.J. & Golding, J.M. (1997) Adaptive forgetting in animals. *Psychonomic Bulletin and Review*, **4**, 480–491.

Laudien, H., Frever, J., Erb, R. & Denzer, D. (1986) Influence of isolation stress and inhibited protein biosynthesis on learning and memory in goldfish. *Physiology and Behaviour*, **38**, 621–628.

Lieberman, D.A. (1990) *Learning: Behavior and Cognition*. Wadsworth, Belmont, California.

Logue, A.W. (1988) A comparison of taste aversion learning in humans and other vertebrates: evolutionary pressures in common. In: R.C. Bolles & M.D. Beecher (eds), *Evolution and Learning*, pp. 97–116. Erlbaum, Hillsdale, New Jersey.

Mackintosh, N.J., Lord, J. & Little, L. (1971) Visual and spatial probability learning in pigeons and goldfish. *Psychonomic Science*, **24**, 221–223.

Mackney, P.A. & Hughes, R.N. (1995) Foraging behaviour and memory window in sticklebacks. *Behaviour*, **132**, 1241–1253.

Magurran, A.E. (1993) Individual differences and alternative behaviours. In: T.J. Pitcher (ed.), *The Behaviour of Teleost Fishes*, pp. 441–477. Chapman and Hall, London.

Marschall, E.A., Chesson, P.L. & Stein, R.A. (1989) Foraging in a patchy environment: prey-encounter rate and residence time distributions. *Animal Behaviour*, **37**, 444–454.

McDowall, R.M. (1996) *Freshwater Fishes of Southeastern Australia*. Reed, Sydney.

Mednick, A.S. & Springer, A.D. (1988) Asymmetric distribution of retinal ganglion cells in goldfish. *Journal of Comparative Neurology*, **268**, 49–59.

Milinski, M. (1993) Predation risk and feeding behaviour. In: T.J. Pitcher (ed.) *The Behaviour of Teleost Fishes*, pp. 285–305. Chapman and Hall, London.

Mills, E.L., Widzowski, D.V. & Jones, S.R. (1987) Food conditioning and prey selection by young yellow perch (*Perca flavescens*). *Canadian Journal of Fisheries and Aquatic Sciences*, **44**, 549–555.

Mittelbach, G.G. (1981) Foraging efficiency and body size a study of optimal diet and habitat use by bluegills. *Ecology*, **62**, 1370–1386.

Murdoch, W.W. (1969) Switching in general predators: experiments on predator specificity and stability of prey populations. *Ecological Monographs*, **39**, 335–354.

Neveu, A. (1999) Feeding strategy of the brown trout (*Salmo trutta* L.) in running water. In: J.L. Bagliniere & G. Maisee (eds), *Biology and Ecology of the Brown and Sea Trout*, pp. 91–113. Praxis, Chichester.

Noda, M., Gushima, K. & Kakuda, S. (1994) Local prey search based on spatial memory and expectation in the planktivorous reef fish, *Chromis chrysurus* (Pomacentridae). *Animal Behaviour*, **47**, 1413–1422.

Norris, A.J. (2003) Sensory modalities, plasticity and prey choice in three sympatric species of whiting (*Pisces*: Sillaginidae). PhD. Thesis, University of Queensland.

Odling-Smee, L. & Braithwaite, V.A. (2003) The role of learning in fish orientation. *Fish and Fisheries*, **4**, 235–246.

Olla, B.L., Davis, M.W. & Ryer, C.H. (1998) Understanding how the hatchery environment represses or promotes the development of behavioural survival skills. *Bulletin of Marine Science*, **62**, 531–550.

Paszkowski, C.A. & Olla, B.L. (1985) Foraging behaviour of hatchery-produced coho salmon (*Oncorhynchus kisutch*) smolts on live prey. *Canadian Journal of Fisheries and Aquatic Sciences*, **42**, 1915–1921.

Pearce, J.M. & Bouton, M.E. (2001) Theories of associative learning in animals. *Annual Reviews Psychology*, **52**, 111–139.

Persson, L. (1985) Optimal foraging: the difficulty of exploiting different feeding strategies simultaneously. *Oecologia (Berlin)*, **67**, 338–341.

Pitcher, T.J. & House, A.C. (1987) Foraging rules for group feeders: area copying depends upon food density in shoaling goldfish. *Ethology*, **76**, 161–167.

Pitcher, T.J. & Magurran, A.E. (1983) Shoal size, patch profitability and information exchange in foraging goldfish. *Animal Behaviour*, **31**, 546–555.

Pitcher, T.J., Lang, S.H. & Turner, J.R. (1988) A risk-balancing trade-off between foraging rewards and predation risk in shoaling fish. *Behavioral Ecology and Sociobiology*, **22**, 225–228.

Ranta, E. & Nuutinen, V. (1986) Experience affects performance of ten-spined sticklebacks foraging on zooplankton. *Hydrobiologia*, **140**, 161–166.

Reader, S.M. & Laland, K.N. (2000) Diffusion of foraging innovations in the guppy. *Animal Behaviour*, **60**, 175–180.

Reebs, S.G. (2000) Can a minority of informed leaders determine the foraging movements of a fish shoal? *Animal Behaviour*, **59**, 403–409.

Regelmann, K. (1984) Competitive resource sharing: a simulation model. *Animal Behaviour*, **32**, 226–232.

Reiriz, L., Nicieza, A.G. & Brana, F. (1998) Prey selection by experienced and naive juvenile Atlantic salmon. *Journal of Fish Biology*, **53**, 100–114.

Rice, M.J. (1989) The sensory physiology of pest fruit flies: conspectus and prospectus. In: A.S. Robinson & G.H.S. Hooper (eds) *Fruit Flies. Their Biology, Natural Enemies and Control*, pp. 249–272. Elsevier, Amsterdam.

Ringler, N.H. (1985) Individual and temporal variation in prey switching by brown trout, *Salmo trutta*. *Copeia*, **1985**, 918–926.

Sanchez-Vazquez, F.J., Madrid, J.A., Zamora, S. & Tabata, M. (1997) Feeding entrainment of locomotor activity rhythms in the goldfish is mediated by a feeding-entrainable circadian oscillator. *Journal of Comparative Physiology*, **181**, 121–132.

Sengpiel, F., Stawinski, P. & Bonhoeffer, T. (1999) Influence of experience on orientation maps in cat visual cortex. *Nature Neuroscience*, **2**, 727–732.

Shepard, R.N. (1987) Toward a universal law of generalisation for psychological science. *Science*, **237**, 1317–1323.

Shepard, R.N. (1988) Time and distance in generalization and discrimination. Reply to Evans (1988). *Journal of Experimental Psychology: General*, **117**, 415–416.

Shettleworth, S.J. (1988) Foraging as operant behavior and operant behavior as foraging: what have we learned? *Psychology of Learning and Motivation*, **22**, 1–49.

Shettleworth, S.J. (1998) *Cognition, Evolution, and Behavior*. Oxford University Press, Oxford.

Smith, M.F.L. & Warburton, K. (1992) Predator shoaling moderates the confusion effect in blue-green chromis, *Chromis viridis*. *Behavioural Ecology and Sociobiology*, **30**, 103–107.

Sosiak, A.J., Randall, R.G. & McKenzie, J.A. (1979) Feeding by hatchery-reared and wild Atlantic salmon (*Salmo salar*) parr in streams. *Journal of the Fisheries Research Board of Canada*, **36**, 1408–1412.

Stradmeyer, L. & Thorpe, J.E. (1987) The response of hatcher-reared Atlantic salmon, *Salmo salar* L., parr to pelletted and wild prey. *Aquaculture and Fisheries Management*, **18**, 51–62.

Suboski, M.D. & Templeton, J.J. (1989) Life skills training for hatchery fish: social learning and survival. *Fisheries Research*, **7**, 343–352.

Sundstrom, L.F. & Johnsson, J.I. (2001) Experience and social environment influence the ability of young brown trout to forage on live novel prey. *Animal Behaviour*, **61**, 249–255.

Switzer, P.V., Stamps, J.A. & Mangel, M. (2001) When should a territory resident attack? *Animal Behaviour*, **62**, 749–759.

Warburton, K. (1990) The use of local landmarks by foraging goldfish. *Animal Behaviour*, **40**, 500–505.

Warburton, K. & Thomson, T. (2006) Costs of learning: the dynamics of mixed-prey exploitation by silver perch, *Bidyanus bidyanus*. *Animal Behaviour*, **71**, 361–370.

Warburton, K., Retif, S. & Hume, D. (1998) Generalists as sequential specialists: diets and prey switching in juvenile silver perch. *Environmental Biology of Fishes*, **51**, 445–454.

Ware, D.M. (1971) Predation by rainbow trout (*Salmo gairdneri*): the effect of experience. *Journal of the Fisheries Research Board of Canada*, **28**, 1847–1852.

Werner, E.E., Mittelbach, G.G. & Hall, D.J. (1981) The role of foraging profitability and experience in habitat use by the bluegill sunfish. *Ecology*, **62**, 116–125.

White, K.G. (2001) Forgetting functions. *Animal Learning and Behavior*, **29**, 193–207.

Wildhaber, M.L., Green, R. & Crowder, L.B. (1994) Bluegills continuously update patch giving-up times based on foraging experience. *Animal Behaviour*, **47**, 501–513.

Williams, B.A. (1994) Reinforcement and choice. In: N.J. Mackintosh (ed.), *Animal Learning and Cognition*, pp. 81–108. Academic Press, San Diego.

Chapter 3
Learned Defences and Counterdefences in Predator–Prey Interactions

Jennifer L. Kelley and Anne E. Magurran

3.1 Introduction

Living with predators is an unavoidable aspect of life for almost all fishes. The activity patterns of predators are highly variable over space and time and consequently prey are faced with the continual need to balance their habitat use (Chapter 7), foraging decisions (Chapter 2) and reproduction (Chapter 5) with the risk of predation (Lima & Dill 1990; Kats & Dill 1998; Lima 1998). Learning is the mechanism by which prey can achieve this important outcome because it allows prey to fine-tune their antipredator responses to variations in predation risk that can occur seasonally, across lunar cycles and on a moment-to-moment basis (Lima & Bednekoff 1999). Through learned predator recognition, prey can respond to novel introduced predators (Kristensen & Closs 2004), changes in community structure (Pollock & Chivers 2004), or learn a response to predators that were previously extinct in the local area (Berger *et al.* 2001).

It was previously thought that there could be little role for learning in the defence against predators because a failure to respond appropriately during the first encounter would result in death. However, predator–prey interactions are far more complex than this scenario implies and there are many ways in which prey can learn about predators while being exposed to relatively low levels of risk. This is particularly the case for remote cues such as odour because prey can obtain information about the predator even when it is not in the vicinity (Smith 1997; Chivers & Smith 1998; Kats & Dill 1998; and see Chapter 4). There is also the potential for prey to acquire information at low risk by observing conspecifics or heterospecifics being attacked, or by observing other prey responding to predator-related cues (i.e. through social learning; see Chapter 10). Predation risk is variable during the predator–prey interaction (see below) and prey must display an antipredator response that reflects the magnitude of the risk posed ('risk-sensitive predator avoidance'; Helfman 1989). This is because prey need to balance their risk of predation against other activities that influence fitness, such as feeding, courtship and habitat use (e.g. Sih 1980, 1988). This balance can be achieved if prey display an antipredator response that is appropriate to the magnitude of the threat, such that greater threats produce stronger avoidance responses (Helfman 1989). Prey that do not show a strong avoidance response have an increased risk of mortality, whereas displaying an overly cautious antipredator response results in a loss of time and energy available for other important activities.

Most studies of predator–prey interactions have addressed the adaptations of prey in their defence against predators. This follows a general trend in behavioural ecology in which the behaviour of predators in the predator–prey interaction appears to have been largely ignored (Lima 2002). In line with this tendency, the majority of studies that have addressed learning in the context of predator–prey interactions have examined the ways in which prey modify their antipredator behaviour as a result of experience with predators. In contrast, few investigators have examined the importance of learning in shaping predator hunting behaviour. Predator learning is a key assumption behind mathematical models of the evolution of aposematism (Speed 2001) but surprisingly few studies have addressed the cognitive abilities of predators that are predicted by these models. Finally, predator learning is an important means by which predators can counteract the behavioural plasticity of their prey and may therefore be a crucial weapon in the predator–prey arms race.

In this chapter, we evaluate the evidence for learning by both predators and prey at each of the five stages of the predator–prey interaction (Fig. 3.1). At each stage of the sequence, predator attack and prey defence may be initiated in response to behavioural, chemical and morphological cues that are associated with the presence of predators and prey. Most experimental studies of predator–prey interactions have examined these cues in isolation and, as a result, many of the examples given in this chapter are based on a particular cue, usually those that are visual or chemical. We consider learning processes that arise through both these stimuli, but the reader is also referred to Chapter 4 for a detailed discussion of chemically mediated learning. This chapter will largely focus on behavioural interactions between predators and prey; however, because it is often the interaction between behaviour and morphology that determines the success of antipredator defences or predator attack strategies,

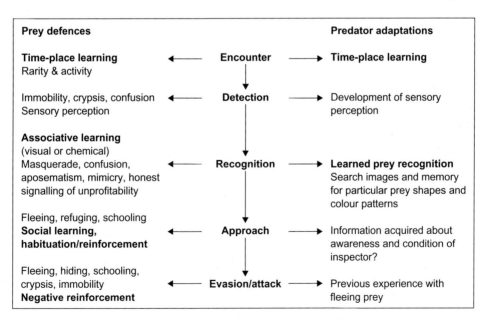

Fig. 3.1 Stages of the predation sequence and opportunities for learning by both predators and prey (adapted from Endler 1986; Lima & Dill 1990; Endler 1991).

morphological adaptations are included in cases where learning is considered to be particularly relevant.

3.2   The predator–prey sequence

The interaction between a predator and its prey can be considered as a sequence of events that begins when predator and prey encounter one another and ends when the prey escapes or is consumed by the predator. As the sequence progresses, individual prey are exposed to increasing levels of predation risk (Endler 1986, 1991). The energetic costs of avoiding predation increase as the interaction progresses and the antipredator behaviours that are performed become less frequent and more specialized (Endler 1986, 1991); for example, evasive manoeuvres such as short bursts of high speed swimming are far more specialized and energetically demanding than avoidance behaviours that reduce the probability of predator encounter (Fuiman & Magurran 1994). However, avoiding risky habitats could be costly in the long term, particularly if alternative habitats are suboptimal or unsuitable.

As in the case of prey, the counterdefences of predators tend to be less frequent, increasingly specialized and more costly to perform as the sequence progresses (Endler 1991); for example, predators tend to use the same sensory and cognitive processes for detecting and recognizing many different prey species (Endler 1991). However, to overcome prey defences at the later stages of the sequence, predators often have costly and specialized adaptations, such as fast attack speeds for capturing their prey and long teeth and venom for subduing them (Endler 1991). At every stage of the interaction, the behavioural defences of prey and the counteradaptations of predators are in conflict. Prey defences are deployed to allow the prey to escape the interaction as early as possible, whereas predator counterdefences aim to continue the interaction until it ends in prey capture (Endler 1991).

Opportunities for learning can arise at any one of the five stages of the predation sequence. However, we predict that learning is likely to play a greater role during the earlier stages of the sequence than during the later stages. This is based on the assumption that behaviours initiated at these early stages are more generalized and less energetically costly to perform than those used at the later stages (see above, Endler 1991). The frequency at which predators and prey experience each stage of the sequence is also important. Prey experience the earlier stages of the interaction more often, for example when encountering a cue from a predator, than at the later stages of the sequence, such as being chased by a predator. Learning in the latter stages will also incur considerable risks to individual prey because these stages of the interaction are associated with higher levels of predation risk; for example, the predation risk of learning to associate a particular habitat with danger is less than learning an appropriate escape response to a pursuing predator. Below, we consider examples of learning by both predators and prey at each of the five stages of the predator–prey sequence.

3.2.1   *Avoidance*

One of the most effective ways for prey to reduce their risk of predation is to adopt behaviours that reduce their probability of encountering a predator. This can be achieved in a variety of ways such as avoiding habitats that are associated with elevated risk and reducing or avoiding activity during the time periods in which predators are active (Lima & Dill 1990). The comprehensive reviews provided by Lima & Dill (1990) and Lima (1998) provide many examples of behavioural decisions that reduce a prey's chance of encountering a predator; for example, the decision of where to feed and for how long is an important one when there is spatial and temporal variability in predator activity (Lima & Bednekoff 1999). In the following sections, we consider studies that have examined whether prey can learn to avoid dangerous habitats and whether predators and prey adjust their activity patterns in order to increase their hunting efficiency or reduce their risk of predation.

3.2.1.1 Avoiding dangerous habitats

Several studies have demonstrated that fishes can learn to avoid dangerous foraging patches following an encounter with a predator; for example, Huntingford & Wright (1992) found that three-spined sticklebacks (*Gasterosteus aculeatus*, Gasterosteidae) learned to avoid a feeding patch that they previously favoured following a simulated predatory attack. Utne-Palm (2001) found that naive two-spotted gobies (*Gobiusculus flavescens*, Gobiidae) subsequently avoided the habitat where they last saw a live cod predator (*Gadus morhua*, Gadidae), but avoided cod odour only after observing cod on three successive occasions. Brown (Brown C. 2003) similarly found that rainbowfish (*Melanotaenia* spp., Melanotaeniidae) avoided habitats where they had previously encountered a model of their natural predator, the mouth almighty (*Glossamia aprion*, Apogonidae). In Brown's experiment (Brown C. 2003), the observation arena was rotated through 90° following predator exposure, suggesting that the fish remembered the location where the predator model appeared using features of the habitat rather than global cues (outside the test arena). Gobies (*Bathygobius soporator*, Gobiidae) living in tidal rock pools also use spatial learning to jump between pools at low tide and avoid contact with fish predators (Aronson 1971).

Studies with fathead minnows (*Pimephales promelas*, Cyprinidae) demonstrate that fish can also learn to recognize dangerous habitats by associating chemical odours from that habitat with 'damage-released alarm cues' (Chivers & Smith 1995a, 1995b). Damage-released alarm cues (hereafter referred to as alarm cues), are chemical cues in the epidermis of the skin that are released if the fish is injured or captured by a predator (von Frisch, cited in Pfeiffer 1974; Smith 1992, 1997; Brown G.E. 2003 and Chapter 4). Detection of alarm cues by conspecifics (or heterospecifics) elicits an unlearned 'fright reaction', characterized by freezing, dashing, hiding, shoaling and/or reduced foraging/mating activity (Chivers & Smith 1998). A large number of studies, particularly those associated with learned predator recognition (see 3.2.3) have provided evidence for associative learning through the association of alarm cues with previously novel stimuli. Fathead minnows were presented with water from either an open habitat or an area with vegetated cover (in the same stream) paired with either alarm cues or distilled water (control). The fathead minnows displayed a learned response to both types of water that were previously presented in conjunction with alarm cues, but not to water that was not paired with alarm cues (Chivers & Smith 1995a, 1995b). Chivers & Smith (1995a) also showed that a learned response to water from a particular habitat can be socially transmitted to naive observers. Collectively these studies demonstrate that spatial memory allows prey to associate visual features of a habitat with increased risk, whereas associative learning is a mechanism by which fish can learn to respond to chemical cues from dangerous habitats.

3.2.1.2   Changing activity patterns

The activity patterns of both marine (reviewed by Neilson & Perry 1990) and freshwater fishes (reviewed by Reebs 2002) are highly plastic and migrations of reef fishes are almost certainly driven by trade-offs between foraging opportunities and risk of predation (Smith 1997); for example, Helfman (1986) found that the timing of migration of juvenile grunts (*Haemulon flavolineatum*, Haemulidae) was delayed following a simulated increase in attack rate by a model predator, the lizardfish (*Synodus intermedus*, Synodontidae). Subsequent studies have demonstrated that social learning allows French grunts (*Haemulon flavolineatum*, Haemulidae) and bluehead wrasse (*Thalassoma bifasciatum*, Labridae) to maintain consistent migration routes over several generations (Helfman & Schultz 1984; Warner 1988; Brown & Laland 2003). Odling-Smee *et al.* provide more information on learned migratory patterns in Chapter 7.

Pettersson *et al.* (2000) showed that crucian carp (*Carassius carassius*, Cyprinidae) that were exposed to water containing a predatory pike exhibited low levels of night time activity, whereas carp exposed to untreated water were nocturnal. The guppy (*Poecilia reticulata*, Poeciliidae), a species that was previously considered to be diurnal (Houde 1997), readily forages at night under release from its nocturnal predator the trahira (*Hoplias malabaricus*, Erythrinidae; Fraser *et al.* 2004). Guppies denied the opportunity of night feeding exhibit low growth rates and reduced daytime courtship intensities relative to their nocturnally foraging counterparts (Fraser *et al.* 2004), illustrating the significance of predator activity on guppy behaviour and life history.

We are only aware of one study that has specifically examined whether prey fishes learn to adjust the timing of their activities in response to predation risk. In this study (Reebs 1999), shoals of inangas (*Galaxias maculates*, Galaxiidae) were presented with either food, a simulated predation threat (a model of a heron's bill), or both treatments. The stimuli were presented twice in the morning in one half of the tank and twice in the afternoon in the other half for a period of 14 days. After this time the stimuli were removed and the activity patterns of the shoals measured. Fish that had received the food treatment continued to anticipate the time and place at which food was delivered, but fish in the predation treatments and in the combined treatments did not display a time-place association (Reebs 1999).

In contrast, a large number of studies have shown that fish predators adjust their activity patterns according to changes in the behaviour of their prey. A well known example of this is the migration of fishes in response to diurnal migrations of their plankton prey. Zooplankton descend into the water column during the day in order to avoid visual-hunting predators. When the risk of predation is lower at night, the zooplankton migrate to the nutrient-rich surface waters. Predator odour has a direct effect on the swimming behaviour of daphnia by increasing the proportion of individuals that perform the migration pattern (Dodson *et al.* 1997). It is most likely that diel patterns of activity are a result of natural selection rather than arising through experience with migrating predators and prey.

Although the rate at which predators encounter prey is influenced by prey abundance, predator learning and experience also plays an important part in this process (Endler 1991). Predator hunting tactics are based on the optimal search rate hypothesis, which states that predators should spend more time searching for prey in patches that have high prey densities and less time in patches in which prey are rare (Gendron & Staddon 1983). The ability of predators to adapt their foraging behaviour according to the availability of different prey types and features of the habitat is discussed under 'recognition' (section 3.2.3).

3.2.2 Detection

One of the most important ways in which prey can avoid being detected by a predator is through being cryptic and matching their colouration with that of their background environment (Edmunds 1974). Although crypsis is usually considered in terms of colouration, prey chemical, auditory and electrical signals can also be difficult for predators to detect at a level above background 'noise' (Smith 1997). Most research on predator–prey interactions in fishes has focused on visual crypsis. The diverse body shapes of species such as pipefishes (*Syngathus* spp., Syngnathidae), the barbeled leaf fish (*Monocirrhus polyacanthus*, Polycentridae), the seadragon (*Phyllopteryx eques*, Syngnathidae) and the frogfish (*Antennarius marmaoratus*, Antennariidae) allow them to be virtually indistinguishable from their surroundings (Keenleyside 1979). Behaviour is an essential component of crypsis and fishes such as the barbeled leaf fish (*M. polyacanthus*) hold themselves vertical and sway back and forth in eel grass (Keenleyside 1979). Because crypsis is a function of the environment, factors that influence habitat selection and the time spent in that habitat will directly affect how conspicuous an animal is in a given environment.

3.2.2.1 Crypsis

Flatfishes (Pleuronectiformes) provide a particularly good example of crypsis as they are able to change their dorsal colouration so that it matches that of the substratum. Burying behaviour also enhances crypsis: sole (*Solea solea*, Soleidae) that are buried in sand react to a predation stimulus at a shorter distance (hence relying on crypsis) than those that are not buried (Ellis *et al.* 1997). The low post-release survival of hatchery-reared flatfishes has been attributed to high predation mortality through poor crypsis, suggesting that some aspect of substratum matching requires experience (Howell 1994). Although flatfish with no experience of burying quickly dig themselves into the substratum when given the opportunity, sole (Ellis *et al.* 1997) and winter flounder (*Pseudopleuronectes americanus*, Pleuronectidae; Fairchild & Howell 2004) with previous experience in a sandy substratum are more efficient buriers (measured as the proportion of sand covering their dorsal surface) than those reared in hard bottomed tanks. In the case of flounder, 5 days of contact with a sandy substratum were required for the burying behaviour of fish reared in hard bottomed tanks to resemble that of fish reared in sandy bottomed tanks (Fairchild & Howell 2004). The ability to change colour in both these species is also influenced by the environment in which fish are reared; the colour (lightness, intensity and hue) of hatchery flatfish reared in hard bottomed tanks took up to 69 days, in the case of sole, to resemble that of wild fish (Ellis *et al.* 1997), and over 90 days for winter flounder (Fairchild & Howell 2004). Although colour change is largely a physiological process rather than a behavioural one, it is the interaction between behaviour and morphology that determines the effectiveness of crypsis (Cott 1940; Edmunds 1974).

3.2.2.2   Sensory perception

Predators are far more successful at capturing prey when the prey animal is unaware of the predator's presence (Dugatkin & Godin 1992). As a result, there is strong selection on predators for rapid detection of prey and vice versa. Movement is the primary means through which predators and prey detect one another (Kislalioglu & Gibson 1976). When detecting the cues from a predator, prey fishes commonly increase schooling, freeze, sink lower into the water column and/or hide in an attempt to avoid being detected. The type of response adopted may depend on the habitat the fish originate from; for example, rainbowfish (*Melanotaenia eachamensis*, Melanotaeniidae) that coexist with predators in an open lake habitat increase their schooling but do not seek refuge when exposed to visual predatory cues (Brown & Warburton 1997). In contrast, rainbowfish from a more structured but predator-free environment were more likely to hide under vegetation when exposed to a predator model (Brown & Warburton 1997).

Exposure of prey to alarm cues results in freezing and/or hiding even in the absence of direct predator cues (Suboski *et al.* 1990; Hall & Suboski 1995; Yunker *et al.* 1999). While also functioning in learned predator recognition and avoidance of dangerous habitats (see Avoidance and Recognition, sections 3.2.1 and 3.2.3, respectively), alarm cues allow prey to respond rapidly to cues that indicate that predators are in the vicinity, allowing them to adopt behaviours that reduce their chance of being detected at an early stage of the encounter. For both predators and prey, detection is a function of the distance over which sensory perception operates and is dependent on the environment. Hartman & Abrahams' sensory compensation model (2000) predicts that fish rely more on alternative cues when the primary source of information is reduced. The relative importance of different sensory cues will fluctuate with both spatial and temporal changes in the environment; in turbid habitats, for instance, both predators and prey are likely to rely disproportionately on non-visual sensory systems such as the lateral line, electroreception and olfaction (Hartman & Abrahams 2000).

3.2.3   Recognition

3.2.3.1   Associative learning

The best known mechanism through which fish learn to recognize predators is associative learning, or releaser-induced recognition learning (Suboski 1990). This learning process is comparable to Pavlovian conditioning and occurs when naive individuals acquire a response to a predator cue (the conditional stimulus) by associating it with an alarm cue (the unconditioned stimulus) (Suboski 1990). It was previously thought that only fishes belonging to the Ostariophysian family display an unlearned response to alarm cues, however, the phenomenon is now known to be widespread among fishes (Chivers & Smith 1998). Although most examples of associative learning have paired alarm cues with predator odour, prey can also be conditioned to respond to visual cues from predators (Chivers & Smith 1994). Furthermore, conditioning cues need not be biologically relevant, as zebra danios (*Brachydanio rerio*, Cyprinidae) and fathead minnows can learn a response to unnatural odours (Suboski *et al.* 1990) and flashing lights (Hall & Suboski 1995; Yunker *et al.* 1999).

In most cases, associative learning occurs after just one simultaneous presentation of the cue and the stimulus (Magurran 1989) and the response can be retained for up to 2 months (Chivers & Smith 1994). If a single exposure to a chemical cue can lead to marked and long-lasting changes in antipredator behaviour, what prevents fishes from learning a response to irrelevant stimuli? Responding to non-biological or irrelevant stimuli would entail significant fitness costs and reduced time available for other activities (Lima & Dill 1990). There is some evidence that fish are predisposed to acquire responses to moving objects (Brown & Warburton 1997; Vogel & Bleckmann 2001; Wisenden & Harter 2001) and to particular predator cues (i.e. 'learning specificity'; Griffin *et al.* 2001); for example, fathead minnows learn to fear the sight of both pike and goldfish, but when retested 2 months later the response to the pike had remained unchanged, whereas the response to the goldfish had diminished (Chivers & Smith 1994). European minnows (*Phoxinus phoxinus*, Cyprinidae), which had acquired a response to the odour of both their natural predator (pike) and a non-piscivorous cichlid, showed a stronger response towards the pike (spent more time schooling and less time foraging (Magurran 1989). Prey may be predisposed to learn a response to generalized cues that indicate a fish is likely to be predatory.

3.2.3.2   Learning specificity

Karplus & Algom (1981) tested the idea of predisposed antipredator responses towards visual cues by conducting a detailed morphometric analysis of the facial features and feeding habitats of 105 species of reef fishes. Although a large number of facial features were associated with a piscivorous habit, fishes with large mouths and eyes that are widely spaced tended to be predatory. Karplus *et al.* (1982) then tested this experimentally, confirming that chromis (*Chromis caeruleus*, Pomacentridae) showed a stronger response to models with these facial characteristics than they did to models with non-predatory features. Similar generalized visual features of predators may allow prey to learn a response to piscivorous species more rapidly than non-predatory species. Csanyi & Doka (1993) tested this idea by presenting paradise fish (*Macropodus opercularis*, Anabantidae) with either a live goldfish or fish models of varying realism in conjunction with an electric shock. Paradise fish showed the strongest learned response towards the goldfish and were more likely to learn a response towards fish models with lateral eye-like spots than models with only one eye or no eyes (Csanyi & Doka 1993).

The predisposition of prey to learn from particular cues in the predator's odour can be explained in terms of the predator's diet (Mathis & Smith 1993a,b; Vilhunen & Hirvonen 2003), and, in particular, of the faeces (Brown *et al.* 1995a,b). This mechanism of learning not only prevents maladaptive conditioning, but also allows the prey the opportunity to learn to respond to the alarm cues of heterospecifics (Chivers & Smith 1995b; Mirza & Chivers 2001; Chivers *et al.* 2002); for instance, Mirza & Chivers (2001) found that fathead minnows learned to respond to brook stickleback (*Culaea inconstans*, Gasterosteidae) through association with conspecific alarm cues in the diet of the predator. Minnows showed a stronger response towards predatory perch that had been fed minnows and sticklebacks than to perch that had been fed swordtails (lacking in alarm cues) or distilled water.

3.2.3.3   Search images

An animal's brain can only process a limited amount of information; therefore predators are predicted to search for only one prey type at any given time, a 'search image' (Tinbergen 1960; Dukas 1998). The development of a search image depends on the predator's experience with particular prey types. Predators develop a search image based on the prey that is encountered most frequently and once this prey type becomes less abundant or less profitable, a new search image will be developed (Dukas 1998). Animals that are able to rapidly adjust their foraging behaviour in response to changes in prey availability will have high fitness.

Ehlinger (1989) investigated the influence of experience on the foraging behaviour of bluegill sunfish (*Lepomis macrochirus*, Centrarchidae) in both open water and vegetated habitats. In both habitat types sunfish adapted their foraging behaviour according to the conspicuousness of the prey. Sunfish that foraged in the open water habitat learned to increase their searching speed, whereas those in the vegetated habitat where prey were more cryptic learned to search more slowly (Ehlinger 1989). If search images for a given prey morphology are acquired more easily and retained in the memory for longer, then selection will be stronger on memorable prey types than on those that are readily forgotten (Endler 1991). In this way, we can envisage how experience with particular prey morphs (which is dependent on the environment, as illustrated by the above example) creates variation in predation pressure that is largely influenced by the cognitive ability of predators (see Chapter 2 for further discussion).

3.2.3.4   Aposematism and mimicry

Prey that are aposematic rely on their conspicuousness to warn potential predators of their unpalatability. Theoretic models suggest that in order for aposematism to evolve, predators must rapidly learn to associate the noxiousness of the prey with its conspicuous colour patterns (Speed 2001; Ruxton *et al.* 2004). If conspicuous colouration is to act as a warning signal, it is predicted that predators learn to associate noxiousness with a conspicuous colour pattern more rapidly than with a cryptic colour pattern (Guildford 1990). There is some evidence for this in birds; chicks learn to avoid conspicuous baits more rapidly than they learn to avoid cryptic distasteful baits (Roper & Wistow 1986; Roper & Redston 1987). Interestingly, aposematism appears to be uncommon in fishes. This is perhaps because less than 3% of teleosts (the majority of which are marine) possess toxic or noxious chemicals in the skin, spines or viscera (Godin 1997). Consequently we know of no studies with fishes that have examined whether predators can learn to associate conspicuous prey morphs with noxiousness.

Batesian mimicry, in which a palatable prey imitates an unpalatable species, relies on predators associating both the mimic and the model species with unpalatability. In fishes, the colour patterns of the leatherjacket (*Paraluteres prionurus*, Monacanthidae) closely resemble those of its putative Batesian mimic, a toxic pufferfish species (*Canthigaster valentine*, Tetraodontidae). In a study designed to investigate the effectiveness of mimicry in relation to the degree of resemblance in colour pattern between the pufferfish species and its mimic, Caley & Schluter (2003) showed that painted model replicas that most resembled the pufferfish received fewer visits from piscivorous fish. It would be interesting to discover whether the avoidance of the colour patterns in the toxic pufferfish is learned or innate. Considering the recent interest in predator learning and memory, it is surprising that few studies with fishes have addressed the role of learning in this context.

3.2.4 *Approach*

3.2.4.1 *Pursuit deterrence*

Fish that detect and recognise a predator frequently perform what is referred to as inspection behaviour, where a small group of fish leave the shoal and slowly approach the predator, often swimming slowly along the length of its body, before returning to the shoal (Pitcher *et al.* 1986). One function of this apparently paradoxical behaviour may be pursuit deterrence (also known as attack inhibition) in which the approach 'informs' the predator that it has been detected (Magurran 1990; Dugatkin & Godin 1992; Godin & Davis 1995; Brown *et al.* 1999). Fin flicking in glowlight tetras (*Hemigrammus erythrozonus*, Characidae; Brown *et al.* 1999) and head bobbing in gobies (*Asterropteryx semipunctatus* and *Gnatholepis anjerensis*, Gobiidae; Smith & Smith 1989) have been proposed to serve a similar function. Predators are more successful at attacking prey that may be less vigilant (e.g. when foraging; Krause & Godin 1996) and consequently the predator may redirect its attention towards fish that have not signalled their awareness (Pitcher & Parrish 1993). Magurran (1990) and Dugatkin & Godin (1992) provide evidence that predators are less likely to attack fish that are inspecting than those that are not inspecting. It has previously been suggested that approach behaviour (and behaviours that are associated with predator detection, e.g. head bobbing and fin flicking) might serve as warning or 'alarm signals' which alert nearby conspecifics of the location of the predator and its potential danger (Smith & Smith 1989). However, this concept is difficult to explain in terms of the benefits accrued by the signal sender, in this case the inspector (reviewed by Smith 1986). A further proposed function of predator inspection behaviour is the preferential selection of bolder males by females (Godin & Dugatkin 1996).

3.2.4.2 *Gaining information about the predator*

Another function of inspection behaviour is that is allows the inspectors to gain information about the predator, such as its condition and motivation to attack. This information may be more reliable than that gained 'secondhand' by observing the inspection behaviour of other individuals (see below). Several studies have suggested that the behaviour of inspectors is changed as a result of inspection, suggesting that information has been acquired about the predator's attack motivation (Pitcher *et al.* 1986; Magurran & Higham 1988; Pitcher 1992). The predator's posture affects reaction distance in three-spot damselfish (*Stegastes planifrons*, Pomacentridae; Helfman 1989) and allows inspecting guppies to differentiate between hungry and satiated predators (Licht 1989). Glowlight tetras and finescale dace (*Phoxinus neogaeus*, Cyprinidae) are more reluctant to inspect and do so from further away and in smaller groups if they detect alarm cues in the odour of the predator (Brown & Godin 1999; Brown *et al.* 2000; Brown *et al.* 2001; Brown & Schwarzbauer 2001; Brown & Dreier 2002). When tetras were subsequently presented with only visual cues from a cichlid, only those tetras that had previously been exposed to the predator and its dietary cues displayed a response (Brown & Godin 1999). 'Attack cone avoidance', whereby inspecting fish tend to avoid the dangerous mouth region of the predator (George 1960), is also mediated by the combined effects of experience and predator diet. Predator-naive tetras directed more inspections towards the tail than the head of a cichlid predator when it was associated with dietary odours that contained alarm cues (Brown & Schwarzbauer 2001; Brown & Dreier 2002). In contrast, tetras that had previously been conditioned to the visual cues from that predator (through alarm cues) displayed a response irrespective of the diet of the predator (Brown & Dreier 2002). Collectively, these studies demonstrate that predator inspection behaviour is mediated both by the effects of predator diet and previous experience (Brown & Godin 1999), even in the absence of visual cues (Brown *et al.* 2000).

3.2.4.3   Social learning

Fish that do not take part in the inspection change their behaviour when the inspector(s) return, suggesting that information is socially transmitted to the shoal (Pitcher *et al.* 1986). Minnows that could not see a pike model reduced their level of activity after observing the 'skittering' behaviour (a startle response) of inspecting fish (Magurran & Higham 1988). Social learning occurs when naive individuals modify their behaviour after observing conspecifics responding fearfully to a particular stimulus, and allows individuals to acquire information about the predator without incurring the potential costs (increased predation risk) associated with gaining the information independently (Box 1984). Note that opportunities for learning from conspecifics (or heterospecifics, see below) could arise at other stages of the interaction, such as during predator detection. However, we discuss it here because of its demonstrated relevance to inspection behaviour. Changes in the behaviour of the inspectors (including signals that may act as pursuit deterrents) could provide clues to the non-inspectors regarding the level of threat posed by the predator. This would parallel the situation in which successful foragers inadvertently disclose information about good foraging sites through changes in their behaviour (Pitcher *et al.* 1982; Pitcher & Parrish 1993). This assumes that the behaviour of the inspector (either during inspection or when returning to the shoal) is a reliable indicator of the behaviour and motivation of the predator. Because inspection behaviour is a risk-sensitive behaviour, balancing the potential risk of being captured against the benefits gained through information acquisition (Murphy & Pitcher 1991), this seems quite likely.

Suboski *et al.* (1990) were the first to demonstrate that conspecific alarm cues can be socially transmitted among naive conspecifics (Brown & Laland 2003). In their study, zebra (*Danio rerio, Cyprinidae*) were conditioned to respond to morpholine (an artificial odour) by presenting it in conjunction with conspecific alarm cues. Naive observers displayed an alarm response after observing (through a clear barrier) conditioned fish responding to morpholine and subsequently retained this response when later tested alone (Suboski *et al.* 1990). Mathis *et al.* (1996) similarly showed that naive fathead minnows can learn to recognize a novel predator odour by observing conspecifics displaying a conditioned fright response. Alarm reactions can also be transmitted among heterospecifics. In Mathis *et al.*'s study (1996), brook stickleback, a species that forms mixed shoals with fathead minnows, acquired a fright reaction after observing minnows displaying a conditioned response. Krause (1993) also reported transmission of information among heterospecifics by demonstrating that sticklebacks exhibited a fright response after observing chub (*Leuciscus cephalus, Cyprinidae*) responding to alarm cues. In all of these experiments, fishes that had never been exposed to the alarm substance were able to learn a fright reaction to a novel odour. This suggests that the response acquired from detecting an alarm cue is similar to that learned by observing the conditioned fright response of conspecifics. It would be interesting to know whether these two sources of information (individual and social) are equivalent, and, specifically, which behaviours are involved in this process. It has been suggested that dashing movements (Chivers & Smith 1994) or the position of the fish relative to the substratum might be important (Griffin 2004). Social learning may partly account for the rapid acquisition of learned predator recognition when novel predators are introduced into a previously predator-free population. Following the introduction of pike into two previously pike-free populations (of approximately 20 000 and 78 000 fish), minnows learned to recognize pike odour within just 14 days (Chivers & Smith 1995c). For a full discussion of the role of social learning in fishes, see chapter 10.

3.2.4.4   Habituation

Importantly, inspection behaviour may play an important role in habituation, a type of learning in which there is a reduced response after repeated exposure to a stimulus (Shettleworth 1988). Inspectors often repeatedly approach the predator, providing an ongoing assessment of the motivation and likely risk posed by it (Pitcher & Parrish 1993). Inspecting fish acquiring information that suggests that a predator is not actively hunting may discontinue inspection if they perceive that there is little threat, or they may continue to inspect the predator but show a less cautious response during subsequent approaches. In this manner, habituation may play an important part in allowing prey to display threat-sensitive antipredator responses. Habituation is less likely to occur towards stimuli that prey fish are predisposed to show a fear response to, but experience also mediates the response. Csanyi (1985) found that paradise fish became habituated to both a goldfish and a satiated pike, but whereas inspection rate rapidly diminished towards the goldfish, paradise fish continued to approach the pike. Magurran & Girling (1986) presented European minnows with pike models that differed in realism. Their results confirmed that minnows showed a stronger response towards the most realistic model and that they habituated most rapidly to the least realistic model.

If predator inspection does indeed function as a pursuit deterrence signal, then predators are also acquiring information during this interaction. This is an intriguing possibility that has so far been little considered. Krause & Godin (1996) reported that subtle differences in prey foraging posture influence the attack success of predators. The predatory cichlid, the blue acara (*Aequidens pulcher*, Cichlidae) preferred to attack foraging guppies rather than non-foraging ones, and guppies that foraged in a 'nose-down' position rather than horizontally (Krause & Godin 1996). This is probably because of the reduced vigilance of foraging prey in a head-down posture. Prey that are foraging or in nose down postures may also have difficulty performing a fast start or c-start escape response, although this explanation is unlikely in this case as the experimental design controlled for different positional effects (Krause & Godin 1996). It would be very interesting to investigate whether the recognition of these and other vigilance cues is based on previous hunting experience.

3.2.5   *Evasion*

Prey fish that are under imminent attack from a predator may either flee, freeze or hide in an attempt to avoid being captured (Edmunds 1974). Acquisition of the latter two responses has been discussed previously in the context of avoiding detection and learned predator recognition. We know of no studies that demonstrate that previous experience with an attacking predator enhances the freezing or hiding responses of prey. This is probably because these responses are used in combination with morphological defences or when the predator is very close and escape seems unlikely (Godin 1997). Fleeing is the more common response to an approaching predator (Edmunds 1974) and several studies demonstrate that the timing, swimming speed and trajectory of the flight response can be improved through experience.

3.2.5.1   *Reactive distance and escape speed*

Dill (1974) found that the reactive distance of zebra danios, the distance between the prey and the moving stimulus when the prey displays an escape response, is increased through repeated exposure to a film of an approaching object or a model predator. Escape velocity also increased with experience of the filmed predator, but not with experience of the predator model (Dill 1974). Huntingford *et al.* (1994) demonstrated that the escape response of three-spined stickleback fry is contingent on both their population of origin and their previous experience. If stickleback fry stray from their nest, the father chases them and carries them back to the nest in his mouth. Huntingford *et al.* (1994) found that the escape speed of fry and the retrieval speed of fathers were highly correlated but that the speeds were significantly greater for fry from high risk populations than low risk ones. Fleeing fish often display a zigzag swimming trajectory (Edmunds 1974), and the angle at which they initially flee and the number of turns executed can affect the success of evading capture (Godin 1997). When exposed to a moving model of a piscivorous fish, stickleback fry from the high risk population were more likely to flee at an angle from the predator than fish from the low risk population (Huntingford *et al.* 1994).

3.2.5.2   Survival benefits

A large number of studies have demonstrated that prior experience with live predators increases their subsequent chance of survival in a predator encounter. Patten (1977) found that the predation mortality of coho salmon (*Oncorhynchus kisutch, Salmonidae*) was lower when fish had previous experience with predatory sculpin or when they had contact with predator-experienced salmon than that for predator-naive salmon. Similar findings were reported by Olla & Davis (1989). Mathis & Smith (1993c) were the first to demonstrate that learned recognition of predators through conditioning with alarm cues confers a survival advantage. Fathead minnows that were conditioned with pike odour and alarm cues survived significantly longer in subsequent trials with live pike. In their study, Mirza and Chivers (2000) found positive correlations between shelter use and survival time and between shoaling behaviour and survival time, suggesting that these behaviours were particularly important (Mathis & Smith 1993c). Parallel results have been presented for brook trout (*Salvelinus fontinalis*, Salmonidae) exposed to live predatory chain pickerel (*Esox niger*; Esocidae); trout conditioned with alarm cues had a higher survival rate than those conditioned with odour not containing alarm cues (Mirza & Chivers 2000). Visual cues can also contribute to learned avoidance behaviour. Berejikian (1995) showed that fry of hatchery-reared steelhead trout (*Oncorhynchus mykiss*, Salmonidae) suffered higher predation mortality than their wild counterparts. After observing sculpin prey on trout through a clear barrier, both wild and hatchery fish showed improved survival skills but wild fish survived for longer. Goodey & Liley (1986) showed that the experience of guppies of being chased by conspecifics when young led to a survival advantage in predator encounters as adults.

There are few studies investigating how predators learn to capture their prey. Domencini & Blake (1993) have suggested that the reason why there is high intraspecific variation in the zigzag trajectories of fleeing fish is to prevent predators from learning and anticipating prey escape movements. This interesting idea is yet to be explored. There are also suggestions in the literature that the hunting behaviour of predators is enhanced through social mechanisms; perch were more successful foraging as a shoal than when foraging alone (Eklov 1992). Again, these ideas remain to be tested.

3.3   Summary and discussion

Predator–prey interactions are complex and learning about predators does not necessarily have to be a life or death experience. There are a variety of mechanisms that allow prey to learn about their predators, some of which are surprisingly subtle. Our understanding of the role of chemical cues, particularly alarm cues, is far ahead of that in other sensory domains and in many cases, the mechanisms by which learning is achieved remain unresolved. The study of animal cognition is still relatively young and has developed through two distinct research disciplines:

psychology and ecology (Kamil 1998). These two perspectives still require integration and we know little of the importance of learning specificity, memory constraints, habituation and reinforcement in shaping the behaviour of predators and prey in the wild.

One of the most striking things to arise from our review of the literature is the lack of research on the learning abilities of predators. In particular, the idea that many of the bright, conspicuous colour patterns of prey have evolved as a result of the memory and learning capabilities of predators is a very exciting notion that requires further investigation. Although predator psychology has been considered in the context of optimal foraging, these investigations often overlook factors affecting prey conspicuousness such as movement and colouration. Experiments are needed to test the learning and memory capability of predators for particular prey colour patterns and movements against a variety of backgrounds (Endler 1986). The use of computer animations is becoming increasingly popular in behavioural ecology and would be one way in which to disentangle the multiple effects of behaviour, colouration, background and movement.

We have suggested that opportunities for learning are likely to be greater at earlier stages of the predator–prey interaction, when the predation risk to individual prey is relatively low and antipredator behaviours are less costly of energy than at the later stages. This review of the literature has revealed examples of learning at every stage of the sequence with no obvious bias of studies towards any particular stage. However, it would be relatively simple to design a set of experiments to test the learning ability of prey at each stage of the sequence. Although there is some evidence for learned avoidance of dangerous habitats, we know little about the influence of predator activity patterns on prey behaviour. Prey can learn to recognize and respond to a novel predator but how do they alter their activity patterns as a result of previous encounters? Can predators learn to 'track' the behaviour of their prey? Studies of learning in predator–prey interactions tend to examine the learning ability of predator or prey in isolation. An exciting area of future research is to investigate how predator–prey interactions proceed when both predators and prey learn; for example, how quickly do predator and prey respond to one another's movements?

The reasons why prey approach predators is still a matter of controversy, despite over 40 years passing since the phenomenon was first described in fishes. In particular, the idea that predators acquire information as a result of the interaction deserves more attention. Furthermore, interactions between predators and prey can involve a diversity of species and more than just one predator and prey. Importantly, the combined effects of multiple predators can be very different to those arising from pairwise predator–prey interactions (reviewed by Sih *et al.* 1998), particularly when a prey's response to one predator increases its predation risk to another (Charnov *et al.* 1976). It would be interesting to investigate how multiple predators influence learning and whether prey can learn the identity of multiple predators simultaneously. It has recently been shown that glowlight tetras can learn to recognize multiple predators through the association of visual predatory cues with alarm cues (Darwish *et al.* 2005). Studies on learned responses to multiple predators will assist in conservation programmes that train captive-bred species to recognize predators (Darwish *et al.* 2005) and increase our understanding of the speed at which populations can respond to changes in species assemblages.

3.4 Acknowledgements

We wish to thank Peter Banks, Culum Brown and John Endler for their valuable comments on this chapter. J.L.K. is funded by a University of New South Wales Vice-Chancellor's Postdoctoral Research Fellowship.

3.5 References

Aronson, L.R. (1971) Further studies on orientation and jumping behaviour in the gobiid fish, *Bathygobius soporator*. *Annals of the New York Academy of Sciences*, **188**, 378–407.

Berejikian, B.A. (1995) The effects of hatchery and wild ancestry and experience on the relative ability of steelhead trout fry (*Oncorhynchus mykiss*) to avoid a benthic predator. *Canadian Journal of Fisheries and Aquatic Sciences*, **52**, 2476–2482.

Berger, J., Swenson, J.E. & Persson, I.-L. (2001) Recolonizing carnivores and naive prey: conservation lessons from Pleistocene extinctions. *Science*, **291**, 1036–1039.

Box, H.O. (1984) *Primate behavior and social ecology*. Chapman & Hall, London.

Brown, C. (2003) Habitat-predator association and avoidance in rainbowfish (*Melanotaenia* spp.). *Ecology of Freshwater Fish*, **12**, 118–126.

Brown, C. & Laland, K.N. (2003) Social learning in fishes: a review. *Fish and Fisheries*, **4**, 280–288.

Brown, C. & Warburton, K. (1997) Predator recognition and antipredator responses in the rainbowfish, *Melanotaenia eachamensis*. *Behavioural Ecology and Sociobiology*, **41**, 61–68.

Brown, G.E. (2003) Learning about danger: chemical alarm cues and local risk assessment in prey fishes. *Fish and Fisheries*, **4**, 227–234.

Brown, G.E. & Dreier, V.M. (2002) Predator inspection behaviour and attack cone avoidance in a characin fish: the effects of predator diet and prey experience. *Animal Behaviour*, **63**, 1175–1181.

Brown, G.E. & Godin, J.-G.J. (1999) Who dares, learns: chemical inspection behaviour and acquired predator recognition in a characin fish. *Animal Behaviour*, **57**, 475–481.

Brown, G.E. & Schwarzbauer, E.M. (2001) Chemical predator inspection and attack cone avoidance in a characin fish: the effects of predator diet. *Behaviour*, **138**, 727–739.

Brown, G.E., Chivers, D.P. & Smith, R.J.F. (1995a) Fathead minnows avoid conspecific and heterospecific alarm pheromone in the faeces of northern pike. *Journal of Fish Biology*, **47**, 387–393.

Brown, G.E., Chivers, D.P. & Smith, R.J.F. (1995b) Localized defecation by pike: a response to labelling by cyprinid alarm pheromone? *Behavioral Ecology and Sociobiology*, **36**, 105–110.

Brown, G.E., Godin, J.-G.J. & Pederson, J. (1999) Fin-flicking behaviour: a visual anti-predator alarm signal in a characin fish, *Hemigrammus erythrozonus*. *Animal Behaviour*, **58**, 469–475.

Brown, G.E., Paige, J.A. & Godin, J.-G.J. (2000) Chemically mediated predator inspection behaviour in the absence of predator visual cues by a characin fish. *Animal Behaviour*, **60**, 315–321.

Brown, G.E., Golub, J.L. & Plata, D.L. (2001) Attack cone avoidance during predator inspection visits by wild finescale dace (*Phoxinus neogaeus*): the effects of predator diet. *Journal of Chemical Ecology*, **27**, 1657–1666.

Caley, M.J. & Schluter, D. (2003) Predators favour mimicry in a tropical reef fish. *Proceedings of the Royal Society of London B*, **270**, 667–672.

Charnov, E.L., Orians, G.H. & Hyatt, K. (1976) Ecological implications of resource depression. *American Naturalist*, **110**, 247–259.

Chivers, D.P. & Smith, R.J.F. (1994) Fathead minnows, *Pimephales promelas*, acquire predator recognition when alarm substance is paired with the sight of an unfamiliar fish. *Animal Behaviour*, **48**, 597–605.

Chivers, D.P. & Smith, R.J.F. (1995a) Chemical recognition of risky habitats is culturally transmitted among fathead minnows, *Pimephales promelas* (Osteichthyes, Cyprinidae). *Ethology*, **99**, 286–296.

Chivers, D.P. & Smith, R.J.F. (1995b) Fathead minnows, *Pimephales promelas*, learn to recognise chemical stimuli from high risk habitats by the presence of alarm substance. *Behavioural Ecology*, **6**, 155–158.

Chivers, D.P. & Smith, R.J.F. (1995c) Free-living minnows rapidly learn to recognize pike as predators. *Journal of Fish Biology*, **46**, 949–954.

Chivers, D.P. & Smith, R.J.F. (1998) Chemical alarm signalling in aquatic predator–prey systems: a review and prospectus. *Ecoscience*, **5**, 338–352.

Chivers, D.P., Mirza, R.S. & Johnston, J.G. (2002) Learned recognition of heterospecific alarm cues enhances survival during encounters with predators. *Behaviour*, **139**, 929–938.

Cott, H.B. (1940) *Adaptive coloration in animals*. Methuen, London.

Csanyi, V. (1985) Ethological analysis of predator avoidance by the paradise fish (*Macropodus opercularis* L.) 1. Recognition and learning of predators. *Behaviour*, **92**, 227–240.

Csanyi, V. & Doka, A. (1993) Learning interactions between prey and predator fish. *Marine Behaviour and Physiology*, **23**, 63–78.

Darwish, T.L., Mirza, R.S., Leduc, A.O.H.C. & Brown, G.E. (2005) Acquired recognition of novel predator odour cocktails by juvenile glowlight tetras. *Animal Behaviour*, **70**, 83–89.

Dill, L.M. (1974) The escape response of the zebra danio (*Brachydanio rerio*) II. The effect of experience. *Animal Behaviour*, **22**, 723–730.

Dodson, S.I., Tollrian, R. & Lampert, W. (1997) Daphnia swimming behaviour and vertical migration. *Journal of Plankton Research*, **19**, 969–978.

Domencini, P. & Blake, R.W. (1993) Escape trajectories in angelfish. *Journal of Experimental Biology*, **177**, 253–272.

Dugatkin, L.A. & Godin, J.-G.J. (1992) Prey approaching predators: a cost-benefit perspective. *Annals Zoologica Fennici*, **29**, 233–252.

Dukas, R. (1998) Constraints on information processing and the effects on behaviour. In: R. Dukas (ed.), *Cognitive Ecology*, pp. 89–127. Chicago University Press, Chicago.

Edmunds, M. (1974) *Defense in animals: a survey of anti-predator defenses*. Longmans, London.

Ehlinger, T.J. (1989) Learning and individual variation in bluegill foraging: habitat-specific techniques. *Animal Behaviour*, **38**, 643–658.

Eklov, P. (1992) Group foraging versus solitary foraging efficiency in piscivorous predators: the perch, *Perca fluviatilis*, and pike, *Esox lucius*, patterns. *Animal Behaviour*, **44**, 313–326.

Ellis, T., Howell, B.R. & Hughes, R.N. (1997) The cryptic responses of hatchery-reared sole to a natural sand substratum. *Journal of Fish Biology*, **51**, 389–401.

Endler, J.A. (1986) Defense against predators. In: J.A. Endler (eds), *Predator–prey relationships: perspectives and approaches from the study of lower vertebrates*, pp. 109–134. University of Chicago Press, Chicago.

Endler, J.A. (1991) Interactions between predators and prey. In: J.R. Krebs & N.B. Davies (eds) *Behavioural ecology: an evolutionary approach*, p. 482. Blackwell Scientific Publications, Oxford.

Fairchild, E.A. & Howell, W.H. (2004) Factors affecting the post-release survival of cultured juvenile *Pseudopleuronectes americanus*. *Journal of Fish Biology*, **65**, 69–87.

Fraser, D.F., Gilliam, J.F., Akkara, J.T., Albanese, B.W. & Snider, S.B. (2004) Night feeding by guppies under predator release: effects on growth and daytime courtship. *Ecology*, **85**, 312–319.

Fuiman, L.A. & Magurran, A.E. (1994) Development of predator defences in fishes. *Reviews in Fish Biology and Fisheries*, **4**, 145–183.

Gendron, R.P. & Staddon, J.E.R. (1983) Searching for cryptic prey: the effect of search rate. *American Naturalist*, **121**, 172–186.

George, C.J.W. (1960) Behavioral interactions of the pickerel (*Esox niger* and *Esox americanus*) and the mosquitofish (*Gambusia patruelis*). PhD thesis. Harvard University.

Godin, J.-G.J. (1997) Evading predators. In: J.-G.J. Godin (ed.), *Behavioural ecology of teleost fishes*, pp. 191–236. Oxford University Press, Oxford.

Godin, J.-G.J. & Davis, S.A. (1995) Who dares, benefits: predator approach behaviour in the guppy (*Poecilia reticulata*) deters predator pursuit. *Proceedings of the Royal Society of London Series B*, **259**, 193–200.

Godin, J.-G.J. & Dugatkin, L.A. (1996) Female mating preferences for bold males in the guppy (*Poecilia reticulata*). *Proceedings of the National Academy of Sciences USA*, **93**, 10262–10267.

Goodey, W. & Liley, N.R. (1986) The influence of early experience in escape behaviour in the guppy (*Poecilia reticulata*). *Canadian Journal of Zoology*, **64**, 885–888.

Griffin, A.S. (2004) Social learning by predators: a review and prospectus. *Learning and Behavior*, **32**, 131–140.

Griffin, A.S., Evans, C.S. & Blumstein, D.T. (2001) Learning specificity in acquired predator recognition. *Animal Behaviour*, **62**, 577–589.

Guildford, T. (1990) The evolution of aposematism. In: D.L. Evans & J.O. Schmidt (eds), *Insect defense: adaptive mechanisms and strategies of prey and predators*, pp. 23–61. State University of New York Press, New York.

Hall, D. & Suboski, M.D. (1995) Visual and olfactory stimuli in learned release of alarm reactions by zebra danio fish (*Brachydanio rerio*). *Neurobiology of learning and memory*, **63**, 229–240.

Hartman, E.J. & Abrahams, M.V. (2000) Sensory compensation and the detection of predators: the interaction between chemical and visual information. *Proceedings of the Royal Society of London B*, **267**, 571–575.

Helfman, G.S. (1986) Behavioral response of prey fishes during predator–prey interactions. In: M.E. Feder & G.V. Lauder (eds), *Predator–prey relationships*, pp. 135–156. Chicago University Press, Chicago.

Helfman, G.S. (1989) Threat-sensitive predator avoidance in damselfish–trumpetfish interactions. *Behavioural Ecology and Sociobiology*, **24**, 47–58.

Helfman, G.S. & Schultz, E.T. (1984) Social transmission of behavioural traditions in a coral reef fish. *Animal Behaviour*, **32**, 379–384.

Houde, A.E. (1997) *Sex, colour, and mate choice in guppies*. Princeton University Press, Princeton, New Jersey.

Howell, B.R. (1994) Fitness of hatchery-reared fish for survival in the sea. *Aquaculture and Fisheries Management*, **25**, 3–17.

Huntingford, F.A. & Wright, P.J. (1992) Inherited population differences in avoidance conditioning in three-spined sticklebacks, *Gasterosteus aculeatus*. *Behaviour*, **122**, 264–273.

Huntingford, F.A., Wright, P.J. & Tierney, J.F. (1994) Adaptive variation in antipredator behaviour in threespine stickleback. In: M.A. Bell & S.A. Foster (eds), *The Evolutionary Biology of the Threespine Stickleback*, pp. 277–296. Oxford University Press, Oxford.

Kamil, A.C. (1998) On the proper definition of cognitive ethology. In: R.P. Balda, I.M. Pepperberg & A.C. Kamil (eds), *Animal cognition in nature*, pp. 1–28. Academic Press, California.

Karplus, I. & Algom, D. (1981) Visual cues for predator face recognition by reef fishes. *Zeitschrift für Tierpsychologie*, **55**, 343–364.

Karplus, I., Goren, M. & Algom, D. (1982) A preliminary experimental analysis of predator face recognition by *Chromis caerulus* (Pisces, Pomacentridae). *Zeitschrift für Tierpsychologie*, **58**, 53–65.

Kats, L.B. & Dill, L.M. (1998) The scent of death: chemosensory assessment of predation risk by prey animals. *Ecoscience*, **5**, 361–394.

Keenleyside, M.H.A. (1979) *Diversity and adaptation in fish behaviour*. Springer Verlag, New York.

Kislalioglu, M. & Gibson, R.N. (1976) Some factors governing prey selection by the 15-spined stickleback, *Spinachia spinachia* (L.). *Journal of Experimental Marine Biology and Ecology*, **25**, 159–169.

Krause, J. (1993) Transmission of fright reaction between different species of fish. *Behaviour*, **127**, 35–48.

Krause, J. & Godin, J.-G.J. (1996) Influence of prey foraging posture on predator detectability and predation risk: predators take advantage of unwary prey. *Behavioural Ecology*, **7**, 264–271.

Kristensen, E.A. & Closs, G.P. (2004) Anti-predator response of naive and experienced common bully to chemical alarm cues. *Journal of Fish Biology*, **64**, 643–652.

Licht, T. (1989) Discrimination between hungry and satiated predators: the response of guppies (*Poecilia reticulata*) from high and low predation sites. *Ethology*, **82**, 238–243.

Lima, S.L. (1998) Stress and decision making under the risk of predation: recent developments from behavioral, reproductive, and ecological perspectives. *Advances in the Study of Animal Behaviour*, **27**, 215–290.

Lima, S.L. (2002) Putting predators back into behavioral predator–prey interactions. *Trends in Ecology and Evolution*, **17**, 70–75.

Lima, S.L. & Bednekoff, P.A. (1999) Temporal variation in danger drives antipredator behaviour: the predation risk allocation hypothesis. *American Naturalist*, **153**, 649–659.

Lima, S.L. & Dill, L.M. (1990) Behavioral decisions made under the risk of predation: a review and prospectus. *Canadian Journal of Zoology*, **68**, 619–640.

Magurran, A.E. (1989) Acquired recognition of predator odour in the european minnow (*Phoxinus phoxinus*). *Ethology*, **82**, 216–223.

Magurran, A.E. (1990) The inheritance and development of minnow anti-predator behaviour. *Animal Behaviour*, **39**, 834–842.

Magurran, A.E. & Girling, S.L. (1986) Predator model recognition and response habituation in shoaling minnows. *Animal Behaviour*, **34**, 510–518.

Magurran, A.E. & Higham, A. (1988) Information transfer across fish shoals under threat. *Ethology*, **78**, 153–158.

Mathis, A. & Smith, R.J.F. (1993a) Fathead minnows (*Pimephales promelas*) learn to recognize pike (*Esox lucius*) as predators on the basis of chemical stimuli in the pike's diet. *Animal Behaviour*, **46**, 645–656.

Mathis, A. & Smith, R.J.F. (1993b) Chemical labelling of northern pike, *Esox lucius*, by the alarm pheromone of fathead minnows, *Pimephales promelas*. *Journal of Chemical Ecology*, **19**, 1967–1979.

Mathis, A. & Smith, R.J.F. (1993c) Chemical alarm signals increase the survival time of fathead minnows (*Pimaphales promelas*) during encounters with northern pike (*Esox lucius*). *Behavioral Ecology*, **4**, 260–265.

Mathis, A., Chivers, D.P. & Smith, J.F. (1996) Cultural transmission of predator recognition in fishes: intraspecific and interspecific learning. *Animal Behaviour*, **51**, 185–201.

Mirza, R.S. & Chivers, D.P. (2000) Predator-recognition training enhances the survival of brook trout: evidence from laboratory and field-enclosure studies. *Canadian Journal of Zoology*, **78**, 2198–2208.

Mirza, R.S. & Chivers, D.P. (2001) Learned recognition of heterospecific alarm signals: the importance of a mixed predator diet. *Ethology*, **107**, 1007–1018.

Murphy, K.E. & Pitcher, T.J. (1991) Individual behavioural strategies associated with predator inspection in minnow shoals. *Ethology*, **88**, 307–319.

Neilson, J.D. & Perry, R.I. (1990) Diel vertical migrations of marine fishes: an obligate or facultative process? *Advances in Marine Biology*, **26**, 115–168.

Olla, B.L. & Davis, M.W. (1989) The role of learning and stress in predator avoidance of hatchery reared coho salmon (*Oncorhynchus kisutch*) juveniles. *Aquaculture*, **76**, 209–214.

Patten, B.G. (1977) Body size and learned avoidance as factors affecting predation on coho salmon fry, *Oncorhynchus kisutch*, by torrent sculpin, *Cottus rhotheus*. *Fisheries Bulletin*, **75**, 457–459.

Pettersson, L.B., Nilsson, P.A. & Bronmark, C. (2000) Predator recognition and defence strategies in crucian carp, *Carassius carassius*. *Oikos*, **88**, 200–212.

Pfeiffer, W. (1974) Pheromones in fish and amphibia. In: M.C. Birch (ed.), *Pheromones*, pp. 269–296. North-Holland, Amsterdam.

Pitcher, T.J. (1992) Who dares wins: the function and evolution of predator inspection behaviour in fish shoals. *Netherlands Journal of Zoology*, **42**, 371–391.

Pitcher, T.J. & Parrish, J.K. (1993) Functions of shoaling behaviour in teleosts. In: T.J. Pitcher (ed.), *Behaviour of Teleost Fishes*, pp. 363–439. Chapman & Hall, London.

Pitcher, T.J., Magurran, A.E. & Winfield, I. (1982) Fish in large shoals find food faster. *Behavioural Ecology and Sociobiology*, **10**, 149–151.

Pitcher, T.J., Green, D.A. & Magurran, A.E. (1986) Dicing with death: predator inspection behaviour in minnow shoals. *Journal of Fish Biology*, **28**, 439–448.

Pollock, M.S. & Chivers, D.P. (2004) The effects of density on the learned recognition of heterospecific alarm cues. *Ethology*, **110**, 341–349.

Reebs, S.G. (1999) Time-place learning based on food but not on predation risk in a fish, the inanga (*Galaxias maculatus*). *Ethology*, **105**, 361–371.

Reebs, S.G. (2002) Plasticity of diel and circadian activity rhythms in fishes. *Reviews in Fish Biology and Fisheries*, **12**, 349–371.

Roper, T.J. & Redston, S. (1987) Conspicuousness of distasteful prey affects the strength and durability of one-trial avoidance learning. *Animal Behaviour*, **35**, 739–747.

Roper, T.J. & Wistow, R. (1986) Aposematic coloration and avoidance learning in chicks. *Quarterly Journal of Experimental Psychology*, **38B**, 141–149.

Ruxton, G.D., Sherratt, T. & Speed, M.P. (2004) *The Evolutionary Ecology of Crypsis, Warning Signals and Mimicry*. Oxford University Press, Oxford.

Shettleworth, S.J. (1988) *Cognition, Evolution and Behavior*. Oxford University Press, New York.

Sih, A. (1980) Optimal behaviour: can foragers balance two conflicting demands? *Science*, **210**, 1041–1043.

Sih, A. (1988) The effects of predators on habitat use, activity and mating behaviour in a semi-aquatic bug. *Animal Behaviour*, **36**, 1846–1848.

Sih, A., Englund, G. & Wooster, D. (1998) Emergent impacts of multiple predators on prey. *Trends in Ecology and Evolution*, **13**, 350–355.

Smith, R.J.F. (1986) Evolution of alarm signals: role of benefits of retaining group members or territorial neighbours. *American Naturalist*, **128**, 604–610.

Smith, R.J.F. (1992) Alarm signals in fishes. *Reviews in Fish Biology and Fisheries*, **2**, 33–63.

Smith, R.J.F. (1997) Avoiding and deterring predators. In: J. -G. J. Godin (ed.) *Behavioural ecology of teleost fishes*, pp. 163–190. Oxford University Press, Oxford.

Smith, R.J.F. & Smith, M.J. (1989) Predator-recognition behaviour in two species of Gobiid fishes, *Asterropteryx semipunctatus* and *Gnatholepis anjerensis*. *Ethology*, **83**, 19–30.

Speed, M.P. (2001) Can receiver psychology explain the evolution of aposematism? *Animal Behaviour*, **61**, 205–216.

Suboski, M.D. (1990) Releaser-induced recognition learning. *Psychological Reviews*, **97**, 271–284.

Suboski, M.D., Bain, S., Carty, A.E., McQuoid, I.M., Seelen, M.I. & Seifert, M. (1990) Alarm reaction in acquisition and social transmission of simulated-predator recognition by zebra danio fish (*Bachydanio rerio*). *Journal of Comparative Psychology*, **104**, 101–112.

Tinbergen, L. (1960) The natural control of insects on pinewoods. I. Factors influencing the intensity of predation by songbirds. *Archives Neerlandaises de Zoologie*, **13**, 265–343.

Utne-Palm, A.C. (2001) Response of naive two-spotted gobies *Gobiusculus flavescens* to visual and chemical stimuli of their natural predator, cod *Gadus morhua*. *Marine Ecology Progress Series*, **218**, 267–274.

Vilhunen, S. & Hirvonen, H. (2003) Innate antipredator responses of Arctic charr depend on predator species and their diet. *Behavioral Ecology & Sociobiology*, **55**, 1–10.

Vogel, D. & Bleckmann, H. (2001) Behavioral discrimination of water motions caused by moving objects. *Journal of Comparative Physiology*, **186**, 1107–1117.

Warner, R.R. (1988) Traditionality of mating site preferences in a coral reef fish. *Nature*, **335**, 719–721.

Wisenden, B.D. & Harter, K.R. (2001) Motion, not shape, facilitates association of predation risk with novel objects by fathead minnows (*Pimephales promelas*). *Ethology*, **107**, 357–364.

Yunker, W.K., Wein, D.E. & Wisenden, B.D. (1999) Conditioned alarm behavior in fathead minnows (*Pimephales promelas*) resulting from association of chemical alarm pheromone with a nonbiological visual stimulus. *Journal of Chemical Ecology*, **25**, 2677–2686.

Chapter 4

Learning About Danger: Chemical Alarm Cues and the Assessment of Predation Risk by Fishes

Grant E. Brown and Douglas P. Chivers

4.1 Introduction

Predation is a constant threat faced by most prey individuals (Lima & Dill 1990), shaping an individual's behaviour, morphology and life history traits. As a result, there exists strong selection pressure for the early detection and avoidance of potential predation threats. However, predator avoidance has the potential to be very costly, as it reduces time and energy available for numerous other fitness-related behaviour patterns such as foraging, mating and territorial defence (Godin & Smith 1988; Sih 1992) or forcing prey to utilize suboptimal habitats (Gotceitas & Brown 1993), leading to a potential reduction in energy intake (Lima & Dill 1990). Thus, an individual's response to predation pressure can be seen as a series of threat-sensitive trade-offs between the benefits of predator avoidance and those of a suite of other fitness related activities (Lima & Dill 1990; Lima & Bednekoff 1999). Prey capable of reliably assessing local predation risk should be at a selective advantage, as they would be able to balance these conflicting benefits.

Why should prey organisms learn to recognize indicators of risk and not respond instead to potential predation threats with a fixed (innate) response? The answer most probably lies in the variability of predation pressure. Prey fishes are typically at risk from a diversity of predators, which undergo changes in population density over time. However, the form of predation and the degree of risk may change dramatically as prey individuals grow (size-dependent predation risk; Brönmark & Miner 1992), and shift habitat preferences with ontogeny (Werner & Gilliam 1984). Prey also move between prey guilds (Olson *et al.* 1995; Olson 1996; Brown *et al.* 2001a) and are subject to seasonal changes in biotic and/or abiotic conditions (Gilliam & Fraser 2001). In addition, predation pressure may change within spaces, as prey move between microhabitat types (Golub *et al.* 2005). Thus, the form and frequency of predation pressure may change from year to year, season to season, day to day or even moment to moment. As a result, individual prey are probably unable to rely solely on genetically inherited antipredator responses. Rather, individuals capable of 'fine-tuning' predator avoidance decisions based on recent experience (i.e. learning) should be better able to optimize such trade-offs. Individuals that are able to alter their behavioural responses, based on learned information, should be expected to have a greater degree of flexibility in their response to predation (Brown & Chivers 2005). Thus, populations exposed to varying predation pressures over time may be selected towards the use of a learned response, rather than a more fixed or innate response.

Learning, in the most general sense, can be defined as the adaptive change in an individual's behavioural response based on experience (Thorpe 1963). By extension, learning of context-appropriate responses to ecologically relevant stimuli (i.e. predators, alarm cues of heterospecific prey guild members) requires reliable information regarding local risks; flexible behaviour patterns, allowing for threat-sensitive responses; and the ability to make lasting associations between relevant cues and context-appropriate responses to predation risk. In this chapter, we will demonstrate that chemosensory cues regarding local predation risk are reliable public information sources, allowing for threat-sensitive response patterns (i.e. flexible predator-avoidance behaviour) and that these cues are indeed used by a variety of prey species to learn important information about local predation risk.

4.2   Chemical alarm cues and flexible responses

The ability to assess local predation risk requires the presence of spatially and temporally reliable information. Of all potential senses capable of conveying such information regarding local predation risk to prey fishes, visual and chemical cues have received the most extensive study (Chivers & Smith 1998; Hartman & Abrahams 2000). Visual and chemical cues, however, provide very different types of information. Visual cues are spatially and temporally reliable. However, because prey and predator must be within close proximity for detection of visual cues, they can represent a high threat situation (Kats & Dill 1998; Brown *et al.* 2000a). Chemosensory information, while spatially and temporally less reliable, may have a lower associated risk, as prey and predator need not be in close proximity for the prey to acquire information (Kats & Dill 1998; Smith 1999). In addition, some visual cues, such as body position or movement (Licht 1989; Murphy & Pitcher 1997), could be readily manipulated by predators. Chemical cues (especially damage-released chemical cues), in comparison, would be less likely to be manipulated, and hence may represent more 'honest' information sources (Brown & Magnavacca 2003).

The use of damage-released chemical alarm cues to assess local predation risk is well documented in a wide variety of taxonomically diverse freshwater, and a growing number of marine, species (Chivers & Smith 1998; Wisenden 2000). Chemical alarm cues have been demonstrated in a wide range of fishes, including ostariophysans, salmonids, gobies, poecilids, sticklebacks (Gasterosteidae), percids, cottids, cichlids (Cichlidae) and centrarchids (Chivers & Smith 1998; Smith 1999). When released, typically following a predation attempt in which a prey individual is injured or captured, these chemical alarm cues can elicit dramatic short-term increases in species-typical antipredator behaviour in nearby conspecifics and some sympatric heterospecifics. Such overt behavioural responses (Smith 1999) have been well documented and include a variety of behaviour patterns including increased shoal cohesion, area avoidance, dashing, freezing and reduced foraging and mating behaviour (Chivers & Smith 1998). The detection of these cues is known to lead to increased survival under laboratory and natural conditions (Mathis & Smith 1993a; Mirza & Chivers 2001a, 2003a).

One persistent argument regarding the potential value of chemosensory cues has been that such cues can only signal the presence of a potential predator and that they are secondary to other sources of sensory information such as visual cues (Hartman & Abrahams 2000). The implication of this is that, based on chemosensory cues alone, prey individuals would be capable of assessing the presence or absence of risk, rather than the relative level of risk. Thus, the assessment of risk and potential for learning based on chemosensory information would be limited. However, several authors have argued that prey fishes rely on the relative concentration of both chemical alarm cues and predator odours to make threat-sensitive decisions regarding local risk (Lawrence & Smith 1989; Dupuch *et al.* 2004; Kusch *et al.* 2004). Given that an individual's risk is inversely related to its distance from a potential predator, and that relative concentration should be proportional to distance between the cue receiver and the cue source, concentration should be a reliable indicator of local risk (Lawrence & Smith 1989; Dupuch *et al.* 2004).

Several researchers have directly tested the hypothesis that overt antipredator response intensity is proportional to the concentration of predator cue or alarm cue detected by prey fishes; for example, Kusch *et al.* (2004) showed that fathead minnows (*Pimephales promelas*, Cyprinidae) exhibited stronger antipredator responses to the odour of predatory pike (*Esox lucius*, Esocidae) as the concentration of pike cue increased. Studies with redbelly dace (*Phoxinus eos*, Cyprinidae; Dupuch *et al.* 2004), roach (*Rutilus rutilus*, Cyprinidae; Jachner & Rydz 2002), and goldfish (*Carassius auratus*, Cyprinidae; Zhao & Chivers 2005) demonstrate that the intensity of overt antipredator responses increases in proportion to the relative concentration of conspecific alarm cues detected. However, studies with juvenile convict cichlids (*Archocentrus nigrofasciatus*, Cichlidae; Roh *et al.* 2004), rainbow trout (*Oncorhynchus mykiss*, Salmonidae; Mirza & Chivers 2003a), and pumpkinseed sunfish (*Lepomis gibbosus*, Centrarchidae; Marcus & Brown 2003) have shown an 'all-or-nothing' response pattern. Studies with fathead minnows show either 'all-or-nothing' responses (Lawrence & Smith 1989; Brown *et al.* 2001b) or responses that increase proportionally with relative concentration (Ferrari *et al.* 2005). More telling is a recent set of experiments that suggests that the response to varying concentrations of conspecific alarm cues is variable within a single species (convict cichlids), depending upon group size. Brown *et al.* (2006) have shown that solitary cichlids exhibit an 'all-or-nothing' response pattern at very low relative concentrations of alarm cues. Small groups of three cichlids exhibit a similar response pattern, although the minimum concentration required to elicit a response is higher. Finally, larger groups of six cichlids exhibited a graded response pattern. These data suggest that the degree of threat-sensitivity in overt behavioural patterns is largely dependent upon immediate perceived risk.

Recent studies have shown that prey fishes can detect conspecific chemical alarm cues well below the concentration threshold required to elicit an overt response (Brown *et al.* 2001c; Mirza & Chivers 2003a). Moreover, these studies demonstrate that prey fishes are gaining valuable information from these low concentration cues and making 'risk aversive' adjustments to their behaviour patterns; for example, juvenile rainbow trout gain significant survival benefits during staged encounters with live predators at concentrations below the overt response threshold (Mirza & Chivers, 2003a). In addition, recent studies suggest that prey individuals may exhibit subtle changes in a suite of behavioural patterns in the absence of a typical overt antipredator response, including changes in foraging posture (Foam *et al.* 2005a), increased inspection and/or exploratory behaviour (Lawrence & Smith 1989; Jachner & Rydz 2002), and increased vigilance towards conspecific visual alarm displays (Brown *et al.* 2004). Thus, the response to conspecific chemical alarm cues may be highly flexible in terms of behavioural patterns and/or intensity. Considered together, these data clearly demonstrate that there is a great deal of information available from chemical alarm cues and that, based on the quality and/or quantity of cue detected, individuals can make threat-sensitive decisions regarding predator avoidance behaviour.

4.3   Temporal variability and the intensity of antipredator behaviour

It is clear that prey fishes can assess relative predation risk based on the quality and/ or quantity of chemosensory cues present. However, recent models have suggested that the temporal variability in predation risk is at least as important in driving the selection for antipredator decisions as is the overall intensity of predation risk (Lima & Bednekoff 1999; Sih *et al.* 2000). Generally, these models predict that, as the frequency of predation risks within a microhabitat increases, individuals should allocate less energy to each predator encounter and should allocate more energy to foraging during the rare times when predation risk is absent (safe periods). Recent studies with aquatic invertebrates and terrestrial vertebrates have provided ambiguous results, but generally support the prediction that temporal variability significantly influences the intensity of antipredator behaviour.

To date, only two studies have examined the role of temporal variability in the intensity of antipredator behaviour of fishes. Foam *et al.* (2005b) exposed juvenile convict cichlids to high (3 times per day) or low (1 time per day) frequency of conspecific chemical alarm cues for 3 days in order to assess the potential effects of recent experience on the intensity of the response to either a conspecific alarm cue or a control [swordtail (*Xiphophorus helleri*, Poeciliidae) skin extract]. They found that cichlids conditioned to the high frequency treatment exhibited significantly lower intensity antipredator responses than did those conditioned to the low frequency treatment. Likewise, they found that cichlids conditioned to the high frequency treatment foraged at an approximately 50% higher rate during safe periods than those conditioned to the low frequency risk treatments. Moreover, the relative concentration of alarm cue used during the conditioning phase did not influence the degree of risk allocation (Foam *et al.* 2005b).

A common criticism of this model is that prey are not actually allocating energy to predator avoidance versus foraging behaviour, but rather are simply habituated to the high encounter rates to risky situations. Foam *et al.* (2005b) demonstrate that this is probably not the case. They found that regardless of the frequency of risk during the conditioning phase, cichlids significantly reduced their foraging rate when exposed to conspecific alarm cues during testing. If simple habituation were the mechanism controlling this, those that were conditioned to the high frequency treatments (i.e. reduced intensity antipredator responses) would not be expected to reduce foraging behaviour.

Mirza *et al.* (2006) have likewise tested the effects of temporal variability in predation risk on the intensity of antipredator behaviour of rainbow trout. They exposed trout to three levels of risk (high, moderate and low) crossed with two levels of temporal variation (exposed to risk three times per day and once a day). Trout exposed to risky situations more frequently responded with significantly less intense antipredator behaviour than those exposed to risk less frequently. Fishes under frequent risk were not more active during periods of safety compared with animals under infrequent risk.

4.4 Predator diet cues and risk assessment during predator inspection

The presence of conspecific alarm cues in the diet of predators is known chemically to label them as dangerous or risky to prey fishes (review in Chivers & Mirza 2001). Fathead minnows, for example, can detect conspecific alarm cues in the diet of northern pike (*Esox lucius*, Esocidae) (Mathis & Smith 1993b, c; Brown *et al.* 1995a). Minnows actively avoid areas labelled with the faeces of northern pike that had been fed a diet of minnows or sympatric brook sticklebacks (*Culaea inconstans*, Gasterosteidae), but not the faeces of pike fed a diet of allopatric swordtails (Brown *et al.* 1995b). Mirza & Chivers (2001b) have shown that juvenile yellow perch (*Perca flavescens*, Percidae) significantly increase their antipredator behaviour in the presence of chemical cues of adult conspecifics, but only when the adult had been fed a diet of juvenile conspecifics or a sympatric prey guild member.

When confronted by potential predators, many prey fishes engage in predator inspection behaviour (Dugatkin & Godin 1992a; Kelley & Magurran 2003). Predator inspection is typically seen as the salutatory approach towards a potential predator, especially novel predators (Brown & Godin 1999), and is an inherently risky behaviour. Inspection behaviour can be seen as a threat-sensitive trade-off between the increased risk of predation and a variety of potential benefits to individuals approaching a potential predator (Brown & Godin 1999; Brown & Schwarzbauer 2001). Several such benefits have been demonstrated, including visual alarm signalling (Smith & Smith 1989; Murphy & Pitcher 1997), predator deterrence (Godin & Davis 1995), and preferential matings (Godin & Dugatkin 1996).

One of the potentially most significant benefits available to inspecting prey is the ability to acquire information regarding local predation risk (Murphy & Pitcher 1997; Brown & Godin 1999). Such dietary labels are reliable sources of information regarding the risk of predation for inspecting individuals. Visual cues, such as the predator's behaviour or posture (Licht 1989; Murphy & Pitcher 1997; Smith & Belk 2001), are also known to provide information to inspecting prey. However, such visual cues may be considered unreliable (see above) and would be of limited use under conditions of low visibility (i.e. at night, or in turbid or highly structured microhabitats). During initial inspection visits, especially for prey inspecting a novel potential predator, chemosensory cues may provide valuable information regarding predator diet, hunger status and/or attack motivation (Brown & Godin 1999; Brown *et al.* 2000a). In addition, prey can potentially inspect a predator based on chemosensory cues at a distance, and continue to approach based on the presence or absence of 'risky' cues in the diet of the predator (Brown & Magnavacca 2003).

Brown & Godin (1999) demonstrated that inspecting individuals could learn to recognize potential predators based on the presence of conspecific alarm cues in the diet of the predator (acquired predator recognition, see section 4.5). Predator naive glowlight tetras (*Hemigrammus erythrozonus*, Characidae) were exposed to live convict cichlid predators fed a diet of tetras (with an alarm cue) or swordtails (lacking a recognizable alarm cue). Two significant findings come from this study. Initially, Brown & Godin (1999) demonstrated that tetras engage in a threat-sensitive trade-off between the potential benefits associated with inspecting predators and the elevated risk of predation. Tetras exposed to a predator in the presence of an alarm cue took longer to initiate an inspection visit, remained further away from the predator and inspected in smaller groups. Brown *et al.* (2000a) have likewise shown that tetras make similar chemically mediated inspection visits in the absence of any visual cues. Additionally, inspecting fishes acquired the recognition of the visual cues of a novel predator, but only when the predator was paired with a dietary cue containing tetra alarm cue (Brown & Godin 1999). Non-inspecting fishes, or those that inspected the predator fed a swordtail diet, did not show any learned predator recognition.

Attack cone avoidance (Kelley & Magurran 2003) has been argued to allow individuals to reduce their overall level of predation risk during predator inspections. Recent reports (Brown & Schwarzbauer 2001; Brown *et al.* 2001d) suggest that inspecting fishes actively avoid the head region of a potential predator only in the presence of conspecific alarm cues in the diet of the predator. However, contradictory reports (Magurran & Seghers 1990; Dugatkin & Godin 1992b) demonstrate attack cone avoidance in the absence of chemical information. Brown & Dreier (2002) examined the role of learned predator recognition in an individual's probability of exhibiting attack cone avoidance. Predator naive tetras exhibited increased attack cone avoidance only in the presence of conspecific alarm cues in the diet of a predator, supporting the role of chemical cues in risk assessment and mediation during inspection visits. However, once tetras had been conditioned to recognize the visual cues of a cichlid predator, Brown & Dreier (2002) found significant attack cone avoidance, independent of predator diet cues. Thus, it appears that chemosensory learning may additionally benefit individuals by allowing them to reduce individual risk during inspections via increased attack cone avoidance.

4.5   Acquired predator recognition

Perhaps the most widely studied example of chemosensory learning is the acquired recognition of predators (Chivers & Smith 1998; Smith 1999; Brown & Chivers 2005). Many prey fishes do not show innate recognition of potential predators. Rather, they must acquire this knowledge based on the pairing of alarm cues with the visual and/ or chemical cues of the predator (Suboski 1990; Smith 1999); for example, European minnows (*Phoxinus phoxinus*, Cyprinidae) and fathead minnows acquire the

recognition of the chemical and/or visual cues of a novel predator after a single exposure to the predator cue paired with conspecific alarm cues (Magurran 1989; Mathis
& Smith 1993c; Chivers & Smith 1994a). In fact, fathead minnow eggs collected from
pike sympatric populations and reared under laboratory conditions exhibited no
'innate' recognition of either the chemical or visual cues of a predatory pike, while
wild-caught individuals (from the same population) of the same size/age did demonstrate recognition (Chivers & Smith 1994b). Similar results have been shown for
juvenile brown trout (*Salmo trutta*, Salmonidae; Alvarez & Nicieza 2003) and common bully (*Gobiomorphus cotidianus*, Eleotridae; Kristensen & Closs 2004).

Recent studies demonstrate that juvenile Chinook salmon (*Oncorhynchus
tshawytscha*, Salmonidae; Berejikian *et al.* 2003), and Arctic charr (*Salvelinus
alpinus*, Salmonidae; Vilhunen & Hirvonen 2003) exhibit increased antipredator
responses when exposed to novel predator odours. Likewise, Hawkins *et al.* (2004)
have shown that juvenile Atlantic salmon (*Salmo salar*, Salmonidae) significantly
increased opercular flap rates upon detection of novel cues, suggesting increased
vigilance or olfactory sampling. Berejikian *et al.* (2003) demonstrated that the
strength of this 'innate' response to novel predators could be enhanced through
acquired recognition learning. An intriguing possibility is that some naive prey
might exhibit 'innate' or neophobic responses (Brown & Chivers 2005) to novel
predators. This initial avoidance response could either be reinforced (strengthened)
or extinguished by learning (Berejikian *et al.* 2003). In fact, Mirza & Chivers (2003b)
demonstrated that while small fathead minnows exhibit increased antipredator
responses to cues from injured sympatric damselfly larvae (*Enallagma boreale*),
larger minnows from the same population do not. While not conclusive, it does
suggest that learned responses are indeed flexible and shaped through repeated
conditioning (see also Ferrari & Chivers 2006a).

Acquired recognition of predators can also result from the pairing of a predator
cue with the alarm cues of heterospecific prey guild members. Chivers *et al.* (1995)
demonstrated that brook sticklebacks acquire recognition of the chemical cues of a
predatory northern pike following exposure to the predator cue paired with either
conspecific or fathead minnow skin extract. Such learned recognition can be culturally transmitted among prey guild members (Mathis *et al.* 1996). Intraspecific and
interspecific cultural transmission of acquired recognition would allow for the rapid
learning of novel predator cues (Chivers & Smith 1995; Brown *et al.* 1997). In fact,
Brown *et al.* (1997) demonstrated that a population of approximately 80 000 fathead
minnows (in a 4-ha surface area pond) learned to recognize the chemical cues of
introduced northern pike within 2–4 days after stocking. This would be particularly
important following the invasion of a habitat by novel predators or early in a growing
season when young-of-the-year prey become more active (Brown & Chivers 2005).

Interestingly, Vilhunen *et al.* (2005) have shown that cultural transmission of
acquired predator recognition is enhanced when the ratio of experienced to naive
conspecifics is low. Juvenile Arctic charr, which had been previously conditioned to
recognize the chemical cues of predatory pikeperch (*Sander lucioperca*, Percidae),
were paired with predator-naive charr at high (16:4), intermediate (10:10) and
low (4:16) ratios of experienced to naive charr and exposed to pikeperch odour.
When the ratio of experienced to naive charr was low, naive prey socially learned
to recognize the pikeperch odour as a predation threat. However, at intermediate

or high experienced to naive ratios, naive charr did not acquire the recognition of pikeperch odour. These results are inconsistent with those of studies demonstrating that socially learned foraging behaviour is enhanced when the number of experienced 'demonstrators' increases (e.g. Laland & Williams 1997; Lachlan *et al.* 1998). Vilhunen *et al.* (2005) suggest that when the ratio of experienced demonstrators is high, the perceived predation risk to demonstrators is relatively low, resulting in less intense antipredator responses to predator odour. However, when experienced demonstrators are in the minority, perceived risk of individual demonstrators may increase and the intensity of the antipredator response is stronger, potentially enhancing the ability of naive observers to acquire the learned recognition of predator odour, suggesting a link between perceived predation risk and the strength of associative learning (see also Chapter 10).

Some studies have shown that chemosensory learning of predator identities is independent of any overt antipredator responses within the prey fish. Brown & Smith (1996) exposed fathead minnows, which had been fed *ad libitum* or which had been deprived of food for up to 48 hours, to paired conspecific alarm cues and pike odour. Only satiated minnows or those food-deprived for 12 hours exhibited an overt antipredator response; those food-deprived for 24 or 48 hours did not differ from a distilled water control. However, when fed *ad libitum* for 4 days and retested for a learned response to the predator odour alone, all minnows exhibited significant antipredator responses, independent of initial food deprivation (Brown & Smith 1996). These results demonstrate that the presence of an overt response is not required for acquired recognition to occur.

Likewise, Brown *et al.* (2001c) demonstrated that fathead minnows learn to recognize a novel predator when exposed to concentrations of conspecific alarm cues well below their minimum behavioural response threshold. Minnows were exposed to hypoxanthine-3-N-oxide (H3NO, an active component of the Ostariophysan alarm cue; G.E. Brown *et al.* 2000b, 2003), at concentrations ranging from 0.4 nM (the population-specific minimum behavioural response threshold; Brown *et al.* 2001b) to 0.05 nM. Although concentrations of H3NO below 0.4 nM did not elicit an overt response during conditioning trials, minnows did learn the chemical cues of a novel predator (yellow perch) when paired with H3NO at concentrations as low as 0.1 nM or 25% of the minimum behavioural response threshold.

Typically, studies examining acquired recognition of predator cues have involved exposing predator-naive prey to a single predator cue paired with an alarm cue. However, under natural conditions, prey may be exposed to multiple predator odours simultaneously (or in close temporal sequence). This is especially true in systems characterized by frequent and variable predation pressure. Darwish *et al.* (2005) directly tested this multiple predator learning hypothesis, exposing predator-naive glowlight tetras to a cocktail consisting of convict cichlid, largemouth bass (*Micropterus salmoides*, Centrarchidae) and goldfish odours simultaneously (paired with a tetra skin extract). When tested for the learned recognition of each of the individual predator odours, tetras exhibited a significant increase in antipredator behaviour, but did not alter their behaviour in response to yellow perch odour (a novel predator control). These results demonstrate that prey can learn multiple predator odours at the same time and that they do not generalize these specific odours to novel predator cues. Moreover, Darwish *et al.* (2005) conditioned naive tetras to recognize pumpkinseed sunfish odour alone or as part of a predator cocktail and tested for potential survival benefits during staged encounters with sunfish predators. They demonstrated that tetras conditioned to either sunfish alone or multiple odours increased survival over unconditioned tetras. In a follow-up study, R.S. Mirza and G.E. Brown (unpublished data) conditioned naive glowlight tetras to recognize the odour of convict cichlids as potential predators. Two days later, they exposed conditioned tetras to a cocktail of predator odours including convict cichlid, pumpkinseed sunfish and goldfish odours and subsequently tested tetras for the learned recognition of the individual predator odours. Tetras were able indirectly (second-order associative learning; Schwartz 1984) to learn to recognize specific predator odours from within the cocktail if a known predator odour was included.

Recent studies have examined the potential link between threat-sensitive responses to chemical alarm cues and acquired predator recognition learning. Ferrari *et al.* (2005) asked whether the strength of the learned response to a novel predator cue is related to the intensity of the initial antipredator response. They exposed predator-naive fathead minnows to low, intermediate or high relative concentrations of conspecific alarm cues, paired with the odour of a novel predator (brook charr, *Salvelinus fontinalis*, Salmonidae). They found that during the initial conditioning phase, minnows exhibited stronger antipredator responses to the high concentration cues. When retested 24 hours later, they found that the learned response to brook charr odour matched the intensity of response during the initial conditioning phase. They likewise found a similar correlation between the intensity of response to brook charr odour between experienced tutors and naive observers in a social learning study (Ferrari *et al.* 2005). Thus, both directly acquired predator recognition and observational (social) learning appears to allow for the acquisition of threat-sensitive response patterns.

Ferrari & Chivers (2006a) demonstrate that recent experience shapes threat-sensitive learning. Prey fishes probably continually update their learned recognition of potential predators. The intensity of antipredator behaviour should, therefore, mimic the most recent learning experience. Ferrari & Chivers (2006a) tested this by exposing fathead minnows to either a high or a low concentration of conspecific alarm cue plus the odour of brook charr. The paired cues were given daily, for six consecutive days in one of four combinations: 6 low; 5 low, 1 high; 1 low, 5 high; or 6 high. The intensity of antipredator behaviour when exposed to brook charr odour alone appeared to match the most recent 'conditioning' experience. This suggests that prey do not simply 'average' the intensity of learning opportunities, but do indeed adjust the level of antipredator response to the most recent experience.

A major gap in our understanding of chemosensory learning is the lack of field studies. To date, only a few studies of chemically mediated learning have been conducted under natural conditions. Chivers & Smith (1995) and Brown *et al.* (1997) demonstrated that fathead minnows from a previously predator-free population rapidly learned to recognize the chemical and/or visual cues of an introduced predator. However, these studies clearly demonstrate learning, but do not elucidate potential mechanisms. A.O.H.C. Leduc and G.E. Brown (unpublished data) exposed marked juvenile Atlantic salmon (*Salmo salar*, Salmonidae) to conspecific alarm cues paired with a novel odour (lemon oil) under natural stream conditions. When individuals were retested 2 days later, they exhibited significant increases in antipredator behaviour (reduced territorial aggression, and foraging attempts and increased time under shelter) when exposed to lemon oil alone. Those that were initially exposed to lemon oil paired with a streamwater control exhibited no such increase in antipredator behaviour. This study represents the first demonstration of a specific mechanism of acquired predator recognition under fully natural conditions.

Mathis & Smith (1993b,c) demonstrated that predator-naive minnows learn to recognize pike as predation threats based on the presence of minnow alarm cues in the predator's diet. Brown *et al.* (1995a,b) have further shown that the prey alarm cue (or some metabolite) is present in the faeces of predators and hence may provide for learned recognition of novel predators, even in the absence of the predator itself. In fact, Brown *et al.* (1997) demonstrated that a previously predator-naive population of wild fathead minnows acquired the recognition of chemical predator cues significantly sooner than visual predator cues. This may be because of the increased opportunities to associate the alarm and predator cues afforded through this 'faecal association' mechanism.

While several studies have demonstrated the potential for increased survival benefits associated with acquired recognition of novel predators, few studies have directly tested for such an effect (Mirza & Chivers 2001c). Mathis & Smith (1993a) and Chivers *et al.* (1996) demonstrated that fathead minnows significantly increase their probability of survival during predator encounters when exposed to conspecific alarm cues, relative to individuals exposed to controls. Similar findings are reported by Mirza & Chivers (2001a) for brook charr under laboratory and field conditions. Mirza & Chivers (2000) conditioned juvenile brook charr to recognize chain pickerel (*Esox niger*, Esocidae) as potential predators. Under laboratory and field conditions, they demonstrated significant survival advantages of trained over naive brook charr. In a similar study, Vilhunen (2006) demonstrated that hatchery-reared juvenile Arctic charr conditioned to recognize the odour of predatory pikeperch that had been fed on Arctic charr exhibited significantly greater survival times (compared to unconditioned charr) when exposed to a live pikeperch, under laboratory conditions. Interestingly, Vilhunen (2006) found that survival times were approximately 50% longer among charr that were repeatedly conditioned (four times over 16 days), compared to those conditioned only once.

Of particular interest is whether hatchery-reared fishes, such as salmonids, retain the learned recognition of novel predators for ecologically relevant time frames (Brown & Day 2002). Shively *et al.* (1996) demonstrated significantly higher predation rates of hatchery-released salmonids (over native individuals) for up to 3 weeks after stocking. They suggest that after this period, hatchery-reared fishes would have had sufficient opportunity to acquire the recognition of novel predators and/or learn context-appropriate avoidance behaviour. Brown & Smith (1998) demonstrated that juvenile rainbow trout retain the acquired recognition of novel predators for at least 21 days, sufficient time for stocked individuals to learn to recognize natural predators. Likewise, Mirza & Chivers (2000) conditioned juvenile brook charr to recognize chain pickerel as a predation threat and found significant increases to predator odour 10 days after conditioning. Thus, individuals conditioned a few days prior to stocking into natural waterways are likely to retain their acquired recognition of novel predators long enough to reinforce this conditioned response (Brown & Smith 1998).

However, while many studies have demonstrated that a single pairing of a chemical alarm cue and a novel predator odour is sufficient for learned recognition to occur, recent studies suggest that the learned recognition of novel predator cues, although initially robust, will wane over time if not reinforced. Both Brown & Smith (1998) and Mirza & Chivers (2000) found that the intensity of learned responses to predator odours decreased with time elapsed since initial conditioning. Similarly, Berejikian *et al.* (1999) found that juvenile Chinook salmon (*Oncorhynchus tshawytscha*, Salmonidae) conditioned to recognize adult cutthroat trout exhibited a significant antipredator response when tested 3 days after conditioning, but not when tested 10 days after conditioning. Under natural conditions, however, it is likely that the recognition of acquired predator odours would be continually reinforced through direct exposure to predators or via social learning mechanisms (Kelley & Magurran 2003; Griffin 2004; Brown & Chivers 2005). The results of Darwish *et al.* (2005) suggest that indirect learning of multiple predator odours may provide an additional mechanism by which the learned recognition could be reinforced, leading to enhanced retention of relevant information.

Recent studies of the potential survival benefits associated with acquired predator recognition have pointed towards an interesting interaction between conditioning and testing environments. Berejikian *et al.* (1999) exposed juvenile Chinook salmon to conspecific skin extracts and the odour of predator cutthroat trout under 'barren' (open raceways) and 'complex' (raceways containing surface and submerged structure) habitat treatments. When tested 2 days later with trout odour alone, conditioned salmon fry, regardless of habitat treatment, significantly increased their antipredator behaviour. When released into natural streams, significantly more conditioned fry were recaptured at a downstream counting fence. However, this significant survival effect was only observed for fishes initially conditioned under the 'complex' treatment. No significant difference was found between control and conditioned fry initially held under 'barren' conditions. Gazdewich & Chivers (2002) also found conditioning-habitat interactions where fathead minnows conditioned to recognize yellow perch as predators exhibited increased survival only when tested in the absence of shelter. C. Brown *et al.* (2003) demonstrated a similar interaction between enriched environments and learning on the ability of hatchery-reared Atlantic salmon to exploit novel food types. Together, these studies suggest that research should be directed towards examining the role of microhabitat conditions on learning.

4.6   Constraints on learning

Recently, chemical alarm signaling systems of several prey fishes have been shown to be significantly impaired by relatively low levels of anthropogenic pollutants. Juvenile Chinook salmon exhibited no significant antipredator response following acute exposure to concentrations of Diazinon (a common neurotoxic insecticide) as low as $0.1\ \mu g\ L^{-1}$, probably because of direct damage to the olfactory epithelium (Schotz *et al.* 2000). Likewise, short-term exposure to low levels of cadmium ($\sim2\ \mu g\ L^{-1}$) impaired the detection of conspecific alarm cues by juvenile rainbow trout (Scott *et al.* 2003). Field studies by McPherson *et al.* (2004) demonstrated a similar impairment in wild Iowa darters (*Etheostoma exile*, Percidae) in lakes contaminated by heavy metal versus 'pristine' lakes. In addition to heavy metal and agrochemical contamination, several recent papers demonstrate that relatively minor changes in ambient pH may likewise impair the detection of chemical alarm cues. Brown *et al.* (2002a) demonstrated that relatively minor reductions in pH (~6.0) result in an inability of fathead minnows and finescale dace to detect and respond to conspecific alarm cues. They further demonstrated that the mechanism responsible for this loss of function is an irreversible covalent bond change to the alarm cue molecule itself, and is not the result of physiological damage or stress. Similar losses of alarm responses have been demonstrated under laboratory conditions for juvenile rainbow trout to conspecific cues (Leduc *et al.* 2004a) and for pumpkinseed sunfish to both conspecific and sympatric heterospecific cues (Leduc *et al.* 2003). Recent field studies have shown that wild populations of brook charr (Leduc *et al.* 2004a) and Atlantic salmon (A.O.H.C. Leduc & G.E. Brown, unpublished data) are impaired in their response to conspecific and heterospecific alarm cues under weakly acidic conditions.

Given the demonstrated survival benefits associated with responding to chemical alarm cues (i.e. Mirza & Chivers 2000, 2001b, 2003a), an inability to detect conspecific and/or heterospecific alarm cues would probably impact significantly on individual fitness and survival. In addition, this inability to detect alarm cues may also impair learning under natural conditions. Leduc *et al.* (2004b) have recently shown that juvenile rainbow trout are unable to learn to recognize the odour of a novel predator following acute exposure to relatively weak acidity levels (~pH 6.0). Trout tested at pH 7.0 were, however, able to learn the predator odour. The impairment of conditioned predator recognition appears to result from the partial or complete degradation of the alarm cue itself, and not from any chemical alteration of the predator odour (Leduc *et al.* 2004b).

One of the outstanding questions regarding acquired predator recognition learning is how individuals avoid acquiring aversive responses towards biologically irrelevant stimuli (Wisenden & Harter 2001). Prey fishes could maximize time and energy devoted to fitness-related behaviour patterns by responding only to risky situations (Lima & Bednekoff 1999). However, learned aversions to non-risky or irrelevant cues have been demonstrated (Magurran 1989; Chivers & Smith 1994a; Yunker *et al.* 1999), and would probably interfere with context-appropriate behavioural responses.

Wisenden & Harter (2001) directly addressed this question by presenting predator-naive fathead minnows with conspecific alarm cues and artificial visual 'predator' cues, which varied in shape and movement. They demonstrated that minnows could be more readily conditioned to avoid a moving versus a stationary stimulus, regardless of its shape. Predators, even ambush predators such as northern pike, must approach, stalk and attack a prey item in order to capture it. Thus, movement rather than shape appears to be the primary visual cue allowing the prey to differentiate between relevant and irrelevant predation threats (Brown & Warburton 1997; Wisenden & Harter 2001).

Another important constraint on the acquired recognition of predator identities is the temporal sequence in which prey perceive the alarm and predator cues. Typically, most studies of acquired predator recognition have presented both the alarm and predator cues simultaneously, or separated by a relatively short period of time. Korpi & Wisenden (2001) exposed zebra danios (*Danio rerio*, Cyprinidae) to conspecific skin extracts in a flow-through aquarium. Five minutes following the introduction of the alarm cue, sufficient time for the cue to be flushed from the test tank, predator odour cues were introduced. When retested with predator odour alone, they found significant antipredator responses, suggesting that acquired recognition can occur, even after relatively long delays between the detection of the alarm and predator cues.

Ferrari & Chivers (2006b) examined how fathead minnows may avoid learning to recognize biologically irrelevant stimuli associated with chemical alarm cues. They showed that minnows exposed to pike odour for 1 hour per day for 5 consecutive days, and then conditioned to recognize pike odour with alarm cues, failed to learn to recognize the predator. In contrast, those individuals exposed to a control of distilled water for the same preconditioning period and then conditioned with pike odour plus alarm cues did learn to respond to the predator. Future studies should examine such latent inhibition effects under field conditions.

4.7   Heterospecific responses

There should be strong selection pressures to respond to the chemical alarm cues from heterospecific prey if the heterospecifics are prey guild members, that is they co-occur and are exposed to the same suite of predators. While many studies have demonstrated strong cross-species responses (Smith 1999), only recently have researchers begun to address the question of how individuals acquire the recognition of heterospecific cues. Cross-species responses can readily be explained among closely related taxonomic groups. Closely related species are likely to have similar chemical alarm cues as a result of similar diets or biochemical production mechanisms (G.E. Brown *et al.* 2003). Alarm cues are highly conserved within ostariophysan fishes (Schutz 1956; G.E. Brown *et al.* 2001e, 2003), salmonids (Mirza & Chivers 2001d), poecilids (Mirza *et al.* 2001) and centrarchids (Brown *et al.* 2002b; Leduc *et al.* 2003). Perhaps more interesting is the observation that taxonomically distant species that share common habitats and predation threats are known to learn to respond to each other's alarm cues with appropriate antipredator responses; for example, sympatric fathead minnows and brook sticklebacks respond to each other's alarm cues, but minnows from populations lacking sticklebacks show no response to stickleback alarm cues (Pollock *et al.* 2003). Pollock *et al.* (2003) directly tested this sympatry-learning hypothesis. They reared minnow eggs collected from stickleback sympatric populations in the laboratory and found no evidence of recognition of stickleback alarm cues.

But how do prey fishes learn to recognize heterospecific alarm cues? Four, non-mutually exclusive learning mechanisms may account for this learned recognition. First, Mathis *et al.* (1996) demonstrated that acquired recognition of predators could be culturally transmitted from interspecific tutors (Brown & Laland 2003; and see Chapter 10). Similar learning of heterospecific alarm cues might also be culturally transmitted. Second, it is also possible that naive individuals could acquire recognition of heterospecific alarm cues through the association with conspecific alarm cues.

There would be ample opportunity for such association in mixed species shoals (Smith 1999). In fact, Pollock & Chivers (2004) demonstrated that fathead minnows reared with high densities of brook sticklebacks were more likely to acquire the learned recognition of stickleback alarm cues than were minnows reared with lower densities of stickleback. Third, individuals may acquire recognition of heterospecific alarm cues through the dietary cues of a shared predator. Mirza & Chivers (2001e) exposed fathead minnows from populations free from brook stickleback to the odour of a predatory yellow perch fed a mixed diet of minnows and sticklebacks or perch-fed sticklebacks and swordtails, and then tested for recognition of stickleback alarm cues. Individuals initially conditioned to the predators fed on minnow and stickleback exhibited a learned response when exposed to stickleback alarm cues alone. Those initially exposed to the odour of a predator fed swordtails and sticklebacks did not respond to stickleback alarm cues alone. Finally, prey may learn to recognize heterospecific alarm cues that they detect in the diet of a known predator. Mirza & Chivers (2003c) exposed fathead minnows to the odour of a known predator (northern pike) fed one of two unknown prey diets (brook sticklebacks or swordtails). They reported that minnows exposed to pike fed on sticklebacks significantly increased their antipredator response when subsequently exposed to stickleback alarm cues, but not when exposed to swordtail alarm cues. Likewise, minnows exposed to the odour of pike fed on swordtails exhibited increased antipredator responses when exposed to swordtail alarm cues, but not when exposed to stickleback alarm cues.

Chivers *et al.* (2002) demonstrated that this learned recognition of heterospecific alarm cues results in increased survival during staged encounters with novel predators. Fathead minnows were exposed to the chemical cues of rainbow trout fed a mixed diet of minnows and sticklebacks (or sticklebacks and swordtails). They were subsequently exposed to live novel predators (yellow perch and northern pike) paired with stickleback alarm cues. Minnows that previously had the opportunity to acquire the recognition of stickleback alarm cues exhibited significantly higher survival compared with those that did not have the opportunity to learn the heterospecific alarm cue.

As with the learned recognition of predator cues, the acquisition of heterospecific cues should be constrained by habitat factors. Pollock & Chivers (2003) stocked experimental ponds containing predatory northern pike with fathead minnows and brook sticklebacks under low structural and high structural complexity conditions. When tested following 8 days of 'conditioning', they found that minnows conditioned in low structure ponds exhibited a significant increase in antipredator response when exposed to stickleback alarm cues (learned recognition of stickleback alarm cue), while those conditioned under high structural complexity conditions did not learn to recognize the stickleback alarm cue. Pollock & Chivers (2003) suggest that this may be the result of one of two non-mutually exclusive mechanisms. Initially, the high degree of structural complexity may limit visual cues and therefore limit the opportunities for observational learning. Second, species-specific use of structured microhabitats might have limited the degree of mixed species shoaling, or constrained shoaling altogether. Either mechanism would function to limit the opportunities for minnows to acquire the recognition of the stickleback alarm cues.

The opportunity to learn that a heterospecific alarm cue represents a potential predation threat should, in theory, be enhanced as the density of heterospecifics within a habitat increases. Pollock & Chivers (2004) tested this hypothesis by stocking experimental ponds with fathead minnows and brook sticklebacks at three densities: 400 minnows to 100 sticklebacks; 475 minnows to 25 sticklebacks; and 500 minnows to 0 sticklebacks. In addition, each pond contained a single predator (northern pike). Following 14 days of 'conditioning', minnows were tested for their behavioural response to the skin extracts of conspecifics, sticklebacks or swordtails (as an allopatric control). Minnows initially stocked with a high density of sticklebacks exhibited a significant increase in antipredator behaviour when exposed to stickleback alarm cues, showing that they had acquired recognition of these cues. However, minnows initially stocked with either a low density of sticklebacks or no sticklebacks, did not show any learned recognition of stickleback cues.

4.8 Conclusions

An individual's behaviour patterns can be conceptualized as a series of threat-sensitive trade-offs between ambient predation pressure and a suite of fitness-related activities, such as resource defence, foraging and mating. Individuals that can reliably assess their current predation risk and adjust their predator avoidance decisions based on experience (i.e. learning) could increase their fitness potential by exhibiting antipredator behaviour only at appropriate times. However, such learned risk assessment requires reliable information regarding current predation risks. A diverse range of prey fishes release chemical alarm cues upon capture by a predator. When detected by conspecifics and some heterospecifics, these chemicals elicit a variety of overt and covert behavioural responses. These chemical cues, alone or as part of a predator's dietary odour, provide reliable information regarding local predation risk for other prey in the vicinity, allowing for the learned recognition of novel predators, risky habitats, and the alarm cues of heterospecific prey guild members. A review of recent literature highlights the role of such chemosensory information in the acquisition of the learned recognition of biologically relevant stimuli. This sophisticated chemosensory learning system allows individuals to assess local predation risk and to modify their behaviour patterns in a threat-sensitive fashion. By doing so, it is possible for prey individuals to reduce their overall level of predation risk, while still gaining benefits associated with a variety of other fitness-related activities.

4.9 Acknowledgements

We thank Maud Ferrari, Antoine Leduc, Mark Harvey and Isabelle Désormeaux for their helpful comments. Financial support was provided by Concordia University and NSERC of Canada to G.E.B. and the University of Saskatchewan and NSERC of Canada to D.P.C.

4.10 References

Alvarez, D. & Nicieza, A.G. (2003) Predator avoidance behaviour in wild and hatchery-reared brown trout: the role of experience and domestication. *Journal of Fish Biology*, **63**, 1565–1577.

Berejikian, B.A., Smith, R.J.F., Tezak, E.P., Schroder, S.L. & Knudsen, C.M. (1999) Chemical alarm signals and complex hatchery rearing habitats affect antipredator behaviour and survival of Chinook salmon (*Oncorhynchus tshawytscha*) juveniles. *Canadian Journal of Fisheries and Aquatic Sciences*, **56**, 830–838.

Berejikian, B.A., Tezak, E.P. & LaRae, A.L. (2003) Innate and enhanced predator recognition in hatchery-reared Chinook salmon. *Environmental Biology of Fishes*, **67**, 241–251.

Brönmark, C. & Miner, J.G. (1992) Predator-induced phenotypic change in body morphology in crucian carp. *Science*, **258**, 1348–1350.

Brown, C. & Day, R. (2002) The future of stock enhancements: bridging the gap between hatchery practice and conservation biology. *Fish and Fisheries*, **3**, 79–94.

Brown, C. & Laland, K.N. (2003) Social learning in fishes: a review. *Fish and Fisheries*, **4**, 280–288.

Brown, C. & Warburton, K. (1997) Predator recognition and anti-predator responses in the rainbowfish, *Melanotaenia eachamensis*. *Behavioral Ecology and Sociobiology*, **41**, 61–68.

Brown, C., Davidson, T. & Laland, K. (2003) Environmental enrichment and prior experience of live prey improve foraging behaviour in hatchery-reared Atlantic salmon. *Journal of Fish Biology*, **63** (**A**), 187–196.

Brown, G.E. & Chivers, D.P. (2005) Learning as an adaptive response to predation. In: P. Barbosa & I. Castellanos (eds), *Ecology of Predator–Prey Interactions*, pp. 34–54. Oxford University Press, Oxford.

Brown, G.E. & Dreier, V.M. (2002) Predator inspection behaviour and attack cone avoidance in a characin fish: the effects of predator diet and prey experience. *Animal Behaviour*, **63**, 1175–1181.

Brown, G.E. & Godin, J.-G.J. (1999) Who dares, learns: chemical inspection behaviour and acquired predator recognition in a characin fish. *Animal Behaviour*, **57**, 475–481.

Brown, G.E. & Magnavacca, G. (2003) Predator inspection behaviour in a characin fish: an interaction between chemical and visual information? *Ethology*, **109**, 739–750.

Brown, G.E. & Schwarzbauer, E.M. (2001) Chemical predator inspection and attack cone avoidance in a characin fish: the effects of predator diet. *Behaviour*, **138**, 727–739.

Brown, G.E. & Smith, R.J.F. (1996) Foraging trade-offs in fathead minnows (*Pimephales promelas*): acquired predator recognition in the absence of an alarm response. *Ethology*, **102**, 776–785.

Brown, G.E. & Smith, R.J.F. (1998) Acquired predator recognition in juvenile rainbow trout (*Oncorhynchus mykiss*): conditioning hatchery-reared fish to recognize chemical cues of a predator. *Canadian Journal of Fisheries and Aquatic Sciences*, **55**, 611–617.

Brown, G.E., Chivers, D.P. & Smith, R.J.F. (1995a) Localized defecation by pike: a response to labeling by cyprinid alarm pheromone? *Behavioral Ecology and Sociobiology*, **36**, 105–110.

Brown, G.E., Chivers, D.P. & Smith, R.J.F. (1995b) Fathead minnows avoid conspecific and heterospecific alarm pheromone in the faeces of northern pike. *Journal of Fish Biology*, **47**, 387–393.

Brown, G.E., Chivers, D.P. & Smith, R.J.F. (1997) Differential learning rates of chemical versus visual cues of a northern pike by fathead minnows in a natural habitat. *Environmental Biology of Fishes*, **49**, 89–96.

Brown, G.E., Paige, J.A. & Godin, J.-G.J. (2000a) Chemically-mediated predator inspection behaviour in the absence of predator visual cues by a characin fish, *Hemigrammus erythrozonus*. *Animal Behaviour*, **60**, 315–321.

Brown, G.E., Adrian, J.C., Jr., Smyth, E., Leet, H. & Brennan, S. (2000b) Ostariophysan alarm pheromones: laboratory and field tests of the functional significance of nitrogen oxides. *Journal of Chemical Ecology*, **26**, 139–154.

Brown, G.E., LeBlanc, V.J. & Porter, L.E. (2001a) Ontogenetic changes in the response of largemouth bass (*Micropterus salmoides*, Centrarchidae, Perciformes) to heterospecific alarm pheromones. *Ethology*, **107**, 401–414.

Brown, G.E., Adrian, J.C., Jr. & Shih, M.L. (2001b) Behavioural responses of fathead minnows to hypoxanthine-3-*N*-oxide at varying concentrations. *Journal of Fish Biology*, **58**, 1465–1470.

Brown, G.E., Adrian, J.C., Jr., Patton, T. & Chivers, D.P. (2001c) Fathead minnows learn to recognize predator odour when exposed to concentrations of artificial alarm pheromone below their behavioural response thresholds. *Canadian Journal of Zoology*, **79**, 2239–2245.

Brown, G.E., Golub, J.L. & Plata, D.L. (2001d) Attack cone avoidance during predator inspection visits by wild finescale dace (*Phoxinus neogaeus*): the effects of predator diet. *Journal of Chemical Ecology*, **27**, 1657–1666.

Brown, G.E., Adrian, J.C., Jr., Kaufman, I.H., Erickson, J.L. & Gershaneck, D. (2001e) Responses to nitrogen-oxides by Characiforme fishes suggests evolutionary conservation in ostariophysan alarm pheromones. In: Marchlewska-Koj, J.J. Lepri & D. Müller-Schwarze (eds), *Chemical Signals in Vertebrates*, Vol. 9, pp. 305–312. Plenum Press, New York.

Brown, G.E., Adrian, J.C. Jr., Lewis, M.G. & Tower, J.M. (2002a) The effects of reduced pH on chemical alarm signaling in ostariophysan fishes. *Canadian Journal of Fisheries and Aquatic Sciences*, **59**, 1331–1338.

Brown, G.E., Gershaneck, D.L., Plata, D.L. & Golub, J.L. (2002b) Ontogenetic changes in response to heterospecific alarm cues by juvenile largemouth bass are phenotypically plastic. *Behaviour*, **139**, 913–927.

Brown, G.E., Adrian, J.C., Jr., Naderi, N.T., Harvey, M.C. & Kelly, J.M. (2003) Nitrogen oxides elicit antipredator responses in juvenile channel catfish, but not in convict cichlids or rainbow trout: Conservation of the Ostariophysan alarm pheromone. *Journal of Chemical Ecology*, **29**, 1781–1796.

Brown, G.E., Poirier, J.-F. & Adrian, J.C., Jr. (2004) Assessment of local predation risk: the role of subthreshold concentrations of chemical alarm cues. *Behavioral Ecology*, **15**, 810–815.

Brown, G.E., Bongiorno, T., DiCapua, D.M., Ivan, L.I. & Roh, E. (2006) Effects of group size on the threat-sensitive response to varying concentrations of chemical alarm cues by juvenile convict cichlids. *Canadian Journal of Zoology*, **84**, 1–8.

Chivers, D.P. & Mirza, R.S. (2001) Predator diet cues and the assessment of predation risk by aquatic vertebrates: a review and prospectus. In: A. Marchlewska-Koj, J.J. Lepri & D. Müller-Schwarze (eds), *Chemical Signals in Vertebrates*, Vol. 9, pp. 227–284. Kluwer Academic, New York.

Chivers, D.P. & Smith, R.J.F. (1994a) Fathead minnows, *Pimephales promelas*, acquire predator recognition when alarm substance is associated with the sight of unfamiliar fish. *Animal Behaviour*, **48**, 597–605.

Chivers, D.P. & Smith, R.J.F. (1994b) The role of experience and chemical alarm signaling in predator recognition by fathead minnows, *Pimephales promelas*. *Journal of Fish Biology*, **44**, 273–285.

Chivers, D.P. & Smith, R.J.F. (1995) Free-living fathead minnows rapidly learn to recognize pike as predators. *Journal of Fish Biology*, **46**, 949–954.

Chivers, D.P. & Smith, R.J.F. (1998) Chemical alarm signalling in aquatic predator–prey systems: a review and prospectus. *Écoscience*, **5**, 338–352.

Chivers, D.P., Brown, G.E. & Smith, R.J.F. (1995) Acquired recognition of chemical stimuli from pike, *Esox lucius*, by brook sticklebacks, *Culaea inconstans* (Osteichthyes, Gasterosteidae). *Ethology*, **99**, 234–242.

Chivers, D.P., Brown, G.E. & Smith, R.J.F. (1996) The evolution of chemical alarm signals: attracting predators benefits alarm signal senders. *American Naturalist*, **148**, 649–659.

Chivers, D.P., Mirza, R.S. & Johnston, J.G. (2002) Learned recognition of heterospecific alarm cues enhances survival during encounters with predators. *Behaviour*, **139**, 929–938.

Darwish, T.L., Mirza, R.S., Leduc, A.O.H.C. & Brown, G.E. (2005) Acquired recognition of novel predator odour cocktails by juvenile glowlight tetras. *Animal Behaviour*, **70**, 83–89.

Dugatkin, L.A. & Godin, J.-G.J. (1992a) Prey approaching predators: a cost-benefit perspective. *Annals Zoologici Fennici*, **29**, 233–252.

Dugatkin, L.A. & Godin, J.-G.J. (1992b) Predator inspection, shoaling and foraging under predation hazard in the Trinidadian guppy, *Poecilia reticulata*. *Environmental Biology of Fishes*, **34**, 265–276.

Dupuch, A., Magnan, P. & Dill, L.M. (2004) Sensitivity of northern redbelly dace, *Phoxinus eos*, to chemical alarm cues. *Canadian Journal of Zoology*, **82**, 407–415.

Ferrari, M.C.O. & Chivers, D.P. (2006a) The role of learning in the development of threat-sensitive predator avoidance: how do fathead minnows incorporate conflicting information? *Animal Behaviour*, **71**, 19–26.

Ferrari, M.C.O. & Chivers, D.P. (2006b) The role of latent inhibition in acquired predator recognition by fathead minnows. *Canadian Journal of Zoology*, **84**, 505–509.

Ferrari, M.C.O., Trowell, J.J., Brown, G.E. & Chivers, D.P. (2005) The role of learning in the development of threat-sensitive predator avoidance by fathead minnows. *Animal Behaviour*, **70**, 777–784.

Foam, P.E., Harvey, M.C., Mirza, R.S. & Brown, G.E. (2005a) Heads up: juvenile convict cichlids rely on chemosensory information to make threat-sensitive foraging decisions. *Animal Behaviour*, **70**, 601–607.

Foam, P.E., Mirza, R.S., Chivers, D.P. & Brown, G.E. (2005b) Juvenile convict cichlids (*Archocentrus nigrofasciatus*) allocate foraging in response to temporal variation in predation risk. *Behaviour*, **142**, 129–144.

Gazdewich, K.J. & Chivers, D.P. (2002) Acquired predator recognition by fathead minnows: influence of habitat characteristics on survival. *Journal of Chemical Ecology*, **28**, 439–445.

Gilliam, J.F. & Fraser, D.F. (2001) Movement in corridors: enhancement by predation threat, disturbance, and habitat structure. *Ecology*, **82**, 124–135.

Godin, J.-G.J. & Davis, S.A. (1995) Who dares, benefits: predator approach behaviour in the guppy (*Poecilia reticulata*) deters predator pursuit. *Proceedings of the Royal Society of London, Series B*, **259**, 193–200.

Godin, J.-G.J. & Dugatkin, L.A. (1996) Female mating preferences for bold males in the guppy, *Poecilia reticulata*. *Proceedings of the National Academy of Sciences*, **93**, 10262–10267.

Godin, J.-G.J. & Smith, S.A. (1988) A fitness cost of foraging in the guppy. *Nature*, **333**, 69–71.

Golub, J.L., Vermette, V. & Brown, G.E. (2005) The response of pumpkinseed sunfish to conspecific and heterospecific chemical alarm cues under natural conditions: the effects of stimulus type, habitat and ontogeny. *Journal of Fish Biology*, **66**, 1073–1081.

Gotceitas, V. & Brown, J.A. (1993) Substrate selection by juvenile Atlantic cod (*Gadus morhua*): effects of predation risk. *Oecologia*, **93**, 31–37.

Griffin, A.S. (2004) Social learning about predators: A review and prospectus. *Learning and Behavior*, **32**, 131–140.

Hartman, E.J. & Abrahams, M.V. (2000) Sensory compensation and the detection of predators: the interaction between chemical and visual information. *Proceedings of the Royal Society of London, Series B – Biological Sciences*, **267**, 571–575.

Hawkins, L.A., Magurran, A.E. & Armstrong, J.D. (2004) Innate predator recognition in newly-hatched Atlantic salmon. *Behaviour*, **141**, 1249–1262.

Jachner, A. & Rydz, M.A. (2002) Behavioural response of roach (Cyprinidae) to different doses of chemical alarm cues (Schreckstoff). *Archieves Hydrobiologica*, **155**, 369–381.

Kats, L.B. & Dill, L.M. (1998) The scent of death: chemosensory assessment of predation risk by prey animals. *Écoscience*, **5**, 361–394.

Kelley, J.L. & Magurran, A.E. (2003) Learned predator recognition and antipredator responses in fishes. *Fish and Fisheries*, **4**, 216–226.

Korpi, N.L. & Wisenden, B.D. (2001) Learned recognition of novel predator odour by zebra danios, *Danio rerio*, following time-shifted presentation of alarm cues and predator odour. *Environmental Biology of Fishes*, **61**, 205–211.

Kristensen, E.A. & Closs, G.P. (2004) Anti-predator response of naive and experienced common bully to chemical alarm cues. *Journal of Fish Biology*, **64**, 643–652.

Kusch, R.C., Mirza, R.S. & Chivers, D.P. (2004) Making sense of predator scents: investigating the sophistication of predator assessment abilities of fathead minnows. *Behavioural Ecology and Sociobiology*, **55**, 551–555.

Lachlan, R.F., Crooks, L. & Laland, K.N. (1998) Who follows whom? Shoaling preferences and social learning of foraging information in guppies. *Animal Behaviour*, **56**, 181–190.

Laland, K.N. & Williams, K. (1997) Shoaling generates social learning of foraging information in guppies. *Animal Behaviour*, **53**, 1161–1169.

Lawrence, B.J. & Smith, R.J.F. (1989) Behavioral response of solitary fathead minnows, *Pimephales promelas*, to alarm substance. *Journal of Chemical Ecology*, **15**, 209–219.

Leduc, A.O.H.C., Noseworthy, M.K., Adrian, J.C., Jr & Brown, G.E. (2003) Detection of conspecific and heterospecific alarm signals by juvenile pumpkinseeds under weakly acidic conditions. *Journal of Fish Biology*, **63**, 1331–1336.

Leduc, A.O.H.C., Kelly, J.M. & Brown, G.E. (2004a) Detection of conspecific chemical alarm cues by juvenile salmonids under neutral and weakly acidic conditions: laboratory and field tests. *Oecologia*, **139**, 318–324.

Leduc, A.O.H.C., Ferrari, M.C.O., Kelly, J.M. & Brown, G.E. (2004b) Learning to recognize novel predators under weakly acidic conditions: the effects of reduced pH on acquired predator recognition by juvenile rainbow trout (*Oncorhynchus mykiss*). *Chemoecology*, **14**, 107–112.

Licht, T. (1989) Discriminating between hungry and satiated predators: the response of guppies (*Poecilia reticulata*) from high and low predation sites. *Ethology*, **82**, 238–243.

Lima, S.L. & Bednekoff, P.A. (1999) Temporal variation in danger drives antipredator behavior: the predation risk allocation hypothesis. *American Naturalist*, **153**, 649–659.

Lima, S.L. & Dill, L.M. (1990) Behavioral decisions made under the risk of predation: a review and prospectus. *Canadian Journal of Zoology*, **68**, 619–640.

Magurran, A.E. (1989) Acquired recognition of predator odour in the European minnow (*Phoxinus phoxinus*). *Ethology*, **82**, 216–233.

Magurran, A.E. & Seghers, B.H. (1990) Population differences in predator recognition and attack cone avoidance in the guppy *Poecilia reticulata*. *Animal Behaviour*, **40**, 443–452.

Marcus, J.P. & Brown, G.E. (2003) Response of pumpkinseed sunfish to conspecific chemical alarm cues: an interaction between ontogeny and stimulus concentration. *Canadian Journal of Zoology*, **81**, 1671–1677.

Mathis, A. & Smith, R.J.F. (1993a) Chemical alarm signals increase the survival time of fathead minnows (*Pimephales promelas*) during encounters with northern pike (*Esox lucius*). *Behavioral Ecology*, **4**, 260–265.

Mathis, A. & Smith, R.J.F. (1993b) Chemical labelling of northern pike, *Esox lucius*, by the alarm pheromone of fathead minnows, *Pimephales promelas*. *Journal of Chemical Ecology*, **19**, 1967–1979.

Mathis, A. & Smith, R.J.F. (1993c) Fathead minnows (*Pimephales promelas*) learn to recognize pike (*Esox lucius*) as predators on the basis of chemical stimuli in the pike's diet. *Animal Behaviour*, **46**, 645–656.

Mathis, A., Chivers, D.P. & Smith, R.J.F. (1996) Cultural transmission of predator recognition in fishes: intraspecific and interspecific learning. *Animal Behaviour*, **51**, 185–201.

McPherson, T.D., Mirza, R.S. & Pyle, G.G. (2004) Responses of wild fishes to alarm chemicals in pristine and metal-contaminated lakes. *Canadian Journal of Zoology*, **82**, 694–700.

Mirza, R.S. & Chivers, D.P. (2000) Predator-recognition training enhances survival of brook trout: evidence from laboratory and field-enclosure studies. *Canadian Journal of Zoology*, **78**, 2198–2208.

Mirza, R.S. & Chivers, D.P. (2001a) Chemical alarm signals enhance survival of brook charr, *Salvelinus fontinalis*, during encounters with predatory chain pickerel, *Esox niger*. *Ethology*, **107**, 989–1006.

Mirza, R.S. & Chivers, D.P. (2001b) Do juvenile yellow perch use diet cues to assess the level of threat posed by intraspecific predators? *Behaviour*, **138**, 1249–1258.

Mirza, R.S. & Chivers, D.P. (2001c) Do chemical alarm signals enhance survival of aquatic vertebrates?: an analysis of the current research paradigm. In: A. Marchlewska-Koj, J.J. Lepri, & D. Müller-Schwarze (eds), *Chemical Signals in Vertebrates*, Vol. 9, pp. 19–26. Kluwer Academic, New York.

Mirza, R.S. & Chivers, D.P. (2001d) Are chemical alarm signals conserved within salmonid fishes? *Journal of Chemical Ecology*, **27**, 1641–1655.

Mirza, R.S. & Chivers, D.P. (2001e) Learned recognition of heterospecific alarm signals: the importance of mixed predator diet. *Ethology*, **107**, 1007–1018.

Mirza, R.S. & Chivers, D.P. (2003a) Response of juvenile rainbow trout to varying concentrations of chemical alarm cue: response thresholds and survival during encounters with predators. *Canadian Journal of Zoology*, **81**, 88–95.

Mirza, R.S. & Chivers, D.P. (2003b) Influence of body size on the responses of fathead minnows, *Pimephales promelas*, to damselfly alarm cues. *Ethology*, **109**, 691–699.

Mirza, R.S. & Chivers, D.P. (2003c) Fathead minnows learn to recognize heterospecific alarm cues they detect in the diet of a known predator. *Behaviour*, **140**, 1359–1369.

Mirza, R.S., Scott, J.J. & Chivers, D.P. (2001) Differential responses of male and female red swordtails to chemical alarm cues. *Journal of Fish Biology*, **59**, 716–728.

Mirza, R.S., Mathis, A. & Chivers, D.P. (2006) Does temporal variation in predation risk influence the intensity of anti-predator responses? A test of the risk allocation hypothesis. *Ethology*, **112**, 44–51.

Murphy, K.E. & Pitcher, T.J. (1997) Predator attack motivation influences the inspection behaviour of European minnows. *Journal of Fish Biology*, **50**, 407–417.

Olson, M.H. (1996) Ontogenetic niche shifts in largemouth bass: variability and consequences for first-year growth. *Ecology*, **77**, 179–190.

Olson, M.H., Mittelback, G.G. & Osenberg, C.W. (1995) Competition between predator and prey: resource-based mechanisms and implication for stage-structured dynamics. *Ecology*, **76**, 1758–1771.

Pollock, M.S. & Chivers, D.P. (2003) Does habitat complexity influence the ability of fathead minnows to learn heterospecific chemical alarm cues? *Canadian Journal of Zoology*, **81**, 923–927.

Pollock, M.S. & Chivers, D.P. (2004) The effects of density on the learned recognition of heterospecific alarm cues. *Ethology*, **110**, 341–349.

Pollock, M.S., Chivers, D.P., Mirza, R.S. & Wisenden, B.D. (2003) Fathead minnows, *Pimephales promelas*, learn to recognize chemical alarm cues of introduced brook stickleback, *Culaea inconstans*. *Environmental Biology of Fishes*, **66**, 313–319.

Roh, E., Mirza, R.S. & Brown, G.E. (2004) Quality or quantity?: the role of donor condition in the production of chemical alarm cues in juvenile convict cichlids. *Behaviour*, **141**, 1235–1248.

Schotz, N.L., Truelove, N.K., French, B.L., Berejikian, B.A., Quinn, T.P., Casillas, E. & Collier, T.K. (2000) Diazinon disrupts antipredator and homing behaviors in Chinook salmon (*Oncorhynchus tshawytscha*). *Canadian Journal of Fisheries and Aquatic Sciences*, **57**, 1911–1918.

Schutz, F. (1956) A comparative study about the fright reaction of fish and their distribution. *Zeitschrift für Vergleichende Physiologie*, **38**, 84–135. (In German.)

Schwartz, B. (1984) *Psychology of Learning and Behavior*, 2nd edn. W.W. Norton, New York.

Scott, G.R., Sloman, K.A., Rouleau, C. & Wood, C.M. (2003) Cadmium disrupts behavioural and physiological responses to alarm substance in juvenile rainbow trout (*Oncorhynchus mykiss*). *Journal of Experimental Biology*, **206**, 1779–1790.

Shively, R.S., Poe, T.P. & Sauter, S.T. (1996) Feeding response by northern squawfish to a hatchery release of juvenile salmonids in the Clearwater River, Idaho. *Transactions of the American Fisheries Society*, **125**, 230–236.

Sih, A. (1992) Prey uncertainty and the balancing of antipredator and foraging needs. *American Naturalist*, **139**, 1052–1069.

Sih, A., Ziemba, R. & Harding, K.C. (2000) New insights on how temporal variation in predation risk shapes prey behavior. *Trends in Ecology and Evolution*, **15**, 3–4.

Smith, M.E. & Belk, M.C. (2001) Risk assessment in western mosquitofish (*Gambusia affinis*): do multiple cues have additive effects? *Behavioral Ecology and Sociobiology*, **51**, 101–107.

Smith, R.J.F. (1999) What good is smelly stuff in the skin? Cross-function and cross-taxa effects in fish 'alarm substances'. In: R.E. Johnston, D. Müller-Schwarze & P.W. Sorenesen (eds), *Advances in Chemical Signals in Vertebrates*, pp 475–488. Kluwer Academic, New York.

Smith, R.J.F. & Smith, M.J. (1989) Predator-recognition behaviour in two species of gobies, *Asterropteryx semipunctatus* and *Gnatholepis anjerensis*. *Ethology*, **81**, 279–290.

Suboski, M.D. (1990) Releaser-induced recognition learning. *Psychological Reviews*, **97**, 271–284.

Thorpe, W.H. (1963) *Learning and Instinct in Animals*, 2nd edn. Methuen, London.

Vilhunen, S. (2006) Repeated antipredator conditioning: a pathway to habituation or to better avoidance? *Journal of Fish Biology*, **68**, 25–43.

Vilhunen, S. & Hirvonen, H. (2003) Innate antipredator response of Arctic charr (*Salvelinus alpinus*) depend on predator species and their diet. *Behavioral Ecology and Sociobiology*, **55**, 1–10.

Vilhunen, S., Hirvonen, H. & Laakkonen, M.V.-M. (2005) Less is more: social learning of predator recognition requires a low demonstrator to observer ratio in Arctic charr (*Salvelinus alpinus*). *Behavioral Ecology and Sociobiology*, **57**, 275–282.

Werner, E.E. & Gilliam, J.F. (1984) The ontogenetic niche and species interactions in size-structured populations. *Annual Reviews in Ecology and Systematics*, **15**, 393–425.

Wisenden, B.D. (2000) Olfactory assessment of predation risk. *Philosophical Transactions of the Royal Society*, **355**, 1205–1208.

Wisenden, B.D. & Harter, K.R. (2001) Motion, not shape, facilitates association of predation risk with novel objects by fathead minnows (*Pimephales promelas*). *Ethology*, **107**, 357–364.

Yunker, W.K., Wein, D.E. & Wisenden, B.D. (1999) Conditioned alarm behavior in fathead minnows (*Pimephales promelas*) resulting from association of chemical alarm pheromone with a non-biological visual stimulus. *Journal of Chemical Ecology*, **25**, 2677–2686.

Zhao, X. & Chivers, D.P. (2005) Response of juvenile goldfish (*Carassius auratus*) to chemical cues: relationship between response intensity, response duration and the level of predation risk. In: R.T. Mason, M. LeMaster & D. Müller-Schwarze (eds), *Chemical Signals in Vertebrates*, Vol. 10, pp. 334–341. Springer, New York.

Chapter 5
Learning and Mate Choice

Klaudia Witte

5.1   Introduction

One of the most fascinating questions within the field of sexual selection is why and how females and males choose particular conspecifics as mates. Sexual selection theories provide different explanations for the origin and evolution of ornamental traits and mate preferences for such traits (overview in Andersson 1994; Jennions *et al.* 2001; Kokko *et al.* 2003). Genes for the ornamental trait and genes for the concomitant preference can co-evolve as a result of a genetic linkage between those genes as a result of the Fisherian runaway process (Fisher 1930; Lande 1981; Kirkpatrick 1982; Brooks 2000) or as a result of a selection for mates advertising 'good genes' (indicator models; Zahavi 1975; Møller & Alatalo 1999). These models generally assume that mate preferences are genetically based (Bakker 1999). Forming mate preferences, however, is a complex process involving not only genetic factors but also non-genetic factors. Increasing evidence suggests that the social environment (Dugatkin 1996a; Westneat *et al.* 2000) and learning are important factors in forming mate preferences. The mate choice of conspecifics influences the mate choice decisions of an individual, who can alter mate preferences through learning processes. Social learning, and using public information (Danchin *et al.* 2004; and see Chapter 10) and other kinds of learning, therefore, significantly influence the process of sexual selection. Forms of social learning have now been recognized as meaningful mechanisms for the non-genetic inheritance (i.e. cultural transmission) of mate preferences, leading to cultural evolution of mate preferences. This chapter illustrates how learning is involved in the mate choices of fishes and emphasizes the important roles that different kinds of learning, particularly social learning, play in sexual selection. It focuses on four different kinds of learning: sexual imprinting; learning after reaching maturity; eavesdropping; and mate-choice copying.

5.2   Sexual imprinting

Sexual imprinting is a learning process restricted to a specific period during early development, which influences mate preferences later on in life (Immelmann 1972). Sexual imprinting is well studied in many bird species (ten Cate & Vos 1999; Witte & Sawka 2003) and occurs also in mammals (Kendrick *et al.* 1998; Penn & Potts 1998; Owens *et al.* 1999), including human beings (Bereczkei *et al.* 2004). A prerequisite for sexual imprinting is that at least one parent, a genetic or social parent, cares for the young to ensure that young have intensive contact with the parent(s) and get the opportunity to learn specific traits of the parent(s). Mate preferences learned by sexual imprinting can be transmitted from one generation to the next generation in a socially inherited way. Thus, sexual imprinting is assumed to be a powerful mechanism for the cultural evolution of mate preferences. Theoretical models show that sexual imprinting potentially plays an important role in sexual selection (Aoki *et al.* 2001) and in sympatric speciation (Laland 1994a).

5.2.1 *Sexual imprinting in fish species*

Is there any evidence for sexual imprinting in fishes? This question was investigated 30–50 years ago but has since been subject to reduced attention within behavioural biology. Early studies provided some evidence that phenotypic traits of parents and/ or siblings are learnt during early development and that this learning process influences mate preferences later on in life. Ideal candidates for studying sexual imprinting are cichlids. Within the family Cichlidae, at least one parent cares intensively for the young; thus, the main prerequisite for sexual imprinting to occur is satisfied. In an early study, Fernö & Sjölander (1976) used the convict cichlid (*Archocentrus nigrofasciatus*, Cichlidae) to investigate whether the colour of parents influenced the mate-choice decision of siblings later on in life. This species is a typical substrate-spawning cichlid with long-lasting parental care in both parents, and both parents have similar appearance. In a cross-fostering experiment, they used two different colour morphs, a commercially available white morph and the natural greyish-blue morph with vertical black stripes. In a pilot study, Fernö & Sjölander (1976) found the first indication that the colour morph of parents seemed to influence the mate preferences in males later on in life. However, their results provided no information on whether an imprinting process occurred because experience with the other colour morph was not restricted to a particular period during early development. Their study was repeated with a more controlled experimental design by Siepen (1984), in which siblings of *C. nigrofasciatum* (*A. nigrofasciatus*) were cross-fostered within and between the two colour morphs (10 males and 10 females in each design). In 7 of 19 cases, fishes reared by parents of the colour morph different from their own morph later mated with a fish of the parents' colour morph. Fishes cross-fostered within the same colour morph did not pair in any of 20 cases with a prospective mate of the other colour morph (Fisher's exact test $P = 0.014$, added by K. Witte). This result provides some indication that the parental colour morph influenced mate-choice decision later in life. Thus, sexual imprinting may play an important role in forming mate preferences in the convict cichlid.

5.2.2 *Does sexual imprinting promote sympatric speciation in fishes?*

In recent years, the topic of sexual imprinting in fishes has again begun to receive attention within behavioural ecology, largely because sexual imprinting is thought to be a meaningful mechanism for promoting sympatric speciation. In an ongoing study, M. Verzijden (Verzijden personal communication) is investigating the effect of sexual imprinting on the mother's phenotype in two sister species of African mouth-breeding cichlids (*Pundamilia nyererei, P. pundamilia*, Cichlidae) living in sympatry in Lake Victoria, Africa. In both species, only females provide brood care to young. In a cross-fostering experiment, Verzijden exchanged eggs between breeding females of the two different species. After reaching maturity, female offspring were allowed to choose between males of both species. Preliminary results show that cross-fostering influenced the mate-choice decisions of females in these species. It seems that female offspring sexually imprinted on the mother's phenotype and preferred heterospecific males over conspecific males when reared by a heterospecific foster mother.

Sexual imprinting might also play a role in forming mate preferences in the three-spined stickleback (*Gasterosteus aculeatus*, Gasterosteidae), a species in which males care intensively for the young. In several lakes in British Columbia, Canada, three-spined sticklebacks occur in sympatric species-pairs. These pairs consist of a large-bodied invertebrate-feeding benthic species and a small-bodied zooplankton-feeding limnetic species (Schluter & McPhail 1992). Albert (2005) investigated sexual imprinting in these two species pairs. She created F_1 hybrid females *in vitro* from wild-caught individuals of both species. Six F_1 families were fostered by limnetic males and five F_1 families were fostered by benthic males. After reaching sexual maturity, F_1 females were tested for their mate preferences in two sequential no-choice tests. In one test, the female could inspect a nesting benthic male; in the other test the female could inspect a nesting limnetic male. Results showed that there was no effect of sexual imprinting on forming mate preferences in stickleback F_1 hybrid females. Females that were raised by benthic males were not more likely to inspect benthic males than females raised by limnetic males. Instead females chose males that were similar to themselves in body length. Size-assortative mating seems to be the underlying mechanism for reproductive isolation in these species, but not sexual imprinting.

Nevertheless, sexual imprinting can potentially be an important learning process in fishes for forming mate preferences. It may be worth investigating which factors facilitate the occurrence of sexual imprinting in a species and its potential role in sympatric speciation.

5.3   Learning after reaching maturity

Whereas sexual imprinting is a learning process that is restricted to a specific sensitive period during early development, there may be other learning processes that are not restricted to a specific sensitive period and that occur in sexually mature individuals. Sexual imprinting is limited to species with intense parental care. Learning processes exhibited later on in life that affect mate choice, however, are observed in a wide range of fish species, including those without any parental care, like live-bearing fishes. In an early study, Haskins & Haskins (1950) showed that when guppy (*Poecilia reticulata*, Poeciliidae) males are reared in isolation until sexual maturity and then exposed to females of a specific colour variant, which differs from their own colour variant, for a month or longer, males preferred females of the colour variant with which males were reared after reaching sexual maturity. Liley (1966) investigated species recognition in four sympatric species within the family Poeciliidae.

Guppy males with female experience restricted to conspecifics did not show a preference for conspecific females when females of three other species were present. He concluded that males must require experience with females of their own species, as well as with females of other species, to learn to discriminate between conspecific and heterospecific females. Haskins and Haskins (1949) investigated whether male guppies learn to discriminate between conspecific and heterospecific females by experience. They presented guppy males with females of three related species (*P. reticulata*, *P. picta* and *P. vivipara*). Male guppies that had had no experience with heterospecifics, initially directed most of their courtship displays towards swamp guppy females (*P. picta*). After about a week, however, males courted mostly conspecific females (*P. reticulata*). Learning may, thus, modify mate preferences in guppy males and may help to prevent them from mating with the wrong species. Such modification of choice via learning may be mediated via feedback emanating from potential mates. Several other studies have shown that male preferences in guppy females are also altered by experience (Breden *et al.* 1995; Rosenqvist & Houde 1997; Jirotkul 1999).

5.3.1   *Learning when living in sympatry or allopatry*

The early study of Haskins & Haskins (1949) left many questions open. Unfortunately, their sample size was small, data from individual males were combined, and important controls were missing. In a recent study, Magurran & Ramnarine (2004) utilized a more controlled design to investigate the role of learning in mate choice among sexually mature Trinidadian guppy (*Poecilia reticulata*) males and swamp guppy (*P. picta*) males living either in sympatry or allopatry. In a baseline mate choice test, two *P. reticulata* males, collected from the same locality, or two *P. picta* males, could physically interact with one *P. reticulata* female and one *P. picta* female, matched for size. One of the two males was the focal male and the authors recorded the number of sneaky matings of that male for a period of 15 min. Males of both species living in allopatry attempted matings with heterospecific females and conspecific females at random. However, males living in sympatry with the other species preferred to attempt matings with females of their own species. In a test for learned preferences, two males, both from the same localities, were housed together with two size-matched females, one of each species. The focal male could interact with different females of both species from day 1 onwards for a further 5 days for 10 min each day. Trinidadian guppy males with no experience with *P. picta* females learnt to discriminate between heterospecific and conspecific females and preferred conspecific females within a few days.

5.3.2   *Learned recognition of colour morphs in mate choice*

Adult learning may also be an important factor influencing male mate-choice decisions in the live-bearing platy fish (*Xiphophorus maculatus*, Poeciliidae). Fernö & Sjölander (1973) investigated how experiences in different stages of life influenced the mate-choice decisions of male platy fishes for females of red and black colour morphs. In one experiment, young platy fishes were reared either with or without the other colour morph. In another condition, platy males were reared together with both colour morphs until they reached sexual maturity. Afterwards they were kept together with fishes of the colour morph different from their own for 2 months. When males were sexually mature for at least 2 months, they were tested in a four-choice arena for their sexual preference and could choose between both sexes of each colour morph. Male platy fishes preferred females of the colour morph they were exposed to during the 2 months after reaching sexual maturity. Exposure to a colour morph during an early stage of development, that is, before reaching maturity, had no influence on mate choice. Thus, learning after reaching maturity is an important process for forming mate preferences in male platy fishes.

   These studies show that learning even after reaching maturity is a significant component in forming mate preferences. Thus, learning is an important factor which should be considered in studies on sexual selection.

5.4   Eavesdropping

Eavesdropping occurs when information from an animal transmitting a signal to another individual is 'overheard' by one or more bystanders towards whom the signal was not directed (McGregor 2005). Eavesdropping is now recognized as representing an important component of animal communication, particularly communication in a network, and has been studied intensively in songbirds and fishes (McGregor 2005). Females can gain information about potential males by assessing their quality on the basis of morphological cues (Endler & Houde 1995; Houde 1997). In addition, however, by observing two males interacting (for instance, fighting) with each other, females gather further reliable information about these males that they can then use to guide mate-choice decisions. Eavesdropping is defined as extracting information from signalling interactions between others (McGregor & Dabelsteen 1996), and it can be an effective way for females to evaluate potential males.

5.4.1   *Eavesdropping and mate choice*

Doutrelant & McGregor (2000) investigated whether female Siamese fighting fish (*Betta splendens*, Osphronemidae), monitor aggressive interactions between two males and whether the information gained by eavesdropping is used to guide mate-choice decisions. In a well-controlled experimental set-up, they found that females that had the opportunity to watch two displaying males subsequently first visited the winner significantly more often, spent significantly more time with the winner, and spent more time looking at, and displaying to the winner, than the other male. Females that had not seen the interaction between two males visited the loser first more often than females under the other condition, and did not behave differently to winner and loser. This experiment shows that females use the information gained from an aggressive interaction between two males in their mate-choice decision.

5.4.2   *The audience effect*

The eavesdropper does not only gain information about the two interacting individuals, but the very presence of the eavesdropper may influence the nature of the interaction. This so-called 'audience effect' or 'bystander effect' has also been invest-igated in Siamese fighting fish. Doutrelant *et al.* (2001) tested whether the presence of a female or male changed the intrasexual interaction between two fighting males. In the experiments, the two fighting males could interact with each other through clear partitions, over two trials. In one trial, both males saw a female prior to interacting with the other male for 3 min. In the other trial, the same males did not see a female before the interaction started. A similar experiment was performed with a male as an audience. The results clearly show that a female audience changes the male–male interactions. With a female audience, significantly more males performed more tail beats, spent more time with gill cover erected, interacted farther away from the other male, and performed fewer bites than without a female audience. Thus, males performed fewer aggressive displays that are used only in male–male interactions and more of the displays that are considered more conspicuous used in the presence of both sexes. Conversely, whether a male audience was previously present or not did not significantly change the characteristics of the male–male interaction. A similar result was found by Matos & McGregor (2002) in the same species. When a male audience was present prior to the encounter, males attempted significantly more bites and spent less time near the opponent than when a female audience was observed prior to the encounter.

5.4.3   *Benefits of eavesdropping*

What are the benefits of eavesdropping? In general, mate choice is costly for females because it requires them to devote time to evaluating males and may expose them to enhanced predation risk (Andersson 1994). Additionally, mate-sampling females may be injured in aggressive courtship displays by males or even suffer from harass-ment by males (Schlupp *et al.* 2001). Eavesdropping females can avoid some of these costs, gaining cheap information about male quality without being directly involved in an interaction with conspecifics. Moreover, they may be able to watch interactions between several conspecifics at the same time thus providing direct comparisons. Females can then use this information about male quality gained from eavesdrop-ping to supplement direct information gained on the basis of male morphological cues. Eavesdropping females gain information on the relative quality of males at little cost and/or risk (McGregor & Peake 2000). Information gained by observing an aggressive interaction between two individuals is assumed to be reliable and not subject to cheating. From this perspective, eavesdropping may be more reliable than mate-choice copying (see below), where a female may copy a 'wrong' choice of the model female. While eavesdropping seems to be a good strategy for mate choice, as yet there is no quantitative evidence for any fitness advantages based on this strategy. Future experiments may focus on this.

5.5   Mate-choice copying

Models of sexual selection assume that females and males choose among poten-tial mates independently of conspecifics. There is, however, strong evidence that females and males sometimes choose a mate non-independently by copying the choices of conspecifics. Mate-choice copying occurs when an observation of a sexual interaction between a male and a female influences the subsequent mate-choice decision making of the observing individual, biasing their decision to favour the observed mating individual. Mate-choice copying is a form of social learning in which individuals gain information and learn about potential mates by observing conspecifics (see Chapter 10 for a review of social learning in fishes). Mate-choice copying is an important mate-choice strategy demonstrating that individuals gather and use social and public information (Danchin *et al.* 2004). The first prerequisite and necessary condition for mate-choice copying to occur is that individuals must be able to observe the mate choices of other (Losey *et al.* 1986). To qualify as mate-choice copying it must be the sexual interaction, and not the consequence of the choice of a female or a male, that influences the mating decision of another (Pruett-Jones 1992). Thus, it is not mate-choice copying when females prefer to lay their eggs in nests that already contain eggs, as in the bullhead goby (*Cottus gobio*, Cottidae; Marconato & Bisazza 1986), fathead minnow (*Pimephales promelas*, Cyprinidae; Unger & Sargent 1988), fantail darter (*Ethiostoma flabellare*, Percidae; Knapp & Sargent 1989), and three-spined sticklebacks (Ridley & Rechten 1981; Goldschmidt *et al.* 1993). This behaviour can be explained by dilution of the risk of egg predation or egg cannibalism (Rohwer 1978), or as resulting when male sticklebacks that have eggs in their nests court more vigorously and are, therefore, preferred by females (Jamieson & Colgan 1989). Patriquin-Meldrum & Godin (1998) provide experimental evidence that three-spined stickleback females do not copy the mate choice of others. Mate-choice copying is most likely to occur in polygynous and promiscuous mating systems with no parental care or with maternal care only, and has been studied intensively in polygynous birds (Höglund *et al.* 1995; White & Galef 1999, 2000) and fishes.

Several theoretical models have investigated how copying could evolve and be maintained in a population (Losey *et al.* 1986; Gibson & Höglund 1992; Pruett-Jones 1992; Laland 1994b; Nordell & Valone 1998; Stöhr 1998; Sirot 2001). Wade & Pruett-Jones (1990) showed that copying is likely to increase the variance in mating success among males, and thus intensify sexual selection. Servedio & Kirkpatrick (1996) showed theoretically that an allele for copying can spread through a population even when there is mild selection against it.

5.5.1   *Mate-choice copying – first experimental evidence and consequence*

The first experimental evidence for mate-choice copying came from Dugatkin's (1992) study of Trinidadian guppies (*Poecilia reticulata*, Poeciliidae). The experiment consisted of an observation period and a preference test. During the observation period, a focal female could observe two males, each in a separate end chamber of the test aquarium, with one male courting another female (the so-called 'model female') who was adjacent to the male but separated by glass. After removing the model female, the focal female was allowed to choose between the two males, and the time the focal female spent within the preference zones adjacent to the male chambers was measured as her mate preference. In 17 out of 20 trials the focal females preferred to associate with the male that they had previously observed interacting with the model female. Although convincing, this result could have been explained by alternative hypotheses, which Dugatkin tested in four other experiments:

1 schooling hypothesis: females prefer the side where they have seen the largest group of fishes;
2 no interaction-hypothesis: females prefer the side where they have seen a male and a female but without sexual interaction between the pair;
3 male activity hypothesis: females prefer a more active male;
4 sexual priming hypothesis: females might become sexually primed after observing an interaction between a male and another female and choose males that behaved as though they have just been involved in an interaction with a female.

Results of these additional experiments provided no support for any of these four alternative hypotheses. By excluding these alternative explanations, Dugatkin's experiment provides the first experimental evidence for mate-choice copying in guppies. Following his study, several other studies regarding mate-choice copying in fishes have been published (see below) and research on this topic is continuing (Godin *et al.* 2005; Uehara *et al.* 2005).

At the present time there is good evidence for mate-choice copying in guppies (Dugatkin & Godin 1992, 1993; Dugatkin 1996a,b, 1998a,b; Godin *et al.* 2005), and in other fish species like the sailfin molly (*Poecilia latipinna*, Poeciliidae; Schlupp *et al.* 1994; Schlupp & Ryan 1997; Witte & Ryan 1998, 2002; Witte & Noltemeier 2002; Witte & Massmann 2003; Witte & Ueding 2003), the humpback limia (*Limia nigrofasciata*, Poeciliidae; Munger *et al.* 2004), and the Japanese medaka (*Oryzias latipes*, Adrianichthyidae (Grant & Green 1996, but see Howard *et al.* 1998).

However, there are other studies that failed to detect mate-choice copying in different fish species, including some of the above. Brooks (1996) could not detect mate-choice copying in guppies from a feral South African population; Lafleur *et al.* (1997) found no indication for mate-choice copying in pet store guppies (but see Dugatkin 1998a for a comment on this study); and Patriquin-Meldrum & Godin (1998) found no evidence for mate-choice copying in the three-spined stickleback. Ambiguous results for the Perugia's limia (*Limia perugiae*, Poeciliidae) were reported by Applebaum & Cruz (2000). It is currently unclear why mate-choice copying should be observed in some situations and not others.

5.5.2 *Mate-choice copying – evidence from the wild*

There is convincing evidence that females and males copy the mate choice of other conspecifics in several fish species. However, until recently all experiments on mate-choice copying in fish species had been performed in the laboratory, rendering it unclear whether males and females copy the choice of others in their natural habitat, and under natural conditions. To show that mate-choice copying is a biologically relevant mate-choice strategy, it is important to demonstrate that this mate-choice strategy occurs in the wild as well. Witte & Ryan (2002) studied this significant issue in the sailfin molly. The sailfin molly is a good candidate for investigating mate-choice copying in the wild because previous studies have shown that males (Schlupp & Ryan 1997) and females (Witte & Ryan 1998) copy the choice of others under standardized conditions in the laboratory. The field study (Witte & Ryan 2002) was performed in the Comal River, near New Braunfels, Texas, USA, where it was easy to observe free-swimming sailfin mollies in a wild sailfin population. Because it was not possible to perform the complicated design for a mate-choice copying experiment in the laboratory in the river, Witte & Ryan (2002) used a simpler design (Fig. 5.1a,b). They performed experiments for male mate-choice copying, female mate-choice copying, and a control for shoaling behaviour in each sex (Fig. 5.1b). In the case of female mate-choice copying (Fig. 5.1a), they presented focal females with two stimulus males, each in a jar with a net on top standing on two upside-down plastic tanks in a natural river. Next to each jar with a stimulus male was a second jar containing a model female or no fish for a symmetrical set-up. The two pairs of jars formed a corridor. Witte & Ryan (2002) counted the number of females

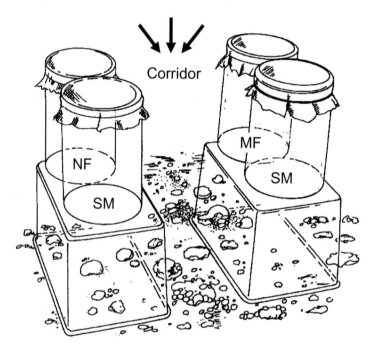

Fig. 5.1   (a) Top view of the set-up of the female mate-choice test in the sailfin molly in a river. Two jars stood on each of two upside-down plastic tanks. Each jar had a net on top and was filled with the water from the river. In the female mate-choice tests, Witte & Ryan (2002) presented stimulus males (SM) in two jars; in one jar next to a stimulus male was the model female (MF) and the fourth jar had no fish (NF). Only females were counted when they entered the set-up from the side with the empty jar and the model female (indicated by the arrows), came through the 'corridor' and stopped within one body length of the jar containing the stimulus males. For the male mate-choice test, stimulus females (SF) replaced stimulus males and a model male the model female. In the male control for shoaling, Witte & Ryan (2002) presented two stimulus males and an extra male. Only males were counted, when they entered the set-up from the side with the empty jar, and the model male (indicated by the arrows) came through the 'corridor' and stopped at the jar within one body length of the jar with the stimulus females.

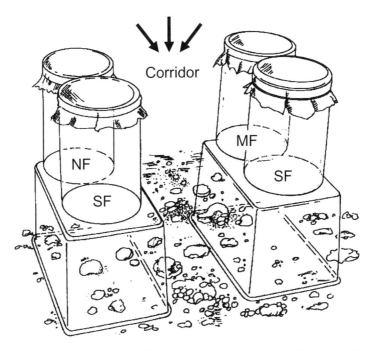

Fig. 5.1 (b) Top view of the set-up of the female control test for shoaling in the river. In the set-up of the female control for shoaling there was a stimulus female (SF) and a jar with a model female (MF) on one tank and a stimulus female next to a jar with no fish (NF) on the other tank. Females were counted only when they entered the set-up from the side with the empty jar and the extra female (indicated by the arrows), came through the 'corridor' and stopped within one body length of the jar with a stimulus female.

swimming into this corridor and interacting with the stimulus male next to the model female or with the lone stimulus male. In the case of male mate-choice, they presented stimulus females and a model male in the same manner and counted the number of males swimming into this corridor and interacting with the stimulus females. These experiments indicated that males and females copy the choices of others; that is, females preferred to associate with the male next to a model female rather than a lone male, and males preferred to associate with a female next to a model male rather than next to a lone female. Neither sex showed shoaling behaviour in this experimental set-up. This field experiment is a convincing indication that mate-choice copying is a biologically relevant mate-choice strategy in sailfin mollies. Witte & Ryan (2002) provide a practicable design for mate-choice copying studies in the natural habitat of a fish species.

   To strengthen the significance of mate-choice copying for sexual selection, mate-choice copying under natural conditions should be investigated in other species as well. The Trinidadian guppy, which lives in different river systems under quite different environmental conditions, would be an ideal system in which to investigate which environmental conditions facilitate or complicate the evolution of mate-choice copying.

5.5.3   *Copying mate rejection*

According to the definition of Pruett-Jones (1992), female mate-choice copying occurs when 'the conditional probability of choice of a given male by a female is either greater or less than the absolute probability of choice depending on whether that male mated previously or was avoided, respectively. The outcome of the female is that if one female mates with or avoids a specific male successively choosing females will be accordingly more or less likely to mate with that male than they would otherwise have been' (Pruett-Jones 1992, pp. 1001–2). Thus, mate-choice copying may decrease the probability that a female mates with a particular male when the female has observed that another female has rejected that male. This idea was investigated experimentally by Witte & Ueding (2003). In their copying experiment, they presented video playbacks of sailfin molly males on two TV monitors, one standing adjacent to each side of the female test tank. First, females were allowed to choose between two male video playbacks and the time females spent within a preference zone in front of the playbacks was used as a measurement of male attractiveness. After the first preference test, females had the opportunity to copy a mate rejection. Females could observe another female escaping from the attractive male when the male tried to court the additional female on one monitor and on the other monitor the previously unattractive male was alone. Afterwards females could choose between the same two single male presentations again. After observing a rejection, females spent significantly less time with the previously attractive male and 8 of 15 females reversed their initial choice and spent more time with the previously unattractive male. In control conditions, sailfin molly females with no opportunity to copy chose consistently. A control condition with the escaping behaviour of a female, but without a.male, did not result in females spending less time with the previously attractive male in the second preference test. Thus, it would seem that the females copied the rejection of a male by another female.

What is the advantage of copying a rejection? During the process of mate choice, females may sample several males (Janetos 1982; Real 1990) and may observe several independent choice events by other females. By observing another female rejecting a male, females might also ignore this male and concentrate on other males as potential mates and by so doing they might be able to save time and energy, and reduce exposure to predation. Additionally, by observing the rejection of a male, females might learn which male phenotype is 'bad' as a potential mate. Thus, in copying the mate rejection of other females, females may protect themselves against 'wrong' choices. According to sexual selection theories, the rejection of a male is as important as the acceptance of a male, because females might suffer low reproductive success when mating with poor quality males. Copying not only mate acceptance but also mate rejection will strongly amplify the skew of reproductive success in males within a population. Copying the rejection of a male is, therefore, potentially as important as copying mate attraction for the process of sexual selection.

5.5.4   *The disruption hypothesis – an alternative explanation to mate-choice copying?*

Mate-choice copying as an alternative mate-choice strategy to independent mate-choice decisions is still under debate, and alternative explanations to some findings have been proposed. Applebaum & Cruz (2000) argued that a disruption effect could better explain mate-choice behaviour in Perugia's limia females than mate-choice copying. They proposed the disruption hypothesis in which 'biotic and/or abiotic events that occur in the perceptible field of a female actively engaged in a mate-choice may disrupt the decision-making and information processing of that female. This may result in a reduction (or increase) in the consistency with which that female selects males with preferred traits'. Applebaum & Cruz stated that previous results showing that guppy females reverse their choices after observing another female interacting with the previously non-preferred male (Dugatkin & Godin 1992) could be interpreted as a disruption effect instead of mate-choice copying. Applebaum & Cruz tested the disruption hypothesis in the live-bearing poeciliid Perugia's limia. They used a design similar to that used by Dugatkin & Godin (1992) and performed three experiments. In each experiment, the focal female could first observe two males placed in the two end chambers of the test tank and separated by glass from the female placed in a clear plastic cylinder. After this observation period, the female was released into the test tank and allowed to choose between the two males. After 30 min with males blocked from view, females could choose a second time between the same males. In Experiment 1, no model females were placed next to a male during the previous observation period. Thus, this experiment tested whether females chose consistently with no opportunity to copy. In Experiment 2, the focal female could view a model female next to the male the focal female had not preferred in the mate choice test before. Afterwards the focal females were allowed to choose a second time between the same males. In Experiment 3, the focal female could view a model female next to her previously preferred male, and was allowed to choose between the same males a second time afterwards. The results indicated that females in Experiment 1 chose consistently when they had not seen a model female next to one male. In Experiment 2, significantly more females than in Experiment 1 reversed their initial preference and preferred the previously non-preferred male in the second mate-choice test. In Experiment 3, a similar fraction of females reversed their initial preference, that is, spent less time with the initially preferred male in the second mate choice test. Applebaum and Cruz interpreted this result as an indication of a disruption effect. In both Experiments 2 and 3, females did not choose consistently and the consistency may decline as a result of disruption during mate-choice decision-making by the presence of the model female.

There are, however, at least two factors that imply that the disruption hypothesis is a less likely explanation than mate-choice copying. First, in all three experiments the absolute time females spent with the preferred male did not decrease between the first and second mate-choice test. If females had been disrupted during mate choice, one might expect that they would have lost interest in the males and spent more time outside the preference zones during the mate-choice test, and thus had spent less time with preferred males. Second, the idea that females may become confused by observing another female interacting with a male seems implausible, especially in fish species living in schools in which school members interact with each other all the time and gather and use social information in the context of foraging behaviour (Laland & Reader 1999), mate choice and predator avoidance. Dugatkin (1992) has shown that guppy females can remember males that they have seen interacting with a female before, and additionally Witte & Massmann (2003) have shown that sailfin molly females remember males they have seen interacting with a female after a day. Thus, the ability of individual recognition in females is much better than assumed by the disruption hypothesis (see Chapter 8 for a review of individual recognition in fishes).

Stimulated by the study of Applebaum & Cruz (2000), Dugatkin *et al.* (2003) repeated their experiment with guppies and tested the two contrary predictions of the disruption hypothesis and mate-choice copying when a model female is placed next to the male that was previously preferred by the focal female. When mate-choice copying is operating, focal females would maintain their initial preference as they would be reinforced by the apparent choice of the model female. If disruption is occurring, focal females would be expected to reverse their initial choice and would not choose consistently. Dugatkin *et al.* (2003) showed that guppy females chose consistently when there was no model female present during the observation period: only 6 of 20 females reversed their choices. When the model female was placed next to the previously non-preferred male, 14 of 20 females reversed their choice. When a model female was place next to the previously preferred male, however, only 6 of 20 females reversed their choice and thus females maintained their initial preference. This result strongly supported mate-choice copying in guppies and the hypothesis that disruption plays no role during mate-choice decision-making in guppy females.

Munger *et al.* (2004) performed the same three experiments with the humpback limia as performed by Applebaum & Cruz (2000) with guppies. Munger *et al.* (2004) found no evidence for the disruption hypothesis. Humpback limia females chose consistently between two males when they had no opportunity to observe another female interacting with a male. Females reversed their initial choices after observing another female interacting with the male the focal female had initially rejected. Females maintained their initial preference for a male when they had observed another female interacting with the initially preferred male. Thus, these experiments indicate that humpback limia females copy the mate-choice of others.

In summary, as the evidence currently stands, the disruption hypothesis is not a compelling alternative explanation to mate-choice copying. It might be possible, however, that females may be influenced differently by different observations of mate choice (Agrawal 2001). It could be that a sexual interaction between an attractive male and a female stimulates the female in a different way than an observation of a sexual interaction between a female and an unattractive male, or a sexual interaction between an attractive male and a female may be perceived by the observing female differently from an interaction between an unattractive male and a female. Future research should focus on how females perceive such sexual interactions and what information extracted from this observation is important for mate-choice decision-making.

5.6   Social mate preferences overriding genetic preferences

5.6.1   *Indications from guppies*

The evolution of mate preference is a complex process in which genetic and non-genetic factors are involved. Several models indicate how genetic factors influence mate choice, and we know how social cues and the environment can influence the mate-choice decision. However, it is less clear how genetic and social factors interact and how this interaction can influence a female's mate-choice decision. Two studies show how a genetically based mate preference is influenced by social learning, that is, by mate-choice copying.

Guppy females generally exhibited a genetically-based preference for males with a higher amount of orange-coloured body surface (Houde 1988, 1992, 1997; Endler & Houde 1995). Guppy females also copy the mate choice of another female when both males presented in a test are matched for size and body colouration, and that is true for guppies of different populations (Dugatkin 1992, 1996a; Dugatkin & Godin 1992; Briggs *et al.* 1996). How do guppy females respond when they are challenged with a conflict between their genetically based mate preference and a socially based mate preference? Dugatkin (1996b) presented guppy females with this dilemma in a mate-choice copying experiment. He varied the difference in male body colouration between males presented in a test. The two males presented simultaneously in a binary choice situation differed by 10%, 25% or 40% in total orange body colouration. In all cases, test females observed model females next to, and interacting with, the less colourful male, for a period of 5 min. Afterwards, the test female was allowed to choose between the two males. When males differed in only 10% or 25% of the amount of orange body colouration, females copied the choice of the model female and preferred to associate with the paler of two males, despite a strong genetic preference for more colourful males. When males differed by 40% in orange body colour, however, test females always preferred the more colourful male, although they observed an interaction between the model female and the paler male. Thus, in this case, the genetic preferences seem to have a stronger influence on the mate-choice decision than the social cues.

In a later study, Dugatkin (1998b) further explored the interaction between genetic and social factors with regard to a preference for orange-coloured males in guppies. In this study, he presented two males simultaneously to females and these males always differed in the amount of orange by on average of 40%. In different experiments, test females could observe either no model female, one model female interacting with the drabber male for 5 min, two different model females interacting with the drabber male each for 5 min, or one model female interacting with the drabber male for 10 min. When females observed no model female or one model female next to the drabber male, they did not copy the choice and preferred the more colourful male, thus females followed their genetic preferences. These results were consistent with the previous findings (Dugatkin 1996b). When females observed two different model females next to the drabber male, 12 of 20 females preferred the drabber male, and 13 of 20 females that had observed one model female next to the drabber male for 10 min preferred the drabber male. Thus, in these two experiments, social cues were shown to override the genetic predisposition and had a stronger influence on mate-choice decision than genetic factors. Thus, it seems that the amount of information a female can gain by observing the sexual interaction between a male and the model female lowers the threshold in favour of social cues having a stronger influence on the mate-choice decision than genetic factors. In these cases, social preference overrides the genetic preference in guppy females.

5.6.2 *Indications from sailfin mollies*

Sailfin mollies also provide evidence for an interaction between genetic and social factors influencing mate-choice decisions. Sailfin molly females show a strong preference for larger over smaller males, which has been documented in different populations of sailfin mollies (Marler & Ryan 1997; Ptacek & Travis 1997; Witte & Ryan 1998; Gabor 1999; Witte & Noltemeier 2002; MacLaren *et al.* 2004; Schlupp *et al.* 1994). Marler & Ryan (1997) provide strong evidence that the preference for larger males in sailfin molly females is genetically based. Witte & Noltemeier (2002) investigated the relative importance of genetic and social cues regarding the female preference for larger males. In a standard mate-choice copying experiment, females could first independently choose between a smaller and a larger male, which differed in standard body length on average by 12 mm. In the first mate-choice test, all females preferred the larger over the smaller male. After this independent mate choice, females had the opportunity to observe a sexual interaction between a model female and the smaller male. Afterwards females were allowed to choose between the same larger and smaller male again. The authors varied the situation during the observing period in three experiments. In the first experiment, females could observe one model female next to the smaller male for 10 min. Afterwards, as predicted by previous experiments (Witte & Ryan 1998), females did not copy and still preferred the larger over the smaller male. In a second experiment, females were allowed to observe two different females interacting with the smaller male each for 5 min. Thereafter, 7 of 15 females reversed their initial preference for the larger of two males and associated significantly more often with the smaller male. The strongest effect was in the experiment in which females could observe one model female interacting with the smaller of two males for 20 min. Here, 13 of 15 females reversed their mate choice in favour of smaller males. Thus, social preference overrides the genetic preference in favour of smaller males. In several control conditions where there was no opportunity to copy, Witte & Noltemeier (2002) found that females consistently preferred the larger over the smaller males, while females in other control conditions exposed solely to stimulus females did not show shoaling behaviour that might explain the experimental findings. This study suggests that genetic factors interact with social cues during mate choice. Depending on the amount of social information received, females may be more influenced by their genetically determined mate preference or social cues. These experiments demonstrate the significance of social learning for mate choice and emphasize the potential of mate-choice copying to precipitate sexual selection.

5.7 Cultural evolution through mate-choice copying

Several studies have demonstrated that females change their initial mate preferences as a result of mate-choice copying. However, for mate-choice copying to be a meaningful mechanism for the cultural inheritance of mate preferences, it is necessary to show that females do not only copy the choice of a particular male, but also acquire and maintain a preference for a particular male phenotype (Brooks 1998). We now have evidence from two studies in fishes that mate-choice copying achieves these criteria for cultural inheritance of female mate preferences.

Females that had previously copied the mate choice of a smaller male after observing one model female interacting with the smaller of two males for 20 min, were retested by Witte & Noltemeier (2002) for a preference for smaller males up to 36 days after copying. In the intervening period, females were kept isolated from males. In a binary choice situation, females could choose between a smaller and a larger male, both of which were unfamiliar males that had not been used in the copying experiment. Thus, the authors tested whether females still exhibited a preference for a specific male phenotype, and not a preference for an individual male. Females that had previously reversed their mate preference in favour of smaller males through mate-choice copying, maintained this preference for smaller males in the binary mate-choice experiment. This was the first evidence in fishes that females copied a mate choice for a male phenotype and that females maintain a mate preference learned by mate-choice copying for a considerable period of time. These females may serve as models for other females and may induce a new mate preference in favour of smaller males within a population. Thus, the prerequisites for mate-choice copying as a mechanism for the cultural inheritance of mate preferences were fulfilled.

More recently, Godin *et al.* (2005) have presented further evidence that guppy females copy the choice for a male phenotype and not just a choice for an individual male and that females maintained their copied mate preference. The authors showed that guppy females copied the choice of other females for less colourful males and that these females still preferred less colourful males the next day when different males were presented in a mate-choice experiment. These two studies provide good evidence that cultural transmission of mate preferences via mate-choice copying is possible in fishes even when the socially induced mate choice conflicts with the genetically based mate preference. These studies, therefore, emphasize mate-choice copying as a powerful influence on sexual selection.

5.8   Does mate-choice copying support the evolution of a novel male trait?

How secondary sexual traits have evolved in males is one of the most fascinating questions in sexual selection. The sensory exploitation hypothesis (Ryan & Keddy-Hector 1992; Ryan 1998) states that females have latent preferences for particular male traits before the evolutionary appearances of these traits in males. These latent mate preferences are shaped by natural selection, mostly acting in the context of foraging behaviour, through pleiotropic effects of genes expressed in both foraging and mate choice. Guppy females prefer males with a higher amount of orange colouration on the body surface (Houde 1988, 1992, 1997; Endler & Houde 1995). Rodd *et al.* (2002) showed that this mate preference probably originated as a pleiotropic effect of a sensory bias for the colour orange, which might have arisen in the context of food detection (e.g. in search of prey items such as crustaceans containing high levels of carotenoids in their exoskeleton). In field and laboratory experiments, Rodd *et al.* (2002) showed that both male and female guppies are more responsive to orange-coloured objects than to objects of other colours, and that was true even outside the reproductive period. The authors assume an innate preference for orange as a cue for rare and high quality food sources in both sexes. Males that developed orange spots, therefore, exploited the pre-existing preference for orange in females and are preferred as mates by females.

5.8.1   *Female preference for swords*

A well known system for a pre-existing sensory bias for a specific male trait has been documented for the genus *Xiphophorus* and the sister group *Priapella*. Male green swordtails (*Xiphophorus helleri*, Poeciliidae) have a yellow-coloured sword with a black border, which is an elongation of certain ventral caudal-fin rays (Basolo 1998), and females base their mate choice largely on the characteristics of this sword (Basolo 1990a). Males of the platyfish (*X. maculatus* and *X. variatus*, Poeciliidae) do not possess a sword. However, females of these species prefer conspecific males with an artificial plastic sword over naturally swordless males (Basolo 1990b, 1995a). A similar preference for artificially sworded males was found in the sister group *Priapella*. Female Olmec priapella (*P. olmaceae*, Poeciliidae) prefer artificially sworded males over naturally non-sworded males (Basolo 1995b). There is, however, no general preference in females of poeciliid fish species for a sword in males. Two-spotted livebearer females (*Heterandria bimaculata*, Poeciliidae) exhibited no pre-existing preference for males with an artificial sword (Basolo 2002a), and results for the sailfin molly regarding a latent preference of females for sworded males are ambiguous. Basolo (2002b) showed a clear preference in sailfin molly females from a Louisiana population for artificially sworded conspecific males of the opposite sex, whereas Witte & Klink (2005) did not find such a preference in sailfin molly females from a Texan population. These population differences in pre-existing biases in females for a sword in males might be explained by differences in female preferences between different populations, and/or by different environmental conditions. The Louisiana sailfin mollies live in murky waters, whereas the Texan sailfin mollies live in clear waters. In murky water, a sword in males would probably enhance the visibility of males for females and also probably enhance their mating success. In clear water, however, a sword may not increase, or may only marginally increase, the visibility of males for females, and may incur high predation risk (Abrahams & Kattenfeld 1997). This may have constrained the evolution of a sword in the Texan male molly and a preference for such a male trait in females.

5.8.2   *Theoretical approaches*

Although the sensory exploitation hypothesis explains how a female preference for male traits has originated, it does not explain how a novel male trait can spread within a population. An interesting question is, therefore, whether mate-choice copying can support the spread of a novel male trait within a population. This fascinating question has been examined theoretically by Kirkpatrick & Dugatkin (1994). They assumed that female mate preferences evolve only through cultural evolution, whereas the male trait on which they act is inherited via a haploid autosomal or a Y-linked locus. In their model, they simulated two different copying situations: 'single mate copying', in which younger females copy the choice of only one older female, and 'mass copying', in which younger females have the opportunity to copy the choices of a large number of older females. Thus, copying females strengthen their mate preference towards the male type they have observed mating. As a result of frequency dependence, females in the 'mass copying' scenario have a stronger preference for the male type they have seen mating most frequently. On the one hand, Kirkpatrick & Dugatkin's model shows that copying can lead to a rapid exaggeration of the male trait and female preference for it. On the other hand, copying seems to make it more difficult for a rare male trait to become established and does not maintain a polymorphism for that trait. Only under specific conditions can copying lead to two alternative evolutionary equilibria for the male trait. Female preference and the male trait can rapidly co-evolve, with a positive frequency-dependent advantage to the more common male trait allele. This is true even for a male trait that lowers male viability, when it has reached a certain threshold in frequency. Both scenarios lead to a positive frequency-dependent advantage to males: the more common a male type, the stronger is the female preference for it. This effect of frequency-dependence is stronger in the 'mass copying' scenario than in the 'single copying' scenario. Thus, according to the model of Kirkpatrick & Dugatkin (1994), mate-choice copying does not favour the spread of a novel male trait within a population. Similar conclusions were reached by Laland (1994b).

Agrawal (2001) has developed another model on evolutionary consequences of mate-choice copying on male traits. In contrast to Kirkpatrick's & Dugatkin's model (1994), Agrawal's model shows that mate-choice copying can cause positive or negative directional selection on male traits, or positive or negative frequency-dependent selection on male traits. Whereas Kirkpatrick & Dugatkin (1994) assume that each copying event influenced the mate-choice decision of the observing female equally, Agrawal (2001) assumed in his model that different observations have differing degrees of influence on the mate-choice decision. Agrawal assumed that females are influenced by the extent to which the male observed mating successfully differs from the population mean regarding the focal male trait. He concluded from his model that mate-choice copying can, first, facilitate the spread of a novel male type through a population, even if there is no inherent preference for the novel male trait, and second, that mate-choice copying can maintain genetic variation for sexually-selected male traits. When a female observes males of different phenotypes successfully mating in proportion to their frequency in the population, her mating preferences are not altered by social cues like mate-choice copying. When a female observes a particular male phenotype mating disproportionately more often than other male phenotypes, her preference is biased towards this type of male. Thus, a female that observes a rare male type mating is more strongly biased towards this rare male phenotype than a female that observes a common male phenotype successfully mating. This assumption is based on the notion that unusual or unexpected stimuli affect individuals more strongly than common stimuli (Cohen 1983). The legitimacy of this assumption has yet to be evaluated within the context of mate-choice behaviour, although see below. Nonetheless, the model of Agrawal (2001) provides some indications that mate-choice copying may favour the spread of a novel rare male trait within a population.

5.8.3 *Experimental approaches*

Is there any experimental evidence for mate-choice copying supporting or preventing the spread of a novel male trait? Sailfin molly females from the Comal River,

Texas, USA, have no pre-existing preference for males with an artificial sword imitating the natural sword of male green swordtails (Witte & Klink 2005), and several previous experiments have demonstrated that sailfin molly females copy the choice of other females (Witte & Ryan 1998; Witte & Noltemeier 2002; Witte & Massmann 2003). To investigate this question Witte et al. (K. Witte, K. Kotten & K.B. Klink, unpublished data submitted) attached an artificial yellow plastic sword with a black border or a transparent plastic sword to the base of the tail fin and created video playbacks of courting males bearing the yellow sword or the transparent sword. In copying experiments, females could first choose between two male videos presented on television monitors at each end of the female test tank. They quantified the time the female spent within a preference zone at each end of the test tank. After this first preference test, females had the opportunity to observe the other with the yellow sword courting another female on a video, whereas the male with the transparent sword was alone. In the second preference test, females were allowed to choose between the two males, one with the coloured sword and the male with the transparent sword, a second time. Fourteen of 23 females that had rejected the male with the coloured sword in the first preference test preferred that male in the second preference test after having observed the male courting another female (McNemar's test, $P < 0.001$). This result seems to indicate that mate-choice copying can support the spread of a novel male trait, because females copy the choice for that novel male type. However, 10 of 17 females, which had preferred the male with the coloured sword in the first preference test, changed their preference and preferred the male with the transparent sword in the second preference test (McNemar, $P = 0.031$). Thus, in this situation mate-choice copying prevented the spread of a novel male trait.

These experiments at least support Agrawal's (2001) assumption that different observations have differing degrees of influence on the female mate-choice decision. The observation of an unattractive male interacting with a female seems to increase the probability that females copy the mate choice, whereas the observation of an attractive male interacting with another female decreases the probability that females copy the mate choice of others. An alternative explanation might be that fishes adopt a 'copy when uncertain' strategy (Laland 2004). An attractive male with familiar characteristics evokes a clear preference and leaves little uncertainty as to courtship behaviour. Conversely a strange-looking male generates uncertainty as to whether he is an appropriate mate, so fishes look to the behaviour of others for guidance. There is good evidence for this strategy being utilized by fishes (Van Bergen et al. 2004). Further experiments are necessary to estimate the evolutionary consequences of mate-choice copying for the evolution of novel traits in males.

5.9 Is mate-choice copying an adaptive mate-choice strategy?

Although mate-choice copying has been studied experimentally in many fish species, there is no clear experimental evidence that mate-choice copying increases the fitness of a copying female. In other words, there is no experimental indication that mate-choice copying is an evolutionary adaptive strategy. In theory, there are several benefits but also some costs associated with mate-choice copying.

5.9.1   *Benefits of mate-choice copying*

Pruett-Jones (1992) demonstrated in a game-theoretical model that the adaptive significance of mate-choice copying depends on the ratio of costs to benefits of independent mate choice. Gibson & Höglund (1992) proposed two important benefits resulting from copying. Copying can serve to increase the accuracy of mate assessment and reduce the costs of mate choice. Increasing the accuracy of mate assessment through mate-choice copying (Losey *et al.* 1986) is especially valid for females inexperienced in mate choice. Dugatkin & Godin (1993) found in guppies that young females, which are assumed to be relatively inexperienced in mate choice, copy the choice of older, presumably more experienced, females in mate choice, but not the reverse. Inexperienced females can learn to recognize a male or male phenotype of good quality by copying the choice of experienced females. Another example of copying facilitating the learning of mate assessment is provided by the sailfin molly. Sailfin molly females copy the choice of others when both males presented in a test are similar in colour and body size. Females do not copy the choice for a smaller male when both males presented in a test differed obviously in size. In the latter case, females prefer the larger of two males, even though a model female is placed next to the smaller male (Witte & Ryan 1998). Thus, when it is difficult to distinguish between males, females are more likely to copy than in a situation when males clearly differ in quality, that is, females are more likely to copy when they are uncertain in their mating strategy.

Another benefit of mate-choice copying is that the observing female is not physically involved in courtship displays with a prospective mate. In some species, courting males behave aggressively towards females, or even harass females, during courtship displays (Schlupp *et al.* 2001). A female that observes how a male courted another female gains information about this male, and may reject an aggressive male, without being physically involved (Witte & Ueding 2003).

Mate-choice copying is assumed to reduce the time costs of searching for a mate. Females often visit several males before they choose one (Forsgren 1997). By observing others' mate-choice decisions, copying females may save time for their own mate choice. Copying females can decrease the time spent on directly assessing potential mates by copying the mate choice of others, and thus can minimize the 'opportunity costs' associated with the assessment of males. Briggs *et al.* (1996) tested this hypothesis. They investigated whether female guppies show a higher tendency for mate-choice copying when a predator is present than with no predator around. Mate-choice copying should reduce the time for mate inspection, and thus should increase the time available for predator vigilance and, therefore, reduce the risk of predation. In both situations, with predator present and predator absent, females copied the mate choice of others, but the authors found no indication of a higher mate-choice copying tendency in females when a predator was present. Copying females may be able to reduce the time for mate assessment and, therefore, increase the time for foraging activities. Hungry females should show a higher tendency to copy the mate choice of others than satiated females. Dugatkin & Godin (1998) tested this assumption in guppy females. However, their results contradicted the expectation; only the well-fed females copied the mate choice of others significantly more often than expected by chance.

5.9.2   *Costs of mate-choice copying*

Mate-choice copying is also likely to entail some costs. Copying females might risk a reduced fertility as a result of sperm depletion in males when the copying female immediately copulates with that male after he has already copulated with several females. Male courtship display is a highly conspicuous behaviour not only to conspecifics but also to predators (Houde 1997); therefore, it might be risky for a copying female to mate with a male immediately after that male has courted another female and might have attracted the attention of a predator. Both disadvantages, however, would be reduced when copying females do not have to copulate with a particular male immediately after observing a sexual interaction between another female and that male. Witte & Massmann (2003) showed that sailfin molly females are able to memorize an observed interaction between a male and a female for at least 1 day. Thus, copying females may copy the choice of others not immediately but rather later, when the male has replenished his sperm supply and at a safer time. It is also possible that the female may acquire outdated, inappropriate or inaccurate information about mate quality through mate-choice copying. Thus females always have to decide whether to base their mate choice rather on social or private information.

5.10   Outlook

In spite of good progress in understanding mate-choice copying, there is, as yet, still little indication that mate-choice copying is an adaptive mate-choice strategy. It would be valuable if future experiments focused on this question, to aid understanding of the function of mate-choice copying. It is also important for future studies to investigate the relative reproductive success of copying and non-copying females. Answers to these questions would strengthen the claim that mate-choice copying plays an important role in sexual selection.

5.11   Conclusions

Learning has an enormous influence on mate choice in fishes and is, therefore, potentially an important influence on sexual selection. Learning during an early phase of development (sexual imprinting) can shape mate preferences later on in life when the individual has reach sexual maturity. Other forms of learning, which involve experience with conspecifics, occur during all phases of life and can form and change mate preferences in adults.

Social learning, which includes observing conspecifics, is arguably the most fascinating kind of learning. Individuals can gather inadvertent social information from conspecifics about the quality of potential mates and use this information for their own mate-choice decision. The evolutionary consequences for social learning by using inadvertent social information is a new expanding research field in evolutionary biology and will provide novel aspects for the intriguing role of socially induced mate preferences in sexual selection.

5.12 Acknowledgements

I thank Machteld Verzijden for allowing me to present some results of her ongoing study. K.W. was supported by the DFG (Wi 1531/4-1). I also thank J.-G.J. Godin and an anonymous referee for comments on a previous version on this manuscript.

5.13 References

Abrahams, M. & Kattenfeld, M. (1997) The role of turbidity as a constraint on predator–prey interactions in aquatic environments. *Behavioral Ecology and Sociobiology*, **40**, 169–174.

Agrawal, A.F. (2001) The evolutionary consequences of mate copying on male traits. *Behavioral Ecology and Sociobiology*, **51**, 33–40.

Albert, A.Y.K. (2005) Mate choice, sexual imprinting, and speciation: a test of a one-allele isolating mechanism in sympatric sticklebacks. *Evolution*, **59**, 927–931.

Andersson, M. (1994) *Sexual Selection*. Princeton University Press, Princeton.

Applebaum, S.L. & Cruz, A. (2000) The role of mate-choice copying and disruption effects in mate preference determination of *Limia perugiae* (Cyprinodontiformes, Poeciliidae). *Ethology*, **106**, 933–944.

Aoki, K., Feldman, M.W. & Kerr, B. (2001) Models of sexual selection on a quantitative genetic trait when preference is acquired by sexual imprinting. *Evolution*, **55**, 25–32.

Bakker, T.C.M. (1999) The study of intersexual selection using quantitative genetics. *Behaviour*, **136**, 1237–1265.

Basolo, A.L. (1990a) Female preference for a male sword length in the green swordtail, *Xiphophorus helleri* (Pisces: Poeciliidae). *Animal Behaviour*, **40**, 332–338.

Basolo, A.L. (1990b) Female preference predates the evolution of the sword in swordtail fish. *Science*, **250**, 808–810.

Basolo, A.L. (1995a) A further examination of a pre-existing bias favouring a sword in the genus *Xiphophorus*. *Animal Behaviour*, **50**, 365–375.

Basolo, A.L. (1995b) Phylogenetic evidence for the role of a pre-existing bias in sexual selection. *Proceedings of the Royal Society of London, Series B*, **259**, 307–311.

Basolo, A.L. (1998) Evolutionary change in a receiver bias: a comparison of female preference functions. *Proceedings of the Royal Society of London, Series B*, **265**, 2223–2228.

Basolo, A.L. (2002a) Female discrimination against sworded males in a poeciliid fish. *Animal Behaviour*, **63**, 463–468.

Basolo, A.L. (2002b) Congruence between the sexes in preexisting receiver responses. *Behavioral Ecology*, **13**, 832–837.

Bereczkei, T., Gyuris, P. & Weisfeld, G.E. (2004) Sexual imprinting in human mate choice. *Proceedings of the Royal Society of London, Series B*, **271**, 1129–1134.

Breden, F., Novinger, D. & Schubert, A. (1995) The effect of experience on mate choice in the Trinidadian guppy, *Poecilia reticulata*. *Environmental Biology of Fishes*, **7**, 323–328.

Briggs, S.E., Godin J.-G.J. & Dugatkin, L.A. (1996) Mate-choice copying under predation risk in the Trinidadian guppy (*Poecilia reticulata*). *Behavioral Ecology*, **7**, 151–157.

Brooks, R. (1996) Copying and the repeatability of mate choice. *Behavioral Ecology and Sociobiology*, **39**, 323–329.

Brooks, R. (1998) The importance of mate copying and cultural inheritance of mating preferences. *Trends in Ecology and Evolution*, **13**, 45–46.

Brooks, R. (2000) Negative genetic correlation between male sexual attractiveness and survival. *Nature*, **406**, 67–70.

ten Cate, C. & Vos, D. (1999) Sexual imprinting and evolutionary processes in birds: a reassessment. *Advances in the Study of Behavior*, **28**, 1–31.

Cohen, J.A. (1983) Sexual selection and the psychophysics of female choice. *Zeitschrift für Tierpsychologie*, **64**, 1–8.

Danchin, E., Giraldeau, L.-A., Valone, T.J. & Wagner, R.H. (2004) Public information: from noisy neighbors to cultural evolution. *Science*, **305**, 487–491.

Doutrelant, C. & McGregor P.K. (2000) Eavesdropping and mate choice in female fighting fish. *Behaviour*, **137**, 1655–1669.

Doutrelant, C., McGregor, P.K. & Oliveira, R.F. (2001) The effect of an audience on intra-sexual communication in male Siamese fighting fish, *Betta splendens*. *Behavioral Ecology*, **12**, 283–286.

Dugatkin, L.A. (1992) Sexual selection and imitation: females copy the mate choice of others. *American Naturalist*, **139**, 1384–1389.

Dugatkin, L.A. (1996a) Copying and mate choice. In: C.M. Heyes & B.G. Galef Jr. (eds), *Social learning in animals: the roots of culture*, pp. 85–105. Academic Press, New York.

Dugatkin, L.A. (1996b) The interface between culturally-based preferences and genetic preferences: female mate choice in *Poecilia reticulata*. *Proceedings of the National Academy of Sciences USA*, **93**, 2770–2773.

Dugatkin, L.A. (1998a) A comment on Lafleur *et al.*'s re-evaluation of mate-choice copying in guppies. *Animal Behaviour*, **56**, 513–514.

Dugatkin, L.A. (1998b) Genes, copying, and female mate choice; shifting thresholds. *Behavioral Ecology*, **9**, 323–327.

Dugatkin, L.A. & Godin, J.-G.J. (1992) Reversal of female mate choice by copying in the guppy (*Poecilia reticulata*). *Proceedings of the Royal Society of London, Series B*, **249**, 179–184.

Dugatkin, L.A. & Godin, J.-G.J. (1993) Female mate copying in the guppy (*Poecilia reticulata*): age-dependent effects. *Behavioral Ecology*, **4**, 289–292.

Dugatkin, L.A. & Godin, J.-G.J. (1998) Effects of hunger on mate-choice copying in the guppy. *Ethology*, **104**, 194–202.

Dugatkin, L.A., Druen, M.W. & Godin, J.-G.J. (2003) The disruption hypothesis does not explain copying in the guppy (*Poecilia reticulata*). *Ethology*, **109**, 67–76.

Endler, J.A. & Houde, A.E. (1995) Geographic variation in female preferences for male traits in *Poecilia reticulata*. *Evolution*, **49**, 456–468.

Fernö, A. & Sjölander, S. (1973) Some imprinting experiments on sexual preferences for colour variants in the platyfish (*Xiphophorus maculates*). *Zeitschrift für Tierpsychologie*, **33**, 417–423.

Fernö, A. & Sjölander, S. (1976) Influence of previous experience on the mate selection of two colour morphs of the convict cichlid, *Cichlasoma nigrofasciatum* (Pisces, Cichlidae). *Behavioural Processes*, **1**, 3–14.

Fisher, R.A. (1930) *The Genetical Theory of Natural Selection*. Clarendon Press, Oxford.

Forsgren, E. (1997) Mate sampling in a population of sand gobies. *Animal Behaviour*, **53**, 267–276.

Gabor, C. (1999) Association patterns of sailfin mollies (*Poecilia latipinna*). Alternative hypotheses. *Behavioral Ecology and Sociobiology*, **46**, 333–340.

Gibson, R.M. & Höglund, J. (1992) Copying and sexual selection. *Trends in Ecology and Evolution*, **7**, 229–232.

Godin, J.-G.J., Herdman, E.J.E. & Dugatkin, L.A. (2005) Social influences on female mate choice in the guppy, *Poecilia reticulata*: generalized and repeatable trait-copying behaviour. *Animal Behaviour*, **69**, 999–1005.

Goldschmidt, T., Bakker, T.C.M. & Feuth-De Bruijn, E. (1993) Selective copying in mate choice of female sticklebacks. *Animal Behaviour*, **45**, 541–547.

Grant, J.W.A. & Green, L.D. (1996) Mate copying versus preference for actively courting males by female Japanese medaka (*Oryzias latipes*). *Behavioral Ecology*, 7, 165–167.

Haskins, C.P. & Haskins, E.F. (1949) The role of sexual selection as an isolating mechanism in three species of poeciliid fishes. *Evolution*, 3, 160–169.

Haskins, C.P. & Haskins, E.F. (1950) Factors governing sexual selection as an isolating mechanism in the poeciliid fish *Lebistes reticulatus*. *Proceedings of the National Academy of Sciences USA*, 36, 464–476.

Höglund, J., Alatalo, R.V., Gibson, R.M. & Lundberg, A. (1995) Mate-choice copying in the black grouse. *Animal Behaviour*, 49, 1627–1633.

Houde, A.E. (1988) Genetic difference in female choice between two guppy populations. *Animal Behaviour*, 36, 510–516.

Houde, A.E. (1992) Sex-linked heritability of sexually selected character in a natural population of *Poecilia reticulata* (Pisces: Poeciliidae) (guppies). *Heredity*, 69, 229–235.

Houde, A.E. (1997) *Sex, Color, and Mate Choice in Guppies*. Princeton University Press, Princeton, New Jersey.

Howard, R.D., Martens, R.S., Innis, S.A., Drenvich, J.M. & Hale, J. (1998) Mate choice and mate competition influence male body size in Japanese medaka. *Animal Behaviour*, 55, 1151–1163.

Immelmann, K. (1972) Sexual and other long-term aspects of imprinting in birds and other species. *Advances in the Study of Behavior*, 4, 147–174.

Jamieson, I.G. & Colgan, P.W. (1989) Eggs in the nest of males and their effect on mate choice in the three-spined stickleback. *Animal Behaviour*, 38, 859–865.

Janetos, A.C. (1982) Strategies of female mate choice: a theoretical analysis. *Behavioral Ecology and Sociobiology*, 7, 107–112.

Jennions, M.D., Møller, A.P. & Petrie, M. (2001) Sexually selected traits and adult survival: a meta-analysis. *Quarterly Review in Biology*, 76, 3–36.

Jirotkul, M. (1999) Operational sex ratio influences female preferences and male–male competition in guppies. *Animal Behaviour*, 58, 287–294.

Kendrick, K.M., Hinton, M.R., Atkins, K., Haupt, M.A. & Skinner, J.D. (1998) Mothers determine sexual preferences. *Nature*, 395, 229–230.

Kirkpatrick, M. (1982) Sexual selection and the evolution of female preferences. *Evolution*, 36, 1–12.

Kirkpatrick, M. & Dugatkin, L.A. (1994) Sexual selection and the evolutionary effects of copying mate choice. *Behavioral Ecology and Sociobiology*, 34, 443–449.

Knapp, R.A. & Sargent, R.C. (1989) Egg mimicry as a mating strategy in the fantail darter, *Ethiostoma flabellare*: females prefer males with eggs. *Behavioral Ecology and Sociobiology*, 25, 321–326.

Kokko, H., Brooks, R., Jennions, M.D. & Morley, J. (2003) The evolution of mate choice and mating biases. *Proceedings of the Royal Society of London, Series B*, 270, 653–664.

Lafleur, D.L., Lozano, G.A. & Sclafini, M. (1997) Female mate-choice copying in guppies, *Poecilia reticulata*: a re-evaluation. *Animal Behaviour*, 54, 579–586.

Laland, K. (1994a) On the evolutionary consequences of sexual imprinting. *Evolution*, 48, 477–489.

Laland, K.N. (1994b) Sexual selection with a culturally transmitted mating preference. *Theoretical Population Biology*, 45, 1–15.

Laland, K.N. (2004) Social learning strategies. *Learning and Behavior*, 32, 4–14.

Laland, K.N. & Reader, S.M. (1999) Foraging innovation in the guppy. *Animal Behaviour*, 57, 331–340.

Lande, R. (1981) Models of speciation by sexual selection on polygenic traits. *Proceedings of the National Academy of Sciences U.S.A.*, 78, 3721–3725.

Liley, N.R. (1966) Ethological isolating mechanisms in four sympatric species of poeciliid fishes. *Behaviour*, **Supplement 13**, 1–197.

Losey, G.S., Jr., Stanton, F.G., Telecky, T.M., Tyler, W.A., III & Zoology 691 Graduate seminar class (1986) Copying others, an evolutionary stable strategy for mate choice: a model. *American Naturalist*, **128**, 653–664.

MacLaren, R.D., Rowland, W.J. & Morgan, N. (2004) Female preference for sailfin and body size in the sailfin molly, *Poecilia latipinna*. *Ethology*, **110**, 363–379.

Magurran, A.E. & Ramnarine, I.W. (2004) Learned mate recognition and reproductive isolation in guppies. *Animal Behaviour*, **67**, 1077–1082.

Marconato, A. & Bisazza, A. (1986) Males whose nest contains eggs are preferred by female *Cottus gobio* L. (Pisces, Cottidae). *Animal Behaviour*, **34**, 1580–1582.

Marler, C.A. & Ryan, M.J. (1997) Origin and maintenance of a female mating preference. *Evolution*, **51**, 1244–1248.

Matos, R.J. & McGregor, P.K. (2002) The effect of the sex of an audience on male–male displays of Siamese fighting fish (*Betta splendens*). *Behaviour*, **139**, 1211–1221.

McGregor, P.K. (2005) *Animal Communication Networks*. Cambridge University Press, Cambridge, UK.

McGregor, P.K. & Dabelsteen, T. (1996) Communication networks. In: D.E. Kroodsma & E.H. Miller (eds), *Ecology and Evolution of Acoustic Communication in Birds*, pp. 409–425. Cornell University Press, Ithaca, New York.

McGregor, P.K. & Peake, T.M. (2000) Communication networks: social environments for receiving and signalling behaviour. *Acta Ethologica*, **2**, 71–81.

Møller, A.P. & Alatalo, R.V. (1999) Good-genes effects in sexual selection. *Proceedings of the Royal Society of London, Series B*, **266**, 85–91.

Munger, L., Cruz, A. & Applebaum, S. (2004) Mate choice copying in female Humpback Limia (*Limia nigrofasciata*, Family Poeciliidae). *Ethology*, **110**, 563–573.

Nordell, S.E. & Valone, T.J. (1998) Mate choice copying as public information. *Ecology Letters*, **1**, 74–76.

Owens, I.P.F., Rowe, C. & Thomas, A.L.R. (1999) Sexual selection, speciation and imprinting: separating the sheep from the goats. *Trends in Ecology and Evolution*, **14**, 131–132.

Patriquin-Meldrum, K.J. & Godin, J.-G.J. (1998) Do female three-spined sticklebacks copy the mate choice of others? *American Naturalist*, **151**, 570–577.

Penn, D. & Potts, W. (1998) MHC-disassortative mating preferences reversed by cross-fostering. *Proceedings of the Royal Society of London, Series B*, **265**, 1299–1306.

Pruett-Jones, S. (1992) Independent versus non-independent mate-choice: do females copy each other? *American Naturalist*, **140**, 1000–1009.

Ptacek, M.B. & Travis, J. (1997) Mate choice in the sailfin molly, *Poecilia latipinna*. *Evolution*, **51**, 1217–1231.

Real, L. (1990) Search theory and mate choice. I. Models of single-sex discrimination. *American Naturalist*, **138**, 901–917.

Ridley, M. & Rechten, C. (1981) Female sticklebacks prefer to spawn with males whose nests contain eggs. *Behaviour*, **76**, 152–161.

Rodd, F.H., Hughes, K.A., Grether, G.F. & Baril, T.C. (2002) A possible non-sexual origin of mate preference: are male guppies mimicking fruit? *Proceedings of the Royal Society of London, Series B*, **269**, 467–481.

Rohwer, S. (1978) Parent cannibalism of offspring and egg raiding as a courtship strategy. *American Naturalist*, **112**, 429–440.

Rosenqvist, G. & Houde, A. (1997) Prior exposure to male phenotypes influences mate choice in the guppy, *Poecilia reticulata*. *Behavioral Ecology*, **8**, 194–198.

Ryan, M.J. & Keddy-Hector, A. (1992) Directional patterns of female mate choice and the role of sensory biases. *American Naturalist*, **139**, S4–35.

Ryan, M.J. (1998) Sexual selection receiver bias, and the evolution of sex differences. *Science*, **281**, 1999–2003.

Schlupp, I. & Ryan, M.J. (1997) Male sailfin mollies (*Poecilia latipinna*) copy the mate choice of other males. *Behavioral Ecology*, **8**, 104–107.

Schlupp, I., Marler, C. & Ryan, M.J. (1994) Benefit to male sailfin mollies of mating with heterospecific females. *Science*, **263**, 373–374.

Schlupp, I., McKnab, R. & Ryan, M.J. (2001) Sexual harassment as a cost for molly females: bigger males cost less. *Behaviour*, **138**, 277–286.

Schluter, D. & McPhail, J.D. (1992) Ecological character displacement and speciation in sticklebacks. *American Naturalist*, **140**, 85–108.

Servedio, M.R. & Kirkpatrick, M. (1996) The evolution of mate choice copying by indirect selection. *American Naturalist*, **148**, 848–867.

Siepen, G. (1984) Der Einfluss der elterlichen Farbmorphe auf die spätere Partnerwahl bei *Cichlasoma nigrofasciatum*. Diploma thesis, Bielefeld.

Sirot, E. (2001) Mate-choice copying by females: the advantages of a prudent strategy. *Journal of Evolutionary Biology*, **14**, 418–423.

Stöhr, S. (1998) Evolution of mate-choice copying: a dynamic model. *Animal Behaviour*, **55**, 893–903.

Uehara, T., Yokomizo, H. & Iwasa, Y. (2005) Mate-choice copying as Bayesian decision making. *American Naturalist*, **165**, 403–410.

Unger, L.M. & Sargent, R.C. (1988) Alloparental care in the fathead minnow, *Pimephales promelas*: Females prefer males with eggs. *Behavioral Ecology and Sociobiology*, **23**, 27–32.

Van Bergen, Y., Coolen, I. & Laland, K.N. (2004) Nine-spined sticklebacks exploit the most reliable source when public and private information conflict. *Proceedings of the Royal Society of London, Series B*, **271**, 957–962.

Wade, M.J. & Pruett-Jones, S.G. (1990) Female copying increases the variance in male mating success. *Proceedings of the National Academy of Sciences USA*, **87**, 5749–5753.

Westneat, D.F., Walters, A., McCarthy, T.M., Hatch, M.I. & Hein, W.K. (2000) Alternative mechanisms of nonindependent mate choice. *Animal Behaviour*, **59**, 467–476.

White, D.J. & Galef, B.G. Jr. (1999) Mate-choice copying and conspecific cueing in Japanese quail (*Cortunix cortunix japonica*). *Animal Behaviour*, **57**, 465–473.

White, D.J. & Galef, B.G. Jr. (2000) 'Culture' in quail: social influences on mate choice in female *Cortunix japonica*. *Animal Behaviour*, **59**, 975–979.

Witte, K. & Klink, K. (2005) No pre-existing bias in sailfin molly females (*Poecilia latipinna*) for a sword in males. *Behaviour*, **142**, 283–303.

Witte, K. & Massmann, R. (2003) Female sailfin mollies, *Poecilia latipinna*, remember males and copy the choice of others after 1 day. *Animal Behaviour*, **65**, 1151–1159.

Witte, K. & Noltemeier, B. (2002) The role of information in mate-choice copying in female sailfin mollies (*Poecilia latipinna*). *Behavioral Ecology and Sociobiology*, **52**, 194–202.

Witte, K. & Ryan, M.J. (1998) Male body length influences mate-choice copying in the sailfin molly *Poecilia latipinna*. *Behavioral Ecology*, **9**, 534–539.

Witte, K. & Ryan, M.J. (2002) Mate-choice copying in the sailfin molly, *Poecilia latipinna*, in the wild. *Animal Behaviour*, **63**, 943–949.

Witte, K. & Sawka, N. (2003) Sexual imprinting on a novel trait in the dimorphic zebra finch: sexes differ. *Animal Behaviour*, **65**, 195–203.

Witte, K. & Ueding, K. (2003) Sailfin molly females (*Poecilia latipinna*) copy the rejection of a male. *Behavioral Ecology*, **14**, 389–395.

Zahavi, A. (1975) Mate selection – a selection for a handicap. *Journal of Theoretical Biology*, **53**, 205–214.

Chapter 6

Modulating Aggression Through Experience

Yuying Hsu, Ryan L. Earley and Larry L. Wolf

6.1 Introduction

Aggressive interactions are a common means of contesting resources for most animals. Considerable variation occurs in whether a specific individual wins a particular aggressive contest. Influences on the behaviour of individuals that might produce this variation include, among many factors, hunger, size, residency and age (Beaugrand *et al.* 1996; Hsu *et al.* 2006). Behavioural ecologists have been quite successful in understanding variation in contest outcomes, employing benefit/cost models to predict such things as contest duration and winner (e.g. Riechert 1998). Benefits are immediate or longer-term positive effects on fitness of the individual, such as gaining access to food or mates as a result of the contest. Costs include time and energy spent in the contest as well as the possibility of being injured or the increased chance of being taken by a predator.

Among fishes and a few other animals, including some insects, experience in a prior contest has been shown to influence the outcome of a later contest (Hsu *et al.* 2006). A recent winning experience tends to increase the chances of winning the current contest, while a losing experience tends to decrease the chance of winning. Among species tested so far considerable variation occurs in how much of an effect prior winning or losing experiences produce and how long the effect lasts. For most organisms winning and losing effects may last for different lengths of time and tend to be asymmetrical, with losing tending to have more influence than winning. Recent experiments indicate, as well, that multiple prior contests, rather than just the most recent experience, can affect behaviour during, and outcome of, the current contest (Hsu & Wolf 2001).

Most of the evidence from fish contests strongly suggests that the effects of prior contest experience influence the individual's perception of its own fighting ability and the accumulation of costs in a subsequent contest (Hsu *et al.* 2006). In contrast to many standard studies of experience effects on learning and memory, the examination of these effects on aggression are complicated by the necessity to consider more than one individual's experience in understanding contest outcomes. It is most likely the combined experience effects of the contestants that influence outcomes. How these combined effects are integrated into the ongoing behaviour of the contesting individuals is still very uncertain.

The effect of prior contest experience wanes over time and has the hallmarks of memory and forgetting. Following the definition of learning offered by Alcock (1993,

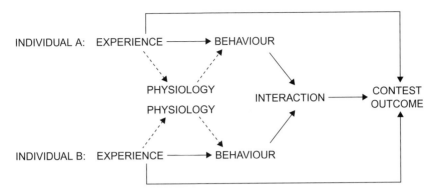

Fig. 6.1 Pathways for individuals A and B from contest experiences to modified contest outcomes based on those experiences. The dashed lines represent internal (i.e. physiological) changes that influence behaviour. The solid lines represent external events, including behaviour, a subsequent interaction, and the outcome of that interaction that are affected by the experience. (Reprinted with permission from Hsu *et al.* 2006.)

p. 29), 'the durable modification of behaviour in response to information acquired from specific experiences', one could argue that these experience effects reflect learning of the potential costs associated with aggressive interactions. Learning in a contest situation does occur in fishes. Classical conditioning can be used to train fishes to cues that signal an impending contest before the opponent is present (Hollis *et al.* 1995). However, if we also consider the mechanisms of learning – modification of neural pathways to facilitate particular behavioural patterns – then whether contest experience actually leads to learning is at present unknown. We will review some of the physiological changes that accompany the experiences, but cannot conclude at present whether learning is involved. Hence, this chapter will refer to experience effects in order not to appear to draw a firm conclusion about whether they actually constitute learning.

Physiological and behavioural modifications are intermediate stages in the translation of experiences into changed contest outcomes (Fig. 6.1). Examining experience effects on aggressive behaviour thus provides a way to investigate how physiological effects are reflected in behavioural variation of each contestant, which can, in turn, affect contest outcomes. These experience effects also influence social behaviour of organisms in as far as it is mediated by aggression. Specifically, contest experience and its effects on subsequent contests might be an important component of dominance hierarchy formation in group-living organisms, with or without individual recognition.

This chapter will review our current knowledge of how experience in contests influences physiology and behaviour and ultimately the outcome of subsequent contests. We also will explore quantitative models of integrating multiple experiences and rates of disappearance of experience effects as well as experience-based algorithms for predicting contest outcomes. Much of this work has been undertaken with fishes as model organisms and adds yet more evidence of the complexity of experience effects in fish behaviour.

6.2   Winner and loser effects in fishes

Many fishes adjust their contest behaviour in accordance with the outcomes of their previous agonistic interactions (see Table 6.1 in Hsu *et al.* 2006), but the changes in fighting behaviour vary among studies. After winning experiences, individuals generally are more likely to initiate contests with more costly (in terms of energy and risk of injury) behaviour (i.e. attack rather than display), to retaliate and escalate contests when attacked, and to have increased probabilities of winning the contests. In contrast, individuals with recent losing experiences become less likely to initiate and participate in any agonistic interactions and more likely to retreat without escalating fights when attacked. In this section, we discuss the influence of methodology on detecting experience effects and the occurrence and variation in winner and loser effects.

6.2.1   *Methodological concerns in detecting experience effects*

In fishes, as in other taxa, loser effects appear to be more commonly detected than winner effects and the importance of experience effects varies among species (Hsu *et al.* 2006). However, caution is needed when interpreting these results because of variation in the experimental procedures employed to examine experience effects. Hsu *et al.* (2006) discussed four methodological concerns in testing experience effects that might have substantial impacts on the interpretation of available empirical results. The first concern is the protocol used to offer fighting experiences to focal individuals: self-selection or random-selection. In self-selection methods, the winner and the loser of a fight are treated as having a winning and a losing experience, respectively. This procedure could confound the effect of fighting experience with intrinsic fighting ability (Chase *et al.* 1994; Bégin *et al.* 1996). Bégin *et al.* (1996) concluded that a self-selected winner would have a 0.67 probability of having intrinsically higher fighting ability than its size-matched, naive opponent (0.83 if compared with self-selected losers), as opposed to a 0.5 probability as usually assumed and tested. In contrast, random selection procedures give predetermined winning or losing experiences to individuals chosen at random by pitting the focal individuals against smaller, habitual losers or larger, habitual winners, respectively. This method attempts to randomize intrinsic differences between contestants in order to focus solely on the effects of the experiences.

The second concern is the frequency and duration of experience training. Experience training can cause energy depletion, bodily injury and physical exhaustion, perhaps differentially affecting winners and losers. Naive opponents do not undergo experience training and are not subject to these effects. Prolonged experience training can temporarily compromise the physical condition of trained winners and losers and lower their probabilities of winning against naive opponents. Consequently, winner/loser effects may appear to be less/more pronounced than they really are. The third concern involves the time interval between the completion of experience training and the subsequent contest. A long time interval provides focal individuals with a chance to recover from the physical exhaustion/injury of experience training. However, because experience effects decay with time (Bakker *et al.* 1989; Chase *et al.* 1994; Hsu & Wolf 1999), the length of this interval will influence the likelihood of detecting any experience effects and the magnitude of the effects detected. The fourth concern relates to isolation of focal individuals before experience training. The purpose of this isolation is to allow focal individuals sufficient time for the effects of their previous fight to disappear. Without sufficient isolation time, the experience effects measured could be influenced by uncontrolled earlier experiences. With isolation procedures, one must also consider the effects of isolation itself on agonistic behaviour (Gomez-Laplaza & Morgan 2000). Bearing in mind the importance of these methodological issues, we will discuss the trends of experience effects.

6.2.2   *Asymmetrical winner and loser effects*

In most species of fishes, a winner effect often is less pronounced and disappears faster than a loser effect (Hsu *et al.* 2006). One exception to this general trend is the mangrove killifish (*Rivulus marmoratus*, Cyprinodontidae), in which winning and losing experiences appear to have opposite but equal effects on contest outcomes (Hsu & Wolf 1999). No studies have explored the possible causes/mechanisms either for the observed asymmetries in winner and loser effects or the variation in the degree of the asymmetry. We will discuss here some hypotheses for the asymmetry, which can be tested in the future. One possible adaptive explanation for this often greater and longer-lasting loser effect is that engaging in contests but losing often incurs more costs (time, energy, injuries) than retreating without confrontation (Neat *et al.* 1998). These high costs of losing could select for individuals that adopt more 'conservative' strategies such that their fighting behaviour is more likely to be modified by losing experiences than by winning experiences. It will be interesting to test whether the cost of losing is correlated with the degree of asymmetry between winner and loser effects. Nonetheless, methodologies for quantifying costs and experience effects probably should first be standardized to facilitate such comparisons.

The asymmetrical winner and loser effects also could be a consequence of loser effects being more easily reflected in the probability of winning than winner effects (Mesterton-Gibbons 1999). Individuals with prior losing experience often voluntarily retreat from a subsequent contest without physically interacting with their naive opponents (e.g. Bakker & Sevenster 1983). Contests between prior winners and naive opponents are likely to escalate into physical fights (e.g. Hsu & Wolf 1999). Prior contest experiences are expected to influence an individual's perception of its own fighting ability and the costs of a future contest without altering the individual's true fighting ability (Hsu *et al.* 2005). In that case, once a contest escalates, the value of information from past fighting experience is greatly diminished and contest outcome should be determined primarily by the intrinsic fighting ability of the two contestants. Thus, with a higher proportion of escalations, the probability of prior winners defeating naive, size-matched opponents will be less likely to deviate from 0.5.

6.2.3   *Interspecific variation in experience effects*

The importance and permanence of experience effects vary among species (Hsu *et al.* 2005); for instance, immediately after experience training, the probability of prior winners winning against size-matched, naive opponents ranges from 0.5 [no effect; e.g. paradise fish, *Macropodus opercularis* (Anabantidae); Francis 1983] to

0.78 [pumpkinseed sunfish, *Lepomis gibbosus* (Centrarchidae); Chase *et al.* 1994] in different species. And, winner effects could decay completely within an hour or last more than 2 days (mangrove killifish, Hsu & Wolf 1999). Loser effects are just as variable among species; the probability of prior losers winning against size-matched naive opponents varies from 0.5 [green swordtail fish, *Xiphophorus helleri* (Poeciliidae); Earley & Dugatkin 2002] to 0.0 [sticklebacks, *Gasterosteus aculeatus* (Gasterosteidae); Bakker *et al.* 1989] and the effect could disappear within 24 h (as suggested by non-analysed preliminary data, Bakker *et al.* 1989) or last more than 3 days (paradise fish; Francis 1983). Part of this variation may be explained by differences in the duration of experience training among different studies. Sticklebacks were trained only for 15 min and their loser effects disappeared fastest (within less than 24 h; Bakker *et al.* 1989); mangrove killifish were trained for an hour and their loser effects were already limited 2 days after experience (Hsu & Wolf 1999); paradise fish were trained for 24 h and their loser effects lasted considerably longer (still significant 3 days later; Francis 1983). This pattern could arise from the duration of experience training influencing the magnitude and/or decay rate of experience effects. The frequency of social encounters also might influence how long information from a social interaction is retained (Schuett 1997; Hsu *et al.* 2006). If, as expected, outcomes of past contests provide individuals with information regarding the costs of engaging in aggressive interactions, individuals in populations with a higher frequency of social encounters will obtain recent and more reliable information more often. Thus, these individuals should preserve information from past interactions to a lesser extent than individuals in populations with a low frequency of social encounters.

6.2.4   *Importance of experience effects in fighting decisions and outcomes*

Although past experience has an important impact on fighting decisions, it is only one of the many factors that influence estimated benefits and/or costs for the contestants. Past experience presumably provides individuals with information regarding the costs of engaging in contests in a similar way to other characteristics related to fighting ability (e.g. size, weaponry). The value and reliability of these indicators of fighting ability increase as asymmetries between contestants increase. Also, as the reliability and importance of other cues increases, the usefulness of prior experience might decrease; for instance, the importance of experience effects on contest outcome is negatively influenced by body size asymmetry (e.g. pumpkinseed sunfish, Beacham 1988; green swordtail fish, Beaugrand *et al.* 1991, 1996). Asymmetries in resource value, prior residency (Beaugrand *et al.* 1996), energy reserves (Marden & Waage 1990), and other factors that influence contest costs (e.g. predation) also should have an impact on the importance of experience effects and should be considered when investigating experience effects on fighting strategies.

6.2.5   *Experience and dominance hierarchies*

The majority of empirical research on experience effects and contest behaviour has been conducted in dyadic contests (but see Chase *et al.* 2003; Dugatkin & Druen

2004). This dyadic method is critical for determining some of the more intricate aspects of experience effects (e.g. decay rates, symmetry, etc.) but essentially removes individuals of many species from their natural social context. Social organisms often are found in groups that interact in ways that range from loosely defined dominance structures to rigid linear hierarchies. These patterns of structural group organization exist for varying periods of time depending usually on the breeding season of the population. Is it possible that changes in individual perceptions of benefits and/or costs via consecutive wins or losses impact the form and/or structure of dominance hierarchies? Early notions held that intrinsic differences between group members in individual attributes such as size were paramount in determining an individual's place in the hierarchy (Collias 1943; Allee *et al.* 1955). Currently, some empirical evidence and a burgeoning theoretical literature indicate that experience effects can predict much of the variation in the observed patterns of dominance-hierarchy formation. Empirical evidence for the role of experience effects dominance-hierarchy formation is slim and stems primarily from tests of Chase's 'jigsaw model' (1980), which posited that linear dominance hierarchies in small groups can emerge from patterns of consecutive wins (double dominance) or losses (double subordination) among group members. These patterns, and subsequent linear hierarchy formation, have been observed in primates (Barchas & Mendoza 1984), birds (Chase & Rohwer 1987), fishes (Chase *et al.* 2002) and invertebrates (Goessmann *et al.* 2000). Evidence also indicates that subjecting one animal to a win or loss can alter its place in a dominance hierarchy (Alexander 1961; Nelissen & Andries 1988). Perhaps the most convincing empirical support for experience effects mediating hierarchy formation comes from a study on swordtail fish (Dugatkin & Druen 2004). Using a random selection procedure, size-matched swordtail males were given winning, losing, or no experience and were then placed together to form a dominance hierarchy. Hierarchies in which previous winners achieved the dominant rank, inexperienced animals the middle rank, and prior losers the lowest rank were significantly more frequent than expected by chance.

Theoretical models provide, at present, the most useful framework for investigating changes in hierarchy structure that result from manipulating factors associated with experience effects. Hsu *et al.* (2006) provide a detailed description of how properties associated with translating experience effects into contest outcomes might affect hierarchy structure. These properties include which decision rules are employed, the symmetry, magnitude and decay of experience effects, how the effects accumulate, and the limits to changes in perceived fighting ability. No consensus exists as to how these properties affect hierarchy structure, either individually or in tandem. This is true even for properties that are commonly manipulated in experiments and simulation models, such as symmetry; for instance, although symmetrical experience effects (equal magnitude, opposite sign) are sufficient to generate a linear hierarchy among a set of initially identical individuals (Bonabeau *et al.* 1999; Hemelrijk 2000; Beacham 2003), it is unclear how hierarchy linearity responds to deviations from symmetry (see Bonabeau *et al.* 1996 and Dugatkin 1997 for opposing predictions). No simulation models to date have systematically explored the influence of different modes of accumulation of experience effects (additive versus multiplicative; see section 6.5, equations 6.1 and 6.2) or limits to increases in perceived fighting ability (i.e. bounds) on emergent hierarchy structure, and very few have manipulated decision rules, interaction frequencies, or decay functions (Bonabeau *et al.* 1996; Hemelrijk 1999; Beacham 2003). It might be difficult to maintain the tractability of simulation models while altering simultaneously all of these properties and introducing variation in intrinsic fighting abilities. However, attempts to do so could provide new insights into the complex influences of experience effects and how and to what degree experience effects shape dominance hierarchies. These modelling approaches should also focus future empirical research.

6.3   Mechanisms of experience effects

Behavioural changes that result from modifications in an individual's perception of costs during and/or after an aggressive contest are likely to be mediated by two somewhat different but overlapping neuroendocrine mechanisms. Changes in neural circuitry facilitate the consolidation and retrieval of information related to contest costs and these types of learning processes are, in turn, influenced to some degree by neuroendocrine factors such as hormone concentrations (e.g. glucocorticoids; Roozendaal 2002). Similarly, integration of information and associated changes in neural circuitry can elicit neuroendocrine changes that allow an animal to respond appropriately to perceived alterations in contest costs. A brief overview of studies that link learning with fighting experience follows below, and it discusses some of the most salient neuroendocrine changes that occur as a result of winning or losing aggressive encounters, keeping in mind that learning itself is a manifestation of changes in neuroendocrine systems.

6.3.1   *Learning*

Psychology has a long history of research into how experiences promote learning of the expectations of benefits and costs of alternative possible behaviours. Pigeons can adjust the frequency of pecking at keys in relation to differential food rewards and gradually improve their performance with repeated trials (Mazur 1995). These changes can be stored in long-term memory for retrieval and used in key-pecking choices at later times. Many memory traces also gradually disappear (forgetting) through time, whether spontaneously or as a result of interference from more recent experiences (Mazur 1996; Devenport 1998).

Recently, more emphasis has been placed on understanding how conditioning, particularly Pavlovian conditioning, might mediate social behaviour (Domjan *et al.* 2000). A series of studies on blue gouramis (*Trichogaster trichopterus*, Osphronemidae) showed that learning to anticipate a rival could reduce the costs of fighting and increase an individual's chances of winning the potential fight through early engagement of the intruder (Hollis 1984, 1999; Hollis *et al.* 1995). Individuals conditioned to anticipate the appearance of a rival, through pairing of a light stimulus with presentation of a rival, were more successful at expelling the intruder from their territory than individuals that had not been conditioned. Success in this initial contest also influenced the probability of winning a future contest (Hollis *et al.* 1995).

Similar effects have been demonstrated in three-spined sticklebacks (*Gasterosteus aculeatus*; Jenkins & Rowland 1996).

Although these studies provide evidence that learning can modulate aggressive behaviour in fishes, identifying the mechanism of learning, if any, that characterizes winner and loser effects presents a considerable challenge. So-called conditioned defeat, which is similar to the loser effect, has been studied extensively in rodents (Huhman *et al.* 2003) but the type of conditioning underlying these effects remains elusive. Changes in behaviour of winners and losers emerge without explicit pairing of conditioned and unconditioned stimuli, thus Pavlovian conditioning is unlikely to be involved. Both winners and losers receive some form of reinforcement, whether positive or negative, and subsequently respond with increases or decreases in aggressive behaviour when presented with a different opponent. These types of behavioural changes appear, superficially, to be the result of operant conditioning but because individuals respond in a more general fashion to similar stimuli (e.g. conspecific opponents), one might qualify the behavioural responses of winner and loser as sensitization (in losers) or stimulus generalization. To establish empirically that experience effects constitute learning might require a reformulation or combination of learning rules coupled with a better understanding of how extinction, forgetting rates, and spontaneous recovery might contribute to intraspecific and interspecific variation in the expression of winner and loser effects. It also will be important to establish whether changes in the behaviour of winners and losers occur as a consequence of, for instance, hormones acting on already established neural pathways or the emergence of new neural pathways that facilitate appropriate behavioural responses in the long term. The experience of both opponents influences the dynamics and outcome of an aggressive contest. Thus, approaching experience effects from a learning perspective will require consideration of learning processes that occur after an initial experience and during the current contest for both contestants (Miklósi *et al.* 1997).

6.3.2 *Neuroendocrine correlates of fighting*

A thorough review of the literature documenting manifold neuroendocrine changes that accompany fighting experience is beyond the scope of this chapter. It will touch briefly on four approaches to studying physiological changes associated with winning or losing aggressive interactions.

The first approach entails measurement of, for instance, hormone or brain neurotransmitter concentrations following contest resolution. In fishes, losers often exhibit increased cortisol concentrations (Hoglund *et al.* 2001; Sloman *et al.* 2001), decreased androgen levels (Cardwell & Liley 1991; Oliveira *et al.* 1996), and increased serotonergic [5-hydroxytryptamine (5-HT)] activity (Winberg & Nilsson 1993) compared to winners. Relative hormonal responses of winners and losers, however, may vary considerably. In some studies, no substantial differences between winners and losers in post-fight cortisol concentrations are found (Correa *et al.* 2003; Buchner *et al.* 2004), and a direct association between winning and elevated androgen levels rarely is observed (e.g. Neat *et al.* 1998; Elofsson *et al.* 2000). An important factor that might explain variation in documented responses across studies is the dynamics of the aggressive contest. The neuroendocrine response of both winners and losers appears to be linked to the intensity and/or length of the contest (e.g. number of escalated interactions or degree of aggressive reinforcement; Winberg & Lepage 1998; Elofsson *et al.* 2000; Sloman *et al.* 2001). Another important factor to consider is the strength of the neuroendocrine response. In particular, hormones associated with the stress response (e.g. cortisol and norepinephrine) will enhance aggression when secreted in low concentrations, but beyond a certain threshold these same hormones will inhibit aggressive behaviour (Haller *et al.* 1998; Øverli *et al.* 2002). Recent evidence also suggests that hormonal responses to fighting, particularly elevations in androgen levels, may be linked to perceptions of contest success rather than the act of fighting alone (Oliveira *et al.* 2005).

The second approach examines changes in the neuroendocrine response of winners and losers over time. This approach is important for understanding the potential for neuroendocrine factors to mediate the differential longevity of winner and loser effects (Hannes *et al.* 1984; Winberg & Lepage 1998; Øverli *et al.* 1999; Summers *et al.* 2003) discussed in Section 6.2.2. Both cortisol concentrations and 5-HT metabolism increase sharply following a contest in both winners and losers, but winners recover baseline concentrations/metabolism much faster than losers. It is tempting to link post-fight temporal decay of plasma hormone or brain neurotransmitter concentrations directly with the persistence of winner and loser effects. However, explicit causal relationships between temporal decay of neuroendocrine factors and similar decay in experience effects have yet to be established.

The third approach involves manipulating hormones, transmitters, or receptor binding capabilities and documenting changes in the expression of behaviour associated with fighting experience. The most common manipulation in fishes is to alter baseline levels of hormone or neurotransmitter via implantation or injection; for instance, androgen supplementation and injections of serotonin synthesis inhibitors facilitate aggressive behaviour in cichlid fishes (Fernald 1976; Adams *et al.* 1996). Beyond this, we know very little about how exogenous manipulations affect the fighting behaviour of fishes. This is due, in part, to the methodological difficulties of conducting such treatments in an aqueous medium on relatively small organisms (which are typically the focus of experimental studies on aggression). Studies on rodents and birds, however, have demonstrated that a variety of neuroendocrine factors can affect aggressive behaviour if applied exogenously (e.g. arginine vasotocin, Goodson 1998 serotonin reuptake inhibitor, Larson & Summers 2001; androgens, Trainor *et al.* 2004). Importantly, some studies in laboratory rodents have successfully regulated the acquisition and/or expression of behaviour associated with defeat by infusing different types of receptor agonists/antagonists into the brain (e.g. γ-amino butyric acid and corticotropin releasing hormone receptors; Jasnow *et al.* 1999; Jasnow & Huhman 2001). These studies provide some compelling directions for exploring the neuroendocrine correlates of the loser effect in fishes.

Lastly, it is important to recognize that aspects of the peripheral and central neuroendocrine systems do not act independently on aggressive behaviour. Changes in aggressive behaviour as a consequence of winning or losing a contest are likely to result from complex interactions between, and correlated changes among hormones, neuropeptides, neurotransmitters, and their receptors (e.g. interactions between the serotonergic system and the neuroendocrine stress axis; Winberg *et al.*

1997). Furthermore, behavioural outcome may depend critically on the site at which neuroendocrine interactions take place (e.g. hypothalamus, forebrain preoptic area; Hayden-Hixson & Ferris 1991; Kruk *et al.* 2004).

6.4   Other types of experience

6.4.1   *Individual recognition*

The behavioural decisions of shoaling fishes (Griffiths 2003) and the dynamics of competitive interactions in territorial species (Miklósi *et al.* 1995; Miklósi *et al.* 1997) are mediated, in part, by the ability of individuals to distinguish familiar conspecifics (see Chapter 8). Individual recognition can reduce fighting costs in situations where the probability of encountering the same individual on a regular basis is high (Pagel & Dawkins 1997). In fishes, memory of past opponents is ascertained by comparing the behavioural response of losers when encountering their former dominant opponent and an unfamiliar opponent. Generally speaking, losers exhibit more pronounced behavioural and physiological (e.g. skin darkening) avoidance responses when faced with familiar opponents (Miklósi *et al.* 1995, 1997; Morris *et al.* 1995; Johnsson 1997; O'Connor *et al.* 2000; Utne-Palm & Hart 2000), suggesting that individual recognition amplifies the loser effect. Contests between familiar opponents seldom escalate and typically are characterized by low levels of aggression (Keeley & Grant 1993; Earley *et al.* 2003). It is thus unclear whether winners fail to escalate because they recognize a former subordinate and/or because the opponent behaves submissively.

6.4.2   *Eavesdropping*

In some fish species, individuals appear to obtain a relatively accurate estimate of possible costs in future contests by watching others engage in aggressive contests. Eavesdropping, the act of extracting information from contest interactions between others (Peake & McGregor 2004; Peake 2005), might be a particularly advantageous assessment strategy when the costs of physical combat are high (Johnstone 2001). In Siamese fighting fish (*Betta splendens*, Osphronemidae) and green swordtail fish, observers appear to update their perception of the fighting abilities of the watched individuals based on the dynamics and/or outcome of the witnessed contest (Oliveira *et al.* 1998; McGregor *et al.* 2001; Earley & Dugatkin 2002; Brown & Laland 2003).

An important consideration for studies on eavesdropping in fishes is whether the observer's response is specific to the watched individuals or more general. Observing an aggressive interaction elevates urinary 11-ketotestosterone levels [Mozambique tilapia, *Oreochromis mossambicus* (Cichlidae); Oliveira *et al.* 2001] and increases the aggressive behaviour of male Siamese fighting fish towards unobserved opponents (Clotfelter & Paolino 2003). It is possible that watching fights elicits behavioural and physiological 'priming' responses (Hollis *et al.* 1995), which cause post-observational changes in agonistic behaviour, independent of integrating information about the observed contestants. Studies on green swordtail fish, however, indicate that eavesdroppers modify predictably their response towards individuals that had been observed to win or lose, but not towards naive animals (Earley & Dugatkin 2002; Earley *et al.* 2005). These results suggest that at least some species are capable of storing information from watched contests for use in future encounters with the observed individuals.

Recent simulation models have addressed whether eavesdropping can act in concert with winner and loser effects to promote linear dominance hierarchies (Dugatkin 2001). Resulting in part from unnecessarily strict assumptions (Earley & Dugatkin 2005), the verdict is uncertain but there is some indication that indirect experience, and subsequent adjustments of perceived relative fighting ability by an eavesdropper, reinforces linear hierarchies (Dugatkin 2001).

6.4.3   *Transitive inference*

Transitive inference in a social context refers to the ability of an individual to combine individual experience with a particular opponent and information obtained through eavesdropping. Fusing these two types of experiences allows an animal to respond more appropriately to its social environment; for instance, if individual A initially loses to B and then witnesses C defeat B, A may also opt not to interact with C so as to avoid the costs of fighting and losing again. Altmann (1981) proposed that non-primate animals were incapable of transitive inference, but pinyon jays (*Gymnorhinus cyanocephalus*; Paz-y-Miño *et al.* 2004), chickens (*Gallus gallus*; Hogue *et al.* 1996), and hyenas (*Crocuta crocuta*; Engh *et al.* 2005) show the capacity for transitive inference, or at least assessment of third-party relationships, in a social context (see also Peake *et al.* 2002 for support in great tits). Combining individual recognition with eavesdropping may reduce errors in assessment of fighting ability and, in turn, further reduce rank-order ambiguity in dominance hierarchies (e.g. Hogue *et al.* 1996; Peake *et al.* 2002). Despite the fact that fishes are well represented in studies on individual recognition and eavesdropping, researchers have yet to test whether species within this taxon are capable of transitive inference. This marks an exciting direction for research on fish cognition over the next decade.

6.5   Integrating experience information

The information derived from previous fighting experience is expected to modify perceptions of fighting costs for individuals in future contests, influence their interactions, and ultimately impact contest outcomes. To understand and model the influence of prior fighting experiences on contest outcomes, we need to consider at least five components:

1   pre-experience expectations;
2   the effect of an experience;
3   how that effect changes with time after the experience;
4   how multiple experiences are integrated for an individual;
5   how the experiences of two individuals interact to determine their probabilities of winning against each other.

Socially naive individuals are aggressive and engage in contests when encountering conspecifics for the first time (Doernberg *et al.* 2001; Chen *et al.* 2002); for instance, socially naive fishes are capable of defeating their competitors to acquire winning experiences (e.g. stickleback, Bakker & Sevenster 1983; mangrove killifish, Hsu & Wolf 1999). These naive individuals are likely to have intrinsic estimates of their own fighting abilities and the possible costs of engaging in contests. Experiences then modify perceived contest costs from what would be expected by naive individuals.

Individuals are likely to participate in more than one aggressive encounter and thus need to process information from these multiple events. Empirical data on the effects of multiple fighting experiences are limited. The contest behaviour of the mangrove killifish is influenced at least by the most recent and the penultimate fighting experiences (Hsu & Wolf 1999, 2001), indicating that the fishes can integrate information from different experiences. The mangrove killifish probably is not the only fish species that uses information from multiple contest experiences. Two mechanisms of experience accumulation often have been considered in the literature on dominance hierarchy formation: additive (e.g. Bonabeau *et al.* 1999; Hemelrijk 2000; Beacham 2003) and multiplicative (e.g. Dugatkin 1997). To allow different experiences to have different magnitudes of effects and decay rates, Hsu *et al.* (2006) proposed a general model for additive effects:

$$F_i = N_i + \sum_{I=1}^{a} (E_I \times W_{I,t}) \tag{6.1}$$

where F_i is individual i's perceived fighting ability (reflected in perceived costs) after a fighting experiences; N_i is the perceived fighting ability of individual i in the naive state; E_I is the information individual i received and incorporated from experience I; and $W_{I,t}$ represents how the effect of experience is weighted with the passage of time or the occurrence of additional experiences. The sign of E_I is positive for a win and negative for a loss. Equation 6.1 can be modified for multiplicative effects:

$$F_i = N_i \times \prod_{I=1}^{a} (1 + E_I \times W_{I,t}) \tag{6.2}$$

These two mechanisms of information integration differ in the relationship between experience effect and perceived fighting ability. In the additive model, the experience effect is E_I and is independent of the naive individual's perceived fighting ability N_i. In the multiplicative model, the experience effect, $N_i^*E_I$, depends on the individual's perceived fighting ability and individuals with higher perceived fighting ability are subjected to stronger experience effects. Consequently, successive winning experiences will result in a more pronounced change in the perceived fighting ability than successive losing experiences. Although the magnitude of an experience effect may be influenced by the state of an individual, the relationship between experience effect and perceived fighting ability has not yet been examined.

If experience effects decay slowly and the frequency of fighting is high, it is theoretically possible that F_i could become large or become negative. However, physiological mechanisms that mediate experience effects may impose a limit

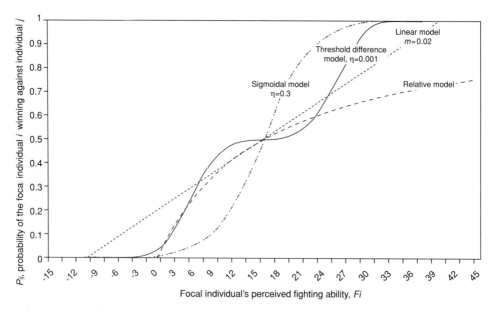

Fig. 6.2   Probability of winning (P_{ij}) predicted from: the Linear model [Eqn 6.3, with m arbitrarily set to 0.02]; the Relative model [Eqn 6.4]; the Sigmoidal model [Eqn 6.5, with η arbitrarily set to 0.3]; and the Threshold difference model [Eqn 6.6, with k set to 3 and η arbitrarily set to 0.001] for a focal individual i. Curves plotted as perceived fighting ability (F_i) of the focal individual i, fighting against an opponent j with a perceived fighting ability (F_j) of 15. (Modified with permission from Hsu *et al.* 2006.)

(minimum/maximum) on the effects; for instance, hormone titres presumably cannot increase indefinitely. However, what regulates the ceilings of experience effects is currently unknown. Different limits could be imposed to allow F_i to vary within a biologically meaningful range of values for different groups/species of animals.

Contest outcomes result from interactions between rivals. Hsu *et al.* (2006) proposed four models that predict fighting outcomes based on the experience-modified perceived fighting abilities of two opponents (Fig. 6.2).

Model 1 is a Linear model:

$$P_{ij} = 0.5 + m(F_i - F_j)   (0 \le P_{ij} \le 1),$$ (6.3)

where P_{ij} is the probability of individual i winning against individual j. The slope, m, scales how important differences in the experience-modified perceived fighting abilities are to contest outcome. In this model, the probability of winning is a linear function only of the difference in the perceived fighting abilities of the opponents.

Model 2, the Relative model:

$$P_{ij} = \frac{F_i}{F_i + F_j}   (0 \le P_{ij} \le 1).$$ (6.4)

In this model, the probability of winning is determined by the relative ratio of the perceived fighting ability of the two opponents. Consequently, the influence of experience is negatively scaled by the sum of the perceived fighting ability of the two opponents. Experience effects are more difficult to detect between bigger contestants or contestants with more winning experiences.

Model 3, the Sigmoidal model (following Bonabeau *et al.* 1999):

$$P_{ij} = \frac{1}{1 + e^{-\eta(F_i - F_j)}} \qquad (6.5)$$

where η scales the rate of approach to the asymptote. A large η produces a deterministic outcome and a small difference in the experience-modified perceived fighting ability would be sufficient to ensure winning by the opponent with a slightly higher perceived fighting ability.

Model 4, the Threshold Difference model:

$$P_{ij} = \frac{1}{1 + e^{-\eta(F_i - F_j)^k}} \qquad (6.6)$$

where k could be any odd number ≥ 3. In this model, small differences in the perceived fighting abilities do not produce detectable effects on fighting outcomes. But once the differences become sufficiently large (i.e. beyond a threshold) then changes occur rapidly toward the asymptote. The parameters η and k influence both: (i) the width of the interval where fighting outcome remains insensitive to the difference in perceived fighting ability; and (ii) how fast the probability of winning approaches the asymptote once the difference reaches the threshold. With a small η and/or k, a big difference in perceived fighting ability is needed to produce a recognizable effect on the contest outcome and the probability of winning approaches the asymptote at a slower rate. However, a change in η produces a bigger impact on how fast the difference reaches the threshold, while a change in k produces a bigger influence on how fast the probability of winning approaches the asymptote once the difference reaches the threshold (Fig. 6.3a,b).

The Linear and the Relative models theoretically could yield $P_{ij} < 0$ or >1, which are empirical impossibilities. However, the theoretical possibility means that even with temporal decay of experience effects, P_{ij} could remain at 0 or 1 for some period of time until the combined experience effects decay to values that yield P_{ij} between 0 and 1. P_{ij} is bounded between 0 and 1 in the Sigmoidal and the Threshold Difference models.

These models make different predictions regarding the relative importance of wins and losses on contest outcomes, even if wins and losses have equivalent effects on an individual's perceived fighting ability. Models 1, 3, and 4 predict that, once integrated, wins and losses have the same magnitude of effects on the probability of winning while the Relative model predicts a more prominent loser than winner effect on contest outcomes. This predicted outcome asymmetry from the Relative model could partially explain why losses have more effect than wins in many contests (see Section 6.2.2) and may explain why the influence of winning or losing may disappear/decay at varying time intervals/rates (Francis 1983; Bakker *et al.* 1989; Chase *et al.* 1994). Empirical data are necessary to understand the applicability of the four models.

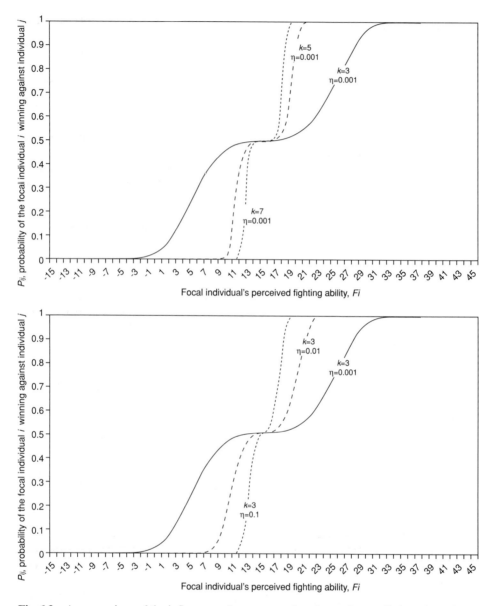

Fig. 6.3   A comparison of the influences of parameters k and η on the predictions from the Threshold difference model [Eqn 6.6]. (a) The impact of parameter k (with η fixed at 0.001 and k set to different values (3, 5 and 7, respectively). As k increases, the function approaches asymptote much faster. Although the interval where the probability of winning (P_{ij}) remains relatively insensitive to the difference in perceived fighting ability, ($F_i - F_j$) becomes narrower as k becomes bigger, the interval does not disappear as fast as when η changes [as in (b)]. (b) The impact of parameter η [with k fixed at 3 and η set to different values (0.001, 0.01 and 0.1, respectively)]. As η becomes bigger, the function also approaches the asymptote faster; and, relative to the curves in (a), the insensitive interval becomes narrower faster.

6.6   Conclusions and future directions

The evidence for fishes and other animals strongly supports an effect of prior contest experience on physiology and behaviour and hence on the outcome of a current contest. These effects vary in their magnitude and with the longevity of earlier winning and losing experiences, and appear to be based on changed perceptions of fighting ability and the costs of the current contest. These changed perceptions can be viewed as at least similar to the phenomena of learning and forgetting of cost expectations based on accumulated experiences.

While we know that these experience effects on aggression occur, we know very little about the details of the effects. We have discussed hormonal changes following contest interactions. Generally one finds considerable variation in hormonal responses but often testosterone is elevated following a win and corticosteroids are elevated following a loss. At the same time, behaviour changes so that winners tend to be more aggressive, willing to initiate a subsequent contest and more willing to attack an opponent, while losers tend to be less likely to participate in a contest. In spite of considerable work in this area, we still know relatively little about the detailed relationships between an experience and its effects on physiology, behaviour and contest outcome. Currently, most studies focus on only one or two stages of the process that transforms experiences into changed contest outcomes (Fig. 6.1). Integrating all stages will be complicated, but ultimately most productive in understanding how experience influences aggression.

Complicating interpretation of changes produced by experiences is the fact that contest outcome is probably based on the relationship of the effects on each contestant, not merely the magnitude of the effects on a single individual. The importance of the interaction in contest outcome requires that models making quantitative predictions of contest outcome include the magnitude of experience effects at any time and how those experience effects of the contestants are integrated to determine contest outcome.

Models integrating multiple experiences assume the effects are additive or multiplicative, but with little supporting evidence. So far experiments have tested a maximum of two prior experiences with only a single interval length between experiences and the subsequent contest. The rate of decay of the effects of individual experiences also is unknown. However, it is clear that the decay rate varies among species and even some populations (e.g. sticklebacks; Mackney & Hughes 1995). Predictions about contest outcomes based on prior experience of the contestants will depend on careful documentation of the symmetry and magnitudes of initial effects and their subsequent decay rates as well as the mode of accumulation of multiple experience effects.

This chapter summarises four different algorithms for predicting contest outcomes from differences in integrated experience effects (Hsu *et al.* 2006). The algorithms differ in the importance of small differences in experiences between the contestants and the degree to which the predicted outcome asymptotes as differences in experience effects become large. The only study of which we are aware that attempts to test predictive power of the algorithms has looked at Equations 6.1 and 6.2. These two models each give good fits to the data on contest outcomes with similar values for additive experience effects at 24 and 48 h following the experiences (Y. Hsu, unpublished data). Carefully designed tests will be required within time limits where experience effects are predicted to be most dissimilar to distinguish among the four current algorithms for their predictive power. One difficulty of this approach is that we do not have estimates of the values of wins and losses at various time periods that are independent of the outcomes data. It will be useful to develop ways to gain independent estimates of these effects.

These algorithms for integrating multiple experience effects and predicting contest outcomes are important for predictions about the role of experience in dominance hierarchy formation. Various models have used additive or multiplicative integration of multiple experiences and have assumed the form of the algorithm predicting contest outcomes. Moving from qualitative to quantitative predictions of contest outcome and hierarchy formation, based on interaction frequencies and experience decay rates, will be an important component of testing recent theoretical models of hierarchy formation.

The experience effects for an individual need not be obtained by direct participation in a contest. Evidence now indicates that observing other individuals fighting can influence future interactions with either fighter, in this case presumably based on the expectation of the costs that could be imposed by the observed fighter. This eavesdropping implies memory of the identity of the individuals and their activities in the fight. Some evidence also suggests that observing a contest can modify the observer's behaviour in subsequent contests (Clotfelter & Paolino 2003). In this case, the bias in behaviour produced by the earlier observations would be with any contestant, not just the ones in the previous contest. This possibility is more problematic and awaits further experimental testing.

Our potential ability to manipulate physiology and behaviour by prior experience also should give us a useful tool for understanding how these two phenomena influence contest outcome. Using experience as an experimental tool to investigate signalling in aggressive interactions also should help us understand the role of self-perception and relative comparisons to the opponent in outcomes of contests. Most game theoretical models assume some comparison to the opponent, but it is possible that self-perception is critical for behaviour in a contest.

Considerable progress has been made in understanding how experience influences subsequent contest behaviour and outcomes, but all aspects of the phenomenon need considerably more research before we approach a reasonable understanding of physiology, behaviour and contest outcomes. As that work moves forward we also need to integrate experience effects into other influences on contest outcomes. Size differences between individuals, residency and differences in perception of possible benefits from the aggression, among other possible factors, would all be expected to be integrated into outcomes of particular contests. Determining the relative importance of each factor has received little investigation and certainly warrants much more effort.

Fishes are very tractable experimental organisms for research on factors, including experience, that influence aggressive behaviour and contest outcomes. In fact, most of the work on this problem so far has been undertaken with fishes and we expect that they will probably increase in importance as model organisms for answering questions in this very fascinating area of behavioural research.

6.7 Acknowledgements

This research was supported in part by Taiwan National Science Council (NSC 93-2311-B-003-004). R.L.E. was supported by an award from the National Institutes of Health (5F32-HD046240-01).

6.8 References

Adams, C.F., Liley, N.R. & Gorzalka, B.B. (1996) PCPA increases aggression in male firemouth cichlids. *Pharmacology*, **53**, 328–330.

Alcock, J. (1993) *Animal Behavior*, 5th edn. Sinauer Associates Inc., Massachusetts.

Alexander, R.D. (1961) Aggressive territoriality and sexual behaviour in field crickets (Orthoptera: Gryllidae). *Behaviour*, **17**, 130–223.

Allee, W.C., Foreman, D., Banks, E.M. & Holabird, C.M. (1955) Effects of an androgen on dominance and subordinance in six common breeds of *Gallus gallus*. *Physiological Zoology*, **28**, 89–115.

Altmann, S.A. (1981) Dominance relationships: the Cheshire cat's grin? *Behavioral and Brain Sciences*, **4**, 430–431.

Bakker, Th.C.M. & Sevenster, P. (1983) Determinants of dominance in male sticklebacks (*Gasterosteus aculeatus* L.). *Behaviour*, **86**, 55–71.

Bakker, Th.C.M., Bruijn, E. & Sevenster, P. (1989) Asymmetrical effects of prior winning and losing on dominance in sticklebacks (*Gasterosteus aculeatus*). *Ethology*, **82**, 224–229.

Barchas, P.R. & Mendoza, S.D. (1984) Emergent hierarchical relationships in rhesus macaques: an application of Chase's model. In: P.R. Barchas (ed.), *Social Hierarchies: essays toward a sociophysiological perspective*, pp. 81–95. Greenwood Press, Westport, Connecticut.

Beacham, J.L. (1988) The relative importance of body size and aggressive experience as determinants of dominance in pumpkinseed sunfish, *Lepomis gibbosus*. *Animal Behaviour*, **36**, 621–623.

Beacham, J.L. (2003) Models of dominance hierarchy formation: effects of prior experience and intrinsic traits. *Behaviour*, **140**, 1275–1303.

Beaugrand, J., Goulet, C. & Payette, D. (1991) Outcome of dyadic conflict in male green swordtail fish, *Xiphophorus helleri*: effects of body size and prior dominance. *Animal Behaviour*, **41**, 417–424.

Beaugrand, J.P., Payette, D. & Goulet, C. (1996) Conflict outcome in male green swordtail fish dyads (*Xiphophorus helleri*): interaction of body size, prior dominance/subordination experience, and prior residency. *Behaviour*, **133**, 303–319.

Bégin, J., Beaugrand, J.P. & Zayan, R. (1996) Selecting dominants and subordinates at conflict outcome can confound the effects of prior dominance or subordination experience. *Behavioural Processes*, **36**, 219–226.

Bonabeau, E., Theraulaz, G. & Deneubourg, J.L. (1996) Mathematical models of self-organizing hierarchies in animal societies. *Bulletin of Mathematical Biology*, **58**, 661–719.

Bonabeau, E., Theraulaz, G. & Deneubourg, J.L. (1999) Dominance orders in animal societies: the self-organization hypothesis revisited. *Bulletin of Mathematical Biology*, **61**, 727–757.

Brown, C. & Laland, K.N. (2003) Social learning in fishes: a review. *Fish and Fisheries*, **4**, 280–288.

Buchner, A., Sloman, K. & Balshine, S. (2004) The physiological effects of social status in the cooperatively breeding cichlid *Neolamprologus pulcher*. *Journal of Fish Biology*, **65**, 1080–1095.

Cardwell, J.R. & Liley, N.R. (1991) Androgen control of social status in males of a wild population of stoplight parrotfish, *Sparisoma viride* (Scaridae). *Hormones and Behavior*, **25**, 1–18.

Chase, I.D. (1980) Social process and hierarchy formation in small groups: a comparative perspective. *American Sociological Review*, **45**, 905–924.

Chase, I.D. & Rohwer, S. (1987) Two methods for quantifying the development of dominance hierarchies in large groups with applications to Harris' Sparrows. *Animal Behaviour*, **35**, 1113–1128.

Chase, I.D., Bartolomeo, C. & Dugatkin, L.A. (1994) Aggressive interactions and inter-contest interval: how long do winners keep winning? *Animal Behaviour*, **48**, 393–400.

Chase, I.D., Tovey, C., Spangler-Martin, D. & Manfredonia, M. (2002) Individual differences versus social dynamics in the formation of animal dominance hierarchies. *Proceedings of the National Academy of Sciences USA*, **99**, 5744–5749.

Chase, I.D., Tovey, C. & Murch, P. (2003) Two's company, three's a crowd: differences in dominance relationships in isolated versus socially embedded pairs of fish. *Behaviour*, **140**, 1193–1217.

Chen, S., Lee, A.Y., Bowens, N.M., Huber, R. & Kravitz, E.A. (2002) Fighting fruit flies: a model system for the study of aggression. *Proceedings of the National Academy of Sciences USA*, **99**, 5664–5668.

Clotfelter, E.D. & Paolino, A.D. (2003) Bystanders to contests between conspecifics are primed for increased aggression in male fighting fish. *Animal Behaviour*, **66**, 343–347.

Collias, N.E. (1943) Statistical analysis of factors which make for success in initial encounters between hens. *American Naturalist*, **77**, 519–538.

Correa, S., Fernandes, M., Iseki, K. & Negrao, J. (2003) Effect of the establishment of dominance relationships on cortisol and other metabolic parameters in Nile tilapia (*Oreochromis niloticus*). *Brazilian Journal of Medical and Biological Research*, **36**, 1725–1731.

Devenport, L.D. (1998) Spontaneous recovery without interference: why remembering is adaptive. *Animal Learning and Behavior*, **26**, 172–181.

Doernberg, S.B., Cromarty, S.I., Heinrich, R., Beltz, B.S. & Kravitz, E.A. (2001) Agonistic behavior in naive juvenile lobsters depleted of serotonin by 5,7-dihydroxytryptamine. *Journal of Comparative Physiology A*, **187**, 91–103.

Domjan, M., Cusato, B. & Villarreal, R. (2000) Pavlovian feed-forward mechanisms in the control of social behavior. *Behavioral and Brain Sciences*, **23**, 235–282.

Dugatkin, L.A. (1997) Winner effects, loser effects, assessment strategies and the structure of dominance hierarchies. *Behavioral Ecology*, **8**, 583–587.

Dugatkin, L.A. (2001) Bystander effects and the structure of dominance hierarchies. *Behavioral Ecology*, **12**, 348–352.

Dugatkin, L.A. & Druen, M. (2004) The social implications of winner and loser effects. *Proceedings of the Royal Society of London, Series B*, **271** (Suppl.), S488–S489.

Earley, R.L. & Dugatkin, L.A. (2002) Eavesdropping on visual cues in green swordtail (*Xiphophorus helleri*) fights: a case for networking. *Proceedings of the Royal Society of London, Series B*, **269**, 943–952.

Earley, R.L. & Dugatkin, L.A. (2005) Fighting, mating and networking: pillars of poeciliid sociality. In: P.K. McGregor (ed.), *Communication Networks*, pp. 84–113. Cambridge University Press, Cambridge.

Earley, R.L., Tinsley, M. & Dugatkin, L.A. (2003) To see or not to see: does previewing a future opponent affect the contest behavior of green swordtail males (*Xiphophorus helleri*)? *Naturwissenschaften*, **90**, 226–230.

Earley, R.L., Druen, M. & Dugatkin, L.A. (2005) Watching fights does not alter a bystander's response towards naive conspecifics in male green swordtail fish, *Xiphohporus helleri*. *Animal Behaviour*, **69**, 1139–1145.

Earley, R.L., Edwards, J.T., Aseem, O., Felton, K., Blumer, L.S., Karom, M. & Grober, M.S. (2006) Social interactions tune aggression and stress responsiveness in a territorial chchlid fish (*Archocentrus nigrofasciatus*). *Physiology & Behaviour* (available online; DOI: 10.1016/j.physbeh.2006.04.002).

Elofsson, U.O.E., Mayer, I., Damsgard, B. & Winberg, S. (2000) Intermale competition in sexually mature Arctic charr: effects on brain monoamines, endocrine stress responses, sex hormone levels, and behavior. *General and Comparative Endocrinology*, **118**, 450–460.

Engh, A.L., Siebert, E.R., Greenberg, D.A. & Holekamp, K.E. (2005) Patterns of alliance formation and postconflict aggression indicate spotted hyaenas recognize third-party relationships. *Animal Behaviour*, **69**, 209–217.

Fernald, R.D. (1976) The effect of testosterone on the behavior and coloration of adult male cichlid fish (*Haplochromis burtoni*, Gunther). *Hormone Research*, **7**, 172–178.

Francis, R.C. (1983) Experiential effects on agnostic behavior in the paradise fish, *Macropodus opercularis*. *Behaviour*, **85**, 292–313.

Goessmann, C., Hemelrijk, C. & Huber, R. (2000) The formation and maintenance of crayfish hierarchies: behavioral and self-structuring properties. *Behavioral Ecology and Sociobiology*, **48**, 418–428.

Gomez-Laplaza, L.M. & Morgan, E. (2000) Laboratory studies of the effects of short-term isolation on aggressive behaviour in fish. *Marine Freshwater Behavior and Physiology*, **33**, 63–102.

Goodson, J.L. (1998) Territorial aggression and dawn song are modulated by septal vasotocin and vasoactive intestinal polypeptide in male field sparrows (*Spizella pusilla*). *Hormones and Behavior*, **34**, 67–77.

Griffiths, S.W. (2003) Learned recognition of conspecifics by fishes. *Fish and Fisheries*, **4**, 256–268.

Haller, J., Makara, G.B. & Kruk, M.R. (1998) Catecholaminergic involvement in the control of aggression: hormones, the peripheral sympathetic, and central noradrenergic systems. *Neuroscience and Biobehavioral Reviews*, **22**, 85–97.

Hannes, R.P., Franck, D. & Liemann, F. (1984) Effects of rank-order fights on whole-body and blood concentrations of androgens and corticosteroids in the male swordtail (*Xiphophorus helleri*). *Zeitschrift für Tierpsychologie*, **65**, 53–65.

Hayden-Hixson, D. & Ferris, C. (1991) Cortisol exerts site-, context- and dose-dependent effects on agonistic responding in hamsters. *Journal of Neuroendocrinology*, **3**, 613–622.

Hemelrijk, C.K. (1999) An individual-oriented model of the emergence of despotic and egalitarian societies. *Proceedings of the Royal Society of London, Series B*, **166**, 361–369.

Hemelrijk, C.K. (2000) Towards the integration of social dominance and spatial structure. *Animal Behaviour*, **5**, 1035–1048.

Hoglund, E., Kolm, N. & Winberg, S. (2001) Stress-induced changes in brain serotonergic activity, plasma cortisol and aggressive behavior in Arctic charr (*Salvelinus alpinus*) is counteracted by L-DOPA. *Physiology & Behavior*, **74**, 381–389.

Hogue, M.E., Beaugrand, J.P. & Lague, P.C. (1996) Coherent use of information by hens observing their former dominant defeating or being defeated by a stranger. *Behavioural Processes*, **38**, 241–252.

Hollis, K.L. (1984) The biological function of Pavlovian conditioning: the best defense is a good offense. *Journal of Experimental Psychology*, **10**, 413–425.

Hollis, K.L. (1999) The role of learning in the aggressive and reproductive behavior of blue gouramis, *Trichogaster trichopterus*. *Environmental Biology of Fishes*, **54**, 355–369.

Hollis, K.L., Dumas, M.J., Singh, P. & Fackelman, P. (1995) Pavlovian conditioning of aggressive behavior in blue gourami fish (*Trichogaster trichopterus*): winners become winners and losers stay losers. *Journal of Comparative Psychology*, **109**, 125–133.

Hsu, Y. & Wolf, L.L. (1999) The winner and loser effect: integrating multiple experiences. *Animal Behaviour*, **57**, 903–910.

Hsu, Y. & Wolf, L.L. (2001) The winner and loser effect: what fighting behaviours are influenced? *Animal Behaviour*, **61**, 777–786.

Hsu, Y., Earley, R.L. & Wolf, L.L. (2006) Modulation of aggressive behaviour by fighting experience: mechanisms and contest outcomes. *Biological Reviews*, **81**, 33–74.

Huhman, K.L., Solomon, M.B., Janicki, M., Harmon, A.C., Lin, S.M., Israel, J.E. & Jasnow, A.M. (2003) Conditioned defeat in male and female Syrian hamsters. *Hormones and Behavior*, **44**, 293–299.

Jasnow, A.M. & Huhman, K.L. (2001) Activation of GABAA receptors in the amygdala blocks the acquisition and expression of conditioned defeat in Syrian hamsters. *Brain Research*, **920**, 142–150.

Jasnow, A.M., Banks, M.C., Owens, E.C. & Huhman, K.L. (1999) Differential effects of two corticotropin-releasing factor antagonists on conditioned defeat in male Syrian hamsters (*Mesocricetus auratus*). *Brain Research*, **846**, 122–128.

Jenkins, J.R. & Rowland, W.J. (1996) Pavlovian conditioning of agonistic behavior in male threespine stickleback (*Gasterosteus aculeatus*). *Journal of Comparative Psychology*, **110**, 396–401.

Johnsson, J.I. (1997) Individual recognition affects aggression and dominance relations in rainbow trout, *Oncorhynchus mykiss*. *Ethology*, **103**, 267–282.

Johnstone, R.A. (2001) Eavesdropping and animal conflict. *Proceedings of the National Academy of Sciences USA*, **98**, 9177–9180.

Keeley, E. & Grant, J. (1993) Visual information, resource value, and sequential assessment in convict cichlid (*Cichlasoma nigrofasciatum*) contests. *Behavioral Ecology*, **4**, 345–349.

Kruk, M.R., Halasz, J., Meelis, W. & Haller, J. (2004) Fast positive feedback between the adrenocortical stress response and a brain mechanism involved in aggressive behavior. *Behavioral Neuroscience*, **118**, 1062–1070.

Larson, E.T. & Summers, C.H. (2001) Serotonin reverses dominant social status. *Behavioural Brain Research*, **121**, 95–102.

Mackney, P.A. & Hughes, R.N. (1995) Foraging behaviour and memory window in sticklebacks. *Behaviour*, **132**, 1241–1253.

Marden, J.H. & Waage, J.K. (1990) Escalated damselfly territorial contests are energetic wars of attrition. *Animal Behaviour*, **39**, 954–959.

Mazur, J.E. (1995) Development of preference and spontaneous recovery in choice behavior with concurrent variable-interval schedules. *Animal Learning and Behavior*, **23**, 93–103.

Mazur, J.E. (1996) Past experience, recency, and spontaneous recovery in choice behavior. *Animal Learning and Behavior*, **24**, 1–10.

McGregor, P.K., Peake, T.M. & Lampe, H.M. (2001) Fighting fish *Betta splendens* extract relative information from apparent interactions: what happens when what you see is not what you get. *Animal Behaviour*, **62**, 1059–1065.

Mesterton-Gibbons, M. (1999) On the evolution of pure winner and loser effects: a game-theoretic model. *Bulletin of Mathematical Biology*, **61**, 1151–1186.

Miklósi, A., Haller, J., & Csanyi, V. (1995) The influence of opponent-related and outcome-related memory on repeated aggressive encounters in the paradise fish (*Macropodus opercularis*). *Biological Bulletin*, **188**, 83–88.

Miklósi, A., Haller, J. & Csanyi, V. (1997) Learning about the opponent during aggressive encounters in paradise fish (*Macropodus opercularis* L.): when it takes place? *Behavioural Processes*, **40**, 97–105.

Morris, M.R., Gass, L. & Ryan, M.J. (1995) Assessment and individual recognition of opponents in the pygmy swordtails *Xiphophorus nigrensis* and *X. multilineatus. Behavioral Ecology and Sociobiology*, **37**, 303–310.

Neat, F.C., Taylor, A.C. & Huntingford, F.A. (1998) Proximate costs of fighting in male cichlid fish: the role of injuries and energy metabolism. *Animal Behaviour*, **55**, 875–882.

Nelissen, M.H.J. & Andries, S. (1988) Does previous experience affect the ranking of cichlid fish in a dominance hierarchy? *Annals of the Royal Zoological Society of Belgium*, **118**, 41–50.

O'Connor, K.I., Metcalfe, N.B. & Taylor, A.C. (2000) Familiarity influences body darkening in territorial disputes between juvenile salmon. *Animal Behaviour*, **59**, 1095–1101.

Oliveira, R.F., Almada, V.C. & Canário, A.V.M. (1996) Social modulation of sex steroid concentrations in the urine of male cichlid fish *Oreochromis mossambicus. Hormones and Behavior*, **30**, 2–12.

Oliveira, R.F., McGregor, P.K. & Latruffe, C. (1998) Know thine enemy: fighting fish gather information from observing conspecific interactions. *Proceedings of the Royal Society of London, Series B*, **265**, 1045–1049.

Oliveira, R.F., Lopes, M., Carneiro, L.A. & Canário, A.V.M. (2001) Watching fights raises fish hormone levels. *Nature*, **409**, 475.

Oliveira, R.F., Carneiro, L.A. & Canário, A.V.M. (2005) No hormonal response in tied fights. *Nature*, **437**, 207–208.

Øverli, Ø., Harris, C.A. & Winberg, S. (1999) Short-term effects of fights for social dominance and the establishment of dominant–subordinate relationships on brain monoamines and cortisol in rainbow trout. *Brain Behavior and Evolution*, **54**, 263–275.

Øverli, Ø., Kotzian, S. & Winberg, S. (2002) Effects of cortisol on aggression and locomotor activity in rainbow trout. *Hormones and Behavior*, **42**, 53–61.

Pagel, M. & Dawkins, M.S. (1997) Peck orders and group size in laying hens: 'future contracts' for non-aggression. *Behavioural Processes*, **40**, 13–25.

Paz-Y-Miño, C.G., Bond, A.B., Kamil, A.C. & Balda, R.P. (2004) Pinyon jays use transitive inference to predict social dominance. *Nature*, **430**, 778–781.

Peake, T.M. (2005) Eavesdropping in communication networks. In: P.K. McGregor (ed.), *Animal Communication Networks*, pp. 13–37. Cambridge University Press, Cambridge.

Peake, T.M. & McGregor, P.K. (2004) Information and aggression in fishes. *Learning and Behavior*, **32**, 114–121.

Peake, T.M., Terry, A.M.R., McGregor, P.K. & Dabelsteen, T. (2002) Do great tits assess rivals by combining direct experience with information gathered by eavesdropping? *Proceedings of the Royal Society of London, Series B*, **269**, 1925–1929.

Riechert, S.E. (1998) Game theory and animal contests. In: L.A. Dugatkin & H.K. Reeves (eds), *Game Theory and Animal Behavior*, pp. 64–92. Oxford University Press, New York.

Roozendaal, B. (2002) Stress and memory: opposing effects of glucocorticoids on memory consolidation and memory retrieval. *Neurobiology of Learning and Memory*, **78**, 578–595.

Schuett, G.W. (1997) Body size and agonistic experience affect dominance and mating success in male copperheads. *Animal Behaviour*, **54**, 213–224.

Sloman, K.A., Metcalfe, N.B., Taylor, A.C. & Gilmour, K.M. (2001) Plasma cortisol concentrations before and after social stress in rainbow trout and brown trout. *Physiological and Biochemical Zoology*, **74**, 383–389.

Summers, C.H., Summers, T.R., Moore, M.C., Korzan, W.J., Woodley, S.K., Ronan, P.J., Hoglund, E., Watt, M.J. & Greenberg, N. (2003) Temporal patterns of limbic monoamine and plasma corticosterone response during social stress. *Neuroscience*, **116**, 553–563.

Trainor, B.C., Bird, I.M. & Marler, C.A. (2004) Opposing hormonal mechanisms of aggression revealed through short-lived testosterone manipulations and multiple winning experiences. *Hormones and Behavior*, **45**, 115–121.

Utne-Palm, A.C. & Hart, P.J.B. (2000) The effects of familiarity on competitive interactions between threespined sticklebacks. *Oikos*, **91**, 225–232.

Winberg, S. & Lepage, O. (1998) Elevation of brain 5-HT activity, POMC expression, and plasma cortisol in socially subordinate rainbow trout. *American Journal of Physiology*, **43 (274)**, R645–R654.

Winberg, S. & Nilsson, G.E. (1993) Time course of changes in brain serotonergic activity and brain tryptophan levels in dominant and subordinate juvenile Arctic charr. *Journal of Experimental Biology*, **179**, 181–195.

Winberg, S., Nilsson, A., Hylland, P., Söderström, V. & Nilsson, G.E. (1997) Serotonin as a regulator of hypothalamic-pituitary-interrenal activity in teleost fish. *Neuroscience Letters*, **230**, 113–116.

Chapter 7

The Role of Learning in Fish Orientation

Lucy Odling-Smee, Stephen D. Simpson, Victoria A. Braithwaite

7.1 Introduction

Until recently, the possibility that fish are capable of flexible, learned orientation strategies has generally been overlooked. However, in most aquatic environments, the physical landscape and the need to relocate various biologically important locations (for example, shelter or the position of a profitable food patch) should favour an ability to learn. Here, we review the evidence from both field and laboratory studies that fish can and do use learning and memory to orientate within their natural environments. Comparisons between different species and populations indicate that fish have a diverse array of sensory systems that they use to encode spatial information; however, what they learn is sometimes constrained by their environment, and/or their genotype.

Much of our knowledge on animal orientation has typically come from terrestrial species such as birds and mammals (Dodson 1988; Healy 1998). However, miniaturization of tracking devices and other telemetry techniques is now allowing us to investigate the movements and spatial behaviours exhibited by fishes (Armstrong *et al.* 1997; Hunter *et al.* 2003). Within the aquatic environment, fishes are exposed to an enormous diversity of potential cues from which they can extract information about their spatial location. Equally diverse is the range of spatial and navigational problems that may be encountered by different species, populations, individuals and even different developmental stages. The aim of this chapter is not only to show that fishes are capable of orienting (i.e. keeping track of their location with respect to an external point of reference) using learning and memory, but also to convey that such flexibility is often constrained or modified by mechanisms that guide learning and associated perceptual processes in response to particular ecological conditions and navigational demands.

7.2 Why keep track of location?

Many behaviours, including foraging, reproduction, competition and predator avoidance, are dependent on a fish being able to plan and execute a route to guide it efficiently between two locations. To achieve this, the fish needs to monitor its location with reference to an external point of reference as it moves through its environment. Resources are often widely separated in space or time such that individuals cannot rely purely on chance encounters to fulfil their energetic or reproductive requirements. Food is often distributed among sites that vary spatially and temporally in profitability, and food patches may differ in the likelihood of renewal after depletion (Warburton 2003; and see Chapter 2). As long as there is some degree of predictability, foraging efficiency will increase if fishes can map the status and renewal rates of individual food patches onto their location and use this information to guide their foraging (Noda *et al.* 1994; Hughes & Blight 1999, 2000). Spatial information may similarly be used to predict the location of receptive mates; for example, in certain reef fishes, spawning aggregations draw fishes to specific locations from disparate areas of the reef (Mazeroll & Montgomery 1998). Several species of fishes guard nest sites or other resources from competitors, displaying territoriality or a tendency to remain in a restricted area or 'home range' (Hallacher 1984; Matthews 1990a,b; Kroon *et al.* 2000). Such familiarity with a particular area may allow the fishes to return to their territory, or re-establish territory boundaries if they become temporarily displaced (Hallacher 1984; Matthews 1990a,b). Furthermore, spatial information may be used to pinpoint the location of shelters, or hiding places within a home range, allowing rapid escape in the event of an attack by a predator (Aronson 1951, 1971; Markel 1994).

In addition to keeping track of site-specific resources, fishes may need to monitor locations associated with risk. Some predators are associated with particular microhabitats or locations that are best avoided by potential prey (Goodyear 1973; Jordan *et al.* 1997; Brown 2003). Others have predictable movements; for example, many reef-based piscivores concentrate on coral reefs at night and move away from the reefs during the day (Mazeroll & Montgomery 1995). Predator avoidance strategies may therefore involve daily migrations requiring fishes to keep track of their location with respect to both feeding areas and protective refuges (Ogden & Buckman 1973; Ogden & Quinn 1984; Mazeroll & Montgomery 1995). Moreover, predators themselves may need to be equipped with information about the spatial structure of their home range in order to avoid being recognized by potential prey (see Brown 2003).

Some species of fish return to their natal area for reproduction, otherwise known as 'homing' (Tsukamoto *et al.* 2003). Homing is likely to enhance reproductive success by allowing mature animals to return to their spawning grounds when conditions are optimal for egg and larval development. In addition, reproductive isolation achieved through homing may facilitate the development of adaptations specific to the particular habitat occupied (Dittman & Quinn 1996). In order to achieve accurate return to their natal sites, fishes must relocate their natal area from impressive distances, in some cases thousands of kilometres (Quinn & Dittman 1990; Dittman & Quinn 1996).

7.3   The use of learning and memory in orientation

Compared to terrestrial groups, considerably less is known about how, or indeed whether, fishes use learning and memory in spatial orientation (Healy 1998). Observations of fishes' movements within their natural habitats suggest impressive spatial abilities without providing hard evidence for the role of learning and memory, or revealing what exactly might be learned. Various studies on fish orientation have used displacement of marked individuals over various distances from the place they were caught to investigate their capacity to home (Green 1971; Carlson &

Haight 1972; Hallacher 1984; Quinn & Ogden 1984; Kolm *et al.* 2005). In the Banggai cardinal fish (*Pterapogon kauderni*, Apogonidae), homing from 50 m was found to be in direct response to the landscape, rather than to conspecifics in the colony (Kolm *et al.* 2005). Some species return home from as far as 22.5 km (Carlson & Haight 1972) or after 6 months in captivity (Green 1971), consistent with the suggestion that they use long-term spatial memory. These observations could, however, also be explained by the fish using reactive mechanisms to home, such as tracking changes in temperature or light levels.

To track the search behaviour of a planktivorous reef damselfish, the stout chromis (*Chromis chrysurus*, Pomacentridae), Noda *et al.* (1994) marked individual fishes with acrylic paint and followed their movements after release. Individual fishes repeatedly visited three distinct foraging areas, swimming slowly and in a stereotypical pattern within each foraging site before swiftly moving off to the next patch. Noda and colleagues suggested that spatial memory may allow the fish to concentrate foraging at relatively high densities of zooplankton, and avoid revisiting depleted areas. Again, alternative explanations cannot be ruled out; for example, *C. chrysurus* may locate food patches simply by responding to olfactory cues.

We owe most of our knowledge about the role of learning and memory in fish orientation and the types of spatial information used to controlled experiments. These involve manipulations of fish sensory systems or spatial cues, or laboratory-based spatial tasks where fish are trained to learn particular associations. Such approaches have revealed that fishes can learn and remember routes and locations by using information from a variety of different sources, ranging from landmarks and compasses to spatially 'informed' conspecifics.

7.4   Learning about landmarks

In most aquatic landscapes, local features (such as visual or olfactory cues) are likely to be changeable components of the environment, requiring that they be stored in memory and updated on the basis of experience. To support this, several species of fish have been successfully trained to use landmark information to solve a range of laboratory-based spatial tasks (Huntingford & Wright 1989; Warburton 1990; Braithwaite *et al.* 1996; Salas *et al.* 1996a; Girvan & Braithwaite 1998; López *et al.* 1999, 2000a,b; Hughes & Blight 2000; Odling-Smee & Braithwaite 2003). However, aquatic environments are very diverse; there are several examples of fish species that still manage to orient even when their visual environment is not heterogeneous (Cain *et al.* 1994; Burt de Perera 2004a). This indicates that certain species of fish are capable of learning about their spatial environment using non-visual senses.

Less is know about the extent to which fishes use landmarks in the wild. One field-based approach has been to manipulate the position of landmarks in the natural migrating paths of fish. Reese (1989) observed that butterflyfish (Chaetodontidae) spend some time searching in an area where coral heads have been removed, before continuing along their original foraging path. Migrating brown surgeonfish (*Acanthurus nigrofuscus*, Acanthuridae) similarly change direction in accordance with the new positions of displaced landmarks (Mazeroll & Montgomery 1998). Moreover, their reliance on particular landmarks is reduced when landmarks are moved more than 6 m from their original location, suggesting that brown surgeonfish may be capable of assessing the stability and thence reliability of visual landmarks (see Biegler & Morris 1996). However, the disruption of migratory routes by shifted landmarks does not prove landmark cues were originally learned to guide orientation. Shifting landmarks could disturb the sediment and increase food availability, promoting exploratory behaviour. Alternatively, fishes may simply be reacting to a change in their local environment.

One successful approach to determining how fishes learn and use landmarks has been to combine laboratory and field work. In laboratory experiments, potentially confounding effects can be more easily controlled; for example, fishes can be trained to use specific landmarks and then additional cues, such as compass or global place cues, can be manipulated or completely removed (by screening or rotating test tanks) to determine whether the fishes used this additional information. The results from these trials can then be used to design and test the spatial cues that the fishes are using in a more natural setting (Braithwaite *et al.* 1996; Armstrong *et al.* 1997; Huntingford *et al.* 1998).

In addition to using landmarks as direct cues or beacons, fishes appear to be capable of learning more complex spatial relationships. Goldfish (*Carassius auratus*, Cyprinidae) can locate a food reward by using landmarks as indirect reference points (Warburton 1990) and remember the spatial positions of three differently rewarded, hidden food patches in a tank (Pitcher & Magurran 1983). They can also learn to use visual cues to locate a goal, even if they approach the goals from a novel direction (Ingle & Sahagian 1973; Warburton 1990; Rodríguez *et al.* 1994). This suggests a capacity to discriminate spatial relationships independently of any particular view of the surroundings, enabling the fishes to take shortcuts or choose between alternative routes to a goal. The level of accuracy achieved by Siamese fighting fish (*Betta splendens*, Osphronemidae), in an eight-arm radial maze similarly suggests that some amount of spatial memory is involved in recognizing which of the eight arms have already been depleted of the food reward (Roitblat *et al.* 1982). The fact that fishes can detect environmental modifications (Welker & Welker 1958) and show an organized pattern of exploration when they are introduced into a novel environment (Kleerokoper *et al.* 1974) also suggests some degree of spatial memory.

A classic demonstration of the ability of fishes to use spatial maps is provided by experiments on the gobiid fish, *Bathygobius soporator* (Gobiidae; Aronson 1951, 1971). When threatened, gobiids jump from their home tidepool to an adjacent pool with impressive accuracy. If required, the gobies can make a series of jumps leading them from one pool to the next and eventually into the open sea. In order to investigate whether gobiids acquire memories of the local topography around their home pools, Aronson constructed three artificial pools and manipulated the water level to simulate low and high tides (Fig. 7.1). Fishes that were given experience of the spatial distribution of the pools at 'high tide' successfully escaped a simulated attack at 'low tide' by jumping into the appropriate pool. There is a clear need for accuracy in this spatial task; a jump in the wrong direction could be the last one that a fish makes. It is surprising that such a simple model of spatial learning has not been investigated further. What types of landmark do the fishes use to recognize which particular pool they are in, local or global? Do the fishes remember a sequence of jumps, or are they able to make an appropriate escape response based on where they are at that point in time? For example, if they were moved to another pool would they be able to make the correct jump from this displaced position.

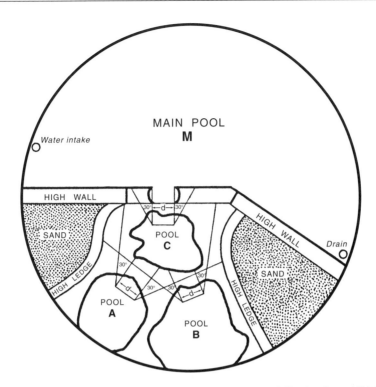

Fig. 7.1 Topography of tide pools used to test spatial learning ability by the gobiid fish (*Bathygobius soporator*, Gobiidae) (adapted from Aronson 1971). Fish placed in the smaller pools A and B were stimulated to jump by being prodded with a stick. Trials were terminated when the fish had reached the main pool.

Markel (1994) used another species of goby to investigate the use of landmarks to find shelter when threatened with a simulated predator attack. Here, the time that blackeye gobies (*Coryphopterus nicholsi*, Gobiidae) took to locate an artificial burrow was recorded. Fishes given additional exploration time in the test tank were quicker at relocating their burrow compared to fishes deprived of this additional experience. When the burrow was shifted to a new location, the experienced group took longer than the less experienced group to find the shifted burrow, suggesting that the fishes had learned and remembered the spatial location of the burrow, as opposed to its appearance.

Where vision is precluded by lifestyle or habitat conditions, information about landmarks may be acquired though alternative sensory channels. The nocturnal African mormyid or elephantnose fish (*Gnathonemus petersii*, Mormyridae) can use electrolocation to learn to locate an opening or aperture in a wall (Cain *et al.* 1994; Cain 1995). Once the fishes have learned the position of an opening, they are able to locate it even after their electric organ has been denervated, suggesting that they develop a stored, internal representation of their environment. The detail that the fishes can perceive with their electroreceptors is impressive; for example, *G. petersii* and another species of electric fish, the longtail knifefish (*Sternopygus macrurus*, Sternopygidae) can recognize the three-dimensional orientations and configurations of landmarks (Graff *et al.* 2004).

Blind Mexican cave fishes (*Astyanax fasciatus*, Characidae) also show some impressive spatial learning abilities. Eye formation in these fishes arrests during development, forcing them to use alternative sensory information to learn about the spatial organization of their environment. The fishes use their lateral line organ (LLO) to provide them with topographical information. As they swim the fishes detect water displacement patterns with these LLOs (Campenhausen *et al.* 1981). When a fish moves forwards it displaces water, and the lateral line detects small differences in water flow patterns as the displaced water is reflected off objects within the environment. The cave fishes move slowly around their environment as they explore a new area, but then increase their swimming velocity once they have learned the position of objects (Teyke 1985, 1989). Teyke (1989) suggests that these fishes develop an internal map that allows them to swim at velocities below this optimal range, once they are familiarized with the environment (but see Bennet 1996). In support of this, *A. fasciatus* have been shown to encode spatial information about configurations of landmarks, including information about their order or sequence, and their three-dimensional relationships (Burt de Perera 2004a,b; Burt de Perera *et al.* 2005). The way in which *A. fasciatus* learns about new landmarks suggests a lateralization in the way its brain processes spatial cues; for example, Burt de Perera & Braithwaite (2005) found that when the fishes were confronted with a new landmark they would preferentially swim past the landmark using the right side of the LLO. This right-side preference persisted whether they swam in a clockwise or anticlockwise direction. However, the preference waned once the fishes became familiar with the landmark. The lateralized use of the right LLO may be related to use of the left hemisphere of the brain, because the neurones associated with the lateral line pass bilaterally from the hindbrain nucleus to the midbrain with contralateral predominance (McCormick 1989). An alternative explanation may be that the fishes react to novelty using their right LLO (Brown *et al.* 2004).

Olfactory cues present an additional type of landmark information, which can be used to relocate home areas or natal streams. Perhaps most acclaimed is the ability of mature salmon to relocate their natal streams based on its unique olfactory composition, a feat described in more detail below. Another sensory modality that has largely been overlooked, but nevertheless provides orientation information, is sound. There is growing evidence that acoustic cues are highly important in the orientation of settlement-stage coral reef fishes when they return to reefs following a pelagic larval phase (Montgomery *et al.* 2006). Studies using light traps coupled with sound systems have shown that temperate (Tolimieri *et al.* 2000) and tropical (Leis *et al.* 2003; Simpson *et al.* 2004) reef fishes are attracted to reef sounds at settlement. Furthermore, Simpson *et al.* (2005a) have demonstrated that this attraction is used by larvae at settlement to locate suitable habitat, demonstrating the ability in larvae, both to orient towards, and to localize source sounds.

7.5   Compass orientation

In common with many terrestrial animals (Able 1993), when learning about locations and routes, fishes are likely to rely on more than one source of spatial information. The use of multiple cues provides back-up points of reference if changes in the environment make one cue unavailable or unreliable. Compasses provide another source of spatial information, which may often be used in combination with landmarks (Goodyear 1973).

Few studies have investigated the role of learning and memory in compass orientation. Early attempts demonstrated that novel sun-compass responses could be learned by training fishes under a clear sky, to seek cover in one of a number of circularly arranged refuges (Hasler *et al.* 1958; Schwassmann & Braemer 1961; Hasler 1971). Goodyear's study (1973) on mosquitofish (*Gambusia affinis*, Poeciliidae) goes a step further by demonstrating the possible adaptive significance of having a flexible compass response (see also Goodyear & Bennet 1979). The directional preferences displayed by mosquitofish when placed in test arenas under an artificial 'sun' suggested that they use a sun compass to move on a course perpendicular to the shore from where they were collected. This movement towards shallow water seems principally a mechanism for predator avoidance as it is absent in mosquitofish captured from predator-free environments (Goodyear & Ferguson 1969). Mosquitofish captured from a pond containing predators and those from a predator-free pond were able to learn a new shoreward direction, after being conditioned to avoid a predatory largemouth bass (*Micropterus salmoides*, Centrachidae) under a clear sky in tanks with a sloped gravel bottom simulating an emergent shoreline (Goodyear 1973). In this case, a modifiable compass response may allow adaptation to changes in predation risk or to shifts in growth of protective vegetation.

In addition to the sun, other cues may be available to fishes for use in compass navigation, such as polarized light and electromagnetic cues. Evidence that fishes learn to orient to polarized light is scant (Hawryshyn *et al.* 1990), although it is a probable cause for fishes that perform daily migrations at sunrise and sunset when polarized light patterns are strongest (Quinn & Ogden 1984; Mazeroll & Montgomery 1998). Laboratory-based training procedures have shown some species of fishes are capable of sensing and responding to magnetic compasses (Walker 1984; Kalmijn 1982; Walker *et al.* 1997) and possible anatomical sites for magnetic detection have been discovered (Kirschvink *et al.* 1985; Walker *et al.* 1988, 1997; Moore *et al.* 1990). As yet little is known about the extent to which learning and memory is involved in the use of polarized light and magnetic compasses. The use of polarized light and the sun for accurate orientation requires reference to an internal clock as the correct heading depends on the time of day (see below).

7.6   Water movements

In habitats where visual and compass cues are unavailable or unreliable, fishes may resort to acquiring spatial information from the body of water that surrounds them. It has already been seen that some species can learn to locate landmarks based on information from water movements occurring between stationary objects in the environment and their own bodies (Teyke 1985, 1989; Burt de Perera 2004a,b). There is also some evidence that three-spined sticklebacks (*Gasterosteus aculeatus*, Gasterosteidae), can learn to use flow direction as an orientation cue (Girvan & Braithwaite 2000; Braithwaite & Girvan 2003). Tidal streams have also been shown to be important in the movements made by migrating plaice (Hunter *et al.* 2004). Electronic data storage tags have been used to follow the movements of individual fishes, and these have shown that some fishes use tidal currents in part of their migration, but that the plaice also show other forms of navigation. This suggests that these fishes, which will perform several migrations in a lifetime, may be learning to use the tidal flows to guide their movements to spawning grounds. Other species appear to learn to sense and orient in response to shifts in hydrostatic pressure; for example, once an internal map-like representation is established in a familiar environment using active electrolocation, the electric elephantnose fish appears to orient using hydrostatic pressure cues (Cain 1995). Similarly, blind Mexican cave fish use hydrostatic pressure cues to learn the location of objects in their environment (Burt de Perera *et al.* 2005).

A more widespread phenomenon is the directional preference or rheotaxis shown by newly hatched juveniles of a number of species when exposed to flowing water (Jonsson *et al.* 1994; Kaya & Jeanes 1995). This response is assumed to develop independently of experience, but a recent study suggests that some species of fishes may learn directional preferences. Smith & Smith (1998) tested the orientation directions of juveniles of two species of amphidromous gobies (*Awaous guamensis* and *Sicyopterus stimpsoni*, Gobiidae), after exposure to a stream of falling water. Juveniles oriented their migratory activity in the direction corresponding to the upstream direction during the preceding period of water flow, indicating a rapid learning of directional cues. This may allow the fishes to maintain oriented movement up rivers, even in areas with no flow and in backwaters where flow is opposite to the general gradient of the river.

7.7   Inertial guidance and internal 'clocks'

Visual or olfactory landmarks are often transient features of the physical landscape and celestial compass cues will be periodically lost under cloud cover or at night. A 'back-up' strategy that may be more resistant to environmental fluctuations is the use of inertial or body-centred information; that is, inertial force and other sensory signals generated by the animal's own movements. Although it is not clear to what extent fishes use this information when orienting in the wild, a diverse array of species solve maze tasks by learning a body-centred pattern of movement; for example, when spatial cues are absent, fifteen-spined sticklebacks (*Spinachia spinachia*, Gasterosteidae) and corkwing wrasse (*Crenilabrus melops*, Labridae) improve their foraging efficiency in an eight-arm radial maze, by developing the algorithm of visiting every third arm (Hughes & Blight 1999; see also Roitblat *et al.* 1982). The use of a body-centred turn response or a sequence of turns has similarly been observed in three-spined sticklebacks and goldfish (Rodríguez *et al.* 1994; Salas *et al.* 1996b; Girvan & Braithwaite 1998; Odling-Smee & Braithwaite 2003).

For some fishes, food availability or the risk of predation may vary spatiotemporally throughout the day but predictably from day to day. In the absence of reliable cues such as sun height, temperature or light intensity, fishes may learn to place themselves in appropriate locations at specific times of the day by consulting an internal circadian clock (Reebs 1999). In a study on golden shiners (*Notemigonus crysoleucas*, Cyprinidae), Reebs (1996) observed time-place learning when fishes were fed in one of two places at specific times of the day. More recently, the inanga (*Galaxias maculates*, Galaxiidae) has been shown to be capable of time-place learning based on feeding but not on predation risk (Reebs 1999). However, the absence of time-place learning in laboratory tests (Reebs 1996, 1999) is difficult to interpret, as fishes may fail to learn because of the relatively low cost of being in the wrong place in space-limited aquaria. More natural set-ups such as outdoor ponds may provide a better test of time-place learning in fishes.

7.8   Social cues

By observing and following the behaviour of 'informed' conspecifics, individuals may acquire spatial knowledge while avoiding many of the costs thought to be associated with individual learning, such as making mistakes or wasting time. Evidence that fishes can learn routes and locations by following conspecifics comes from both field and laboratory-based studies in which naive fishes are allowed to follow informed 'demonstrator' fishes before being tested for route acquisition (Brown & Laland 2003; and see Chapter 10). Spatial information about the nature of certain sites – the profitability of one patch over another, for example (Coolen *et al.* 2003) – can also be learned.

7.9   How flexible is orientation behaviour?

Fishes are clearly exposed to an enormous diversity of potential cues from which they can extract information about their spatial location. Across the diverse array of aquatic habitats occupied by fishes, different sources of spatial information will differ in their availability and/or reliability. Equally diverse may be the range of spatial problems encountered. Individuals might therefore be expected to pay preferential attention to those cues that are most reliable within their particular habitats and invest time and energy in learning during developmental stages that require flexibility. The following section discusses how learned orientation behaviour differs between species, populations and individuals exposed to different ecological conditions.

7.9.1   *When to learn?*

Fishes that remain in a restricted area or home range may need to update stored representations of local topography throughout life. However, in species that return to their natal site to breed, particularly those with an anadromous lifestyle (fishes that migrate from salt to fresh water to spawn), sensory contact with the home site may be lost for prolonged periods of life. In these fishes, 'imprinting' may be the mechanism by which young learn characteristics of the home site, allowing recognition later in life (Morin *et al.* 1989a). Although the concept of imprinting and the validity of some of the criteria are controversial issues, imprinting is thought to be a specialized type of learning which takes place during a restricted period known as a sensitive period, and results in relatively long-lasting memory (Immelmann & Suomi 1981). In migratory fishes, long-term memory that is resistant to change is likely to be essential if the natal site is to be successfully recognized at the end of the return migration (Dodson 1988).

Recent studies have shown that the larval stages of coral reef fishes do not undergo random dispersal. Despite a potentially dispersive larval phase, genetic heterogeneity over small spatial scales is often high (Planes 2002), retention of larvae in near-shore waters is prevalent (Swearer *et al.* 1999), and self-recruitment back to the natal reef can occur (Jones *et al.* 1999). Imprinting of chemical cues from host anemones on embryonic clownfish (*Amphiprion melanopus*, Pomacentridae) has been observed, where later larval orientation behaviour is related to earlier experiences (Arvedlund 1999). Because many reef fishes spawn their eggs demersally on the reef, there is also potential for acoustic imprinting. A study on the embryonic stages of clownfish found that for several days during development and prior to hatching the sounds of coral reefs would be audible (Simpson *et al.* 2005b). This possibility is the focus of ongoing research.

7.9.2   *What to learn?*

Control over when to learn may be accompanied by mechanisms that predispose fishes to use specific types of information or to learn certain associations in preference to others. Recent experiments suggest that three-spined sticklebacks originating from ponds and rivers differ in their propensity to use landmarks during spatial learning (Girvan & Braithwaite 1998; Odling-Smee & Braithwaite 2003). Odling-Smee & Braithwaite (2003) trained sticklebacks from five rivers and five ponds to locate a goal in one arm of a T-maze, either by learning a turn direction out of the start box, or by using plant landmarks as signposts indicating the rewarded end (Fig. 7.2). Pond fishes appear to use both turn direction and landmarks, while river fishes show a strong preference for using turn direction. In fast-flowing rivers, the visual landscape is likely to be continually disrupted by flow and turbulence making landmarks unreliable indicators of location. Therefore, even within a species, learned orientation behaviour may be adapted in response to specific habitat conditions. Huntingford & Wright (1989) observed similar population differences in the

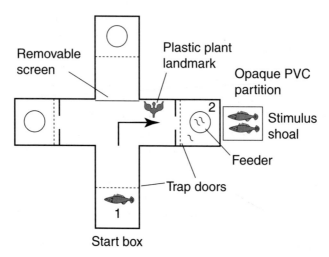

Fig. 7.2   Diagrammatic representation of the spatial task used to compare cue use by pond and river three-spined sticklebacks (*Gasterosteus aculeatus*, Gasterosteidae). The arrow indicates the correct route a right-trained fish had to take to reach the goal (food and shoal mates). Landmarks were always placed in the rewarded arm. The numbers indicate the sequence of start box positions for a run of three consecutive trials starting at position 1.

use of local visual cues by three-spined sticklebacks collected from two sites of high and low predation risk. Fishes from the high risk site used local landmark cues to learn an avoidance task more often than fishes from the low risk site. However, other factors are likely to vary between the two sites in addition to predation pressure, which could affect the use of visual cues.

Many fishes experience conditions where visual cues do not vary greatly. This is true throughout the lives of cave-dwelling or deep-water species. It is also true for fishes that inhabit 'blue-water' environments, for all, or part of their lives. For the pelagic larvae of coral reef fishes, cues that are relevant at spatial scales of hundreds of metres, such as auditory and olfactory signals, are important (Montgomery *et al.* 2001; Kingsford *et al.* 2002). Chemical cues can provide information about the environment they emanate from, while biological sounds such as snapping shrimp and fish vocalizations can indicate the proximity and direction of the local reef inhabitants. As such, they make ideal candidates for orientation in blue water. How and whether fishes learn to interpret these cues as they develop in the plankton and so improve their orientation is unknown, but an ability to discriminate between different habitats would provide valuable spatial information both at settlement and in later life.

7.9.3 *Spatial learning capacity*

Numerous studies have suggested that the telencephalon is involved in spatial learning in fishes (Salas *et al.* 2003; and see Chapter 13). Preliminary evidence suggests that differences in telencephalon morphology may correlate with differences in the ecological demand for spatial learning ability. Male Azorean rock-pool blennies (*Parablennius sanguinolentus*, Blenniidae) establish nests. In this species, males establish nests in crevices and almost never leave their nest area during the entire breeding season, while females must travel relatively long distances in order to visit different nests and spawn with males. Females may need to retain a spatial map of the area and remember the location of previously visited nest sites, a requirement that may explain why the dorsolateral region of the telencephalon is larger in females (Carneiro *et al.* 2001). Similarly in cichlids, variation in telencephalon size appears to relate closely to the challenges of spatial, environmental complexity (Kotrschal *et al.* 1998). Van Staaden *et al.* (1994) and Huber *et al.* (1997) examined the brains of 189 species of cichlids from the East African Lakes and Madagascar, and found that species living in complex habitats created by shallow rock and vegetation had comparatively larger telencephalons than those living in pelagic zones. However, the telencephalon is likely to govern many cognitive abilities from spatial learning, and in none of these studies have spatial learning abilities of fishes been tested. Thus, until the relationship between spatial learning ability and specific features of the telencephalon is better understood, the results of these studies present suggestive, but inconclusive, evidence for a relationship between spatial learning ability and ecological demand.

Ecologically driven differences in the demand for spatial learning may explain why two lake-dwelling sympatric species of three-spined stickleback differ in the rate at which they learn a spatial task (L. Odling-Smee & V. Braithwaite, unpublished data). 'Benthic' and 'limnetic' sticklebacks live in reproductive isolation in several lakes in south-western British Columbia, USA (Schluter & McPhail 1992). The benthic species lives in the vegetated littoral zone, feeding predominantly on mud-dwelling invertebrates, while the limnetic species lives in a comparatively homogeneous environment in terms of spatial structure, out in the open water column, where they feed mainly on plankton. Two independently derived populations of benthics and limnetics from two lakes were trained to locate a goal arm in a T-maze using either plant landmarks or a turn direction out of the start box (Fig. 7.2). Although both species use both types of spatial cue, benthics learned the task almost twice as quickly as limnetics, consistent with the suggestion that pelagic and benthic lifestyles make differential demands on spatial learning ability. However, a classic problem in comparative studies of learning is the possibility that contextual variables rather than differences in learning ability are responsible for species differences in performance (Shettleworth 1993); for example, species may differ in motivation or in adaptability to laboratory conditions. Although attempts were made to minimize this possibility, further experiments are needed for it to be ruled out.

Other ecological variables have also been found to contribute to differences in fishes' spatial behaviour; for instance, observations from the Panamanian bishop (*Brachyraphis episcopi*, Poeciliidae) indicate the environment the fishes inhabit has a large effect on their spatial learning skills. Fishes from low predation areas learned the location of a hidden food patch more quickly than those from high predation areas (Brown & Braithwaite 2005). This result can partly be explained by differences in the strength of cerebral lateralization found in fishes from high or low predation sites, which generated a strong turning preference that influenced maze performance (Brown *et al.* 2004). Additionally, spatial ability may be more important to fishes from the low predation areas (where the species dominates the fauna) as they will need to use spatial cues to help them maintain territories, whereas in high predation areas the fishes are displaced into marginal habitats and do not appear to utilize territories.

There is enormous potential for future research investigating the relationship between learned orientation strategies in fishes and their habitat ecology. So far, preliminary but suggestive evidence indicates that in fishes, as appears to be the case in terrestrial vertebrates (Sherry 1998), spatial learning may be modified or fine-tuned in response to particular ecological conditions. Fishes appear predisposed to 'know' when to learn or what stimuli to attend to. Furthermore, preliminary evidence suggests that fishes may invest only as much into spatial learning capacity as their ecologies and lifestyles demand.

7.10   Salmon homing – a case study

In this final section, we investigate homing in salmon, paying particular attention to the final return by reproductive adults to freshwater streams, a behaviour governed by olfactory recognition of home-stream water. There are several excellent reviews that deal in depth with salmonid migrations (Stabell 1984; Quinn & Dittman 1990; Dittman & Quinn 1996) and it is not our intention to make an exhaustive coverage of the literature. Rather, salmon homing presents a useful case study that illustrates the complex role of genetic, developmental and environmental influences in shaping orientation behaviour in response to particular ecological conditions.

Although many species of fish migrate at some point in their life, the study of fish migration has largely focused on the migratory behaviour of salmonids (salmon, trout and charr species). The interest in understanding the migratory behaviour of this group is, in part, driven by their economic importance. However, the relative ease with which their migratory behaviour can be researched as they move up or downstream on their homeward or outward journey has also made them an amenable model species. Salmon therefore provide an ideal case study in which to determine what they learn as they undergo their migration. We now know, for example, that both genetic and environmental influences shape how and when the fishes learn information relevant to their migration. Typically, salmon spawn in fresh water and after a variable period of residence the offspring migrate to the ocean, to take advantage of the higher productivity of the marine environment (Gross *et al.* 1988). Once the fishes reach coastal waters, some remain close to the mouth of their natal river over the summer, but many others migrate thousands of kilometres to feeding grounds. Virtually all the surviving fishes will at some point return to their natal stream to spawn. Homing is in many cases so precise that it has led to reproductive isolation of spawning populations and specialized adaptations for the natal habitat (Quinn & Dittman 1990; Dittman & Quinn 1996).

Although a diverse array of sensory mechanisms and cues have been proposed, there is still remarkably little empirical information on how the salmon navigate once they are at sea (Quinn & Dittman 1990; Ueda *et al.* 1995; Dittman & Quinn 1996). It is likely that the fishes rely on a compass to migrate the long distances once they leave their natal coastal area (Dittman & Quinn 1996); one of the possible compasses is magnetic (Walker *et al.* 1997). Tracking work has shown that when the fishes are at sea they will swim both night and day, and magnetite, biogenically produced iron oxide has been found in the lateral line of Atlantic salmon (*Salmo salar*, Salmonidae; Moore *et al.* 1990, 1991).

The freshwater phase of the spawning migration is primarily governed by olfactory discrimination of home-stream waters (Hasler & Scholz 1983). The fishes learn the olfactory characteristics of their home site through imprinting during their outward migration. Recognition of the natal stream by returning adults has been investigated using two approaches. First, salmonids are exposed as juveniles to artificial odorants and these fishes are subsequently decoyed to unfamiliar streams scented with the odorants during their homing migrations (Scholz *et al.* 1976; Hasler & Scholz 1983). Second, hatchery fishes are transported from a rearing site and adult return patterns are monitored (Hansen *et al.* 1993). Monitoring the stray-patterns of returning marked hatchery-reared salmon has also given clues about the imprinting process (Unwin & Quinn 1993; Pascual & Quinn 1994, 1995; Heard 1996; Hard & Heard 1999).

One unresolved issue relates to when salmon imprint (Dittman *et al.* 1996). Experiments with artificial odours indicate that imprinting occurs at the time of peak thyroid hormone levels during smolt transformation, the physiological processes that prepare salmon for oceanic residence (Hasler & Scholz 1983; Morin *et al.* 1989a,b; Morin & Døving 1992). However, several species of salmon commonly move from their natal site and develop elsewhere in freshwater prior to smolt transformation, suggesting that olfactory imprinting and smolt transformation are not inextricably linked (Quinn 1985). Furthermore, kokanee (*Oncorhynchus* spp., Salmonidae) are able to imprint on artificial odorants as alevins and emergent fry as well as at the smolt stage (Dittman & Quinn 1996). A likely resolution to this controversy is that species and populations will differ in the timing of imprinting, according to the nature of their ecologies and migratory patterns. There is extraordinary diversity of freshwater habitats and migratory patterns both within and among salmonid species, and individuals may need to be sensitive to odours at different stages of their development (Dittman *et al.* 1996).

Salmon may also learn odours at more than one time and place and then use the sequence of learned odours to guide their homeward migration (Quinn 1985; Hansen *et al.* 1987; Heard 1996). One possibility is that thyroxine surges trigger transient neural changes in the peripheral olfactory system, permitting windows of sensitivity that enable olfactory imprinting. Short-term increases in plasma thyroid hormones may enable wild salmon to learn key freshwater landmarks, not only at developmentally regulated periods but also during migrations in response to novel environmental stimuli (Dittman & Quinn 1996). Elevated levels of thyroxine that last for only a few hours have also been shown for migrating chum salmon (*Oncorhynchus keta*, Salmonidae) when the salmon encounter different habitats (Iwata *et al.* 2003). Animals typically go through a phase of learning when they encounter new environments; it is possible then that thyroxine may in some way prepare the fishes for a period of learning. Further evidence for a link between thyroxine and learning is that structural reorganization in certain brain circuits has been shown to occur during the downstream migration of Atlantic salmon when levels of thyroxine are high (Ebbesson *et al.* 2003).

Overall, these studies provide a convincing illustration of the extent to which developmental and genetic influences are likely to regulate and constrain flexible learned behaviour in response to species-typical and even population-typical habitat conditions and migratory patterns.

7.11   Conclusions

Most species of fishes experience environments in which biologically important locations, as well as the physical landscape from which spatial information is extracted, will change at rates that cannot be tracked by adaptive changes in the gene pool. A capacity to learn provides flexibility, allowing fishes to match their orientation strategy to a variable environment on the basis of experience. A combined approach of field- and laboratory-based experiments have shown that fishes are indeed capable of spatial learning and can use information from a vast array of different sources. Flexible learning appears to act in concert with genetic and developmental influences such that orientation behaviour is adapted for particular ecological conditions and navigational demands. Future studies, that attempt to assess how and to what extent fishes learn to use different types of spatial information in orientation, should involve experiments with a design that is based on a detailed understanding of the habitat ecology and spatial problems likely to be encountered by fishes in their natural habitats.

7.12   Acknowledgements

We would like to thank the editors for inviting us to make this contribution and for their helpful comments on the earlier versions of this chapter.

7.13   References

Able, K.P. (1993) Orientation cues used by migratory birds – a review of cue-conflict experiments. *Trends In Ecology & Evolution*, **8**, 367–371.

Armstrong, J.D., Braithwaite, V.A. & Huntingford, F.A. (1997) Spatial strategies of wild Atlantic salmon parr: exploration and settlement in unfamiliar areas. *Journal of Animal Ecology*, **66**, 203–211.

Aronson, L.R. (1951) Orientation and jumping behaviour in the Gobiid fish *Bathygobius soporator*. *American Museum Novitates*, **1486**, 1–22.

Aronson, L.R. (1971) Further studies on orientation and jumping behaviour in the Gobiid fish *Bathygobius sporator*. *Annals of the New York Academy of Science*, **188**, 378–392.

Arvedlund, M., McCormick, M.I., Fautin, D.G. & Bildsøe, M. (1999) Host recognition and possible imprinting in the anemonefish *Amphiprion melanopus* (Pisces: Pomacentridae). *Marine Ecology Progress Series*, **188**, 207–218.

Bennet, A.T.D. (1996) Do animals have cognitive maps? *Journal of Experimental Biology*, **199**, 219–224.

Biegler, R. & Morris, R.G.M. (1996) Landmark stability: studies exploring whether perceived stability of the environment influences spatial representation. *Journal of Experimental Biology*, **199**, 187–193.

Braithwaite, V.A. & Girvan, J.R. (2003) Use of waterflow to provide spatial information in a small-scale orientation task. *Journal of Fish Biology*, **63**, 74–83.

Braithwaite, V.A., Armstrong, J.D., McAdam, H.M. & Huntingford, F.A. (1996) Can juvenile Atlantic salmon use multiple cue systems in spatial learning? *Animal Behaviour*, **51**, 1409–1415.

Brown, C. (2003) Habitat-predator association and avoidance in rainbowfish (*Melanotaenia spp.*). *Ecology of Freshwater Fishes*, **12**, 118–126.

Brown, C. & Braithwaite, V.A. (2005) Effects of predation pressure on the cognitive ability of the poeciliid *Brachyraphis episcopi*. *Behavioural Ecology*, **16**, 482–497.

Brown, C. & Laland, K.N. (2003) Social learning in fishes: a review. *Fish and Fisheries*, **4**, 280–288.

Brown, C., Gardner, C. & Braithwaite, V.A. (2004) Population variation in lateralised eye use in the poeciliid *Brachyraphis episcopi*. *Proceedings of the Royal Society of London, Series B*, **271 (6)**, S455–S457.

Burt de Perera, T. (2004a) Spatial parameters encoded in the spatial map of blind Mexican cave fish, *Astynax fasciatus*. *Animal Behaviour*, **68**, 291–295.

Burt de Perera, T. (2004b) Fish can encode order in their spatial map. *Proceedings of the Royal Society of London, Series B*, **271**, 2131–2134.

Burt de Perera, T. & Braithwaite, V.A. (2005) Laterality in a non-visual sensory modality – the lateral line of fish. *Current Biology*, **15**, R241–R242.

Burt de Perera, T., de Vos, A. & Guilford, T. (2005) The vertical component of a fish's spatial map. *Animal Behaviour*, **70**, 405–409.

Cain, P. (1995) Navigation in familiar environments by the weakly electric elephantnose fish, *Gnathonemus petersii* L. (Mormyriformes, Teleostei). *Ethology*, **99**, 332–349.

Cain, P., Gerin, W. & Moller, P. (1994) Short-range navigation of the weakly electric fish, *Gnathonemus petersii* L. (Mormyridae, Teleostei), in novel and familiar environments. *Ethology*, **96**, 33–45.

Campenhausen, C.V., Reiss, I. & Weissert, R. (1981) Detection of stationary objects by the blind cave fish *Anoptichthys jordani* (Characidae). *Journal of Comparative Physiology A*, **143**, 369–374.

Carlson, H.R. & Haight, R.E. (1972) Evidence for a home site and homing of adult yellowtail rockfish, *Sebastes flavidus*. *Journal of the Fisheries Research Board of Canada*, **29**, 1011–1014.

Carneiro, L.A., Andrade, R.P., Oliveira, R.F. & Kotrschal, K. (2001) Sex differences in home range and dorso-lateral telencephalon in the Azorean rock-pool blenny. *Social Neuroscience Abstract*, **27**, Program No. 535.4.

Coolen, I., van Bergen, Y., Day, R.L. & Laland, K.N. (2003) Species differences in adaptive use of public information in sticklebacks. *Proceedings of the Royal Society of London, Series B*, **270**, 2413–2419.

Dittman, A.H. & Quinn, T.P. (1996) Homing in pacific salmon: mechanisms and ecological basis. *Journal of Experimental Biology*, **199**, 83–91.

Dittman, A.H., Quinn, T.P. & Nevitt, G.A. (1996) Timing of imprinting to natural and artificial odors by coho salmon (*Oncorhynchus kisutch*). *Canadian Journal of Fisheries and Aquatic Sciences*, **53**, 434–442.

Dodson, J.J. (1988) The nature and role of learning in the orientation and migratory behaviour of fishes. *Environmental Biology of Fishes*, **23**, 161–182.

Ebbesson, L.O.E., Ekstrom, P., Ebbesson, S.O.E., Stefansson, S.O. & Holmqvist, B. (2003) Neural circuits and their structural and chemical reorganisation in the light-brain-pituitary axis during parr–smolt transformation. *Aquaculture*, **222**, 59–70.

Girvan, J.R. & Braithwaite, V.A. (1998) Population differences in spatial learning in three-spined sticklebacks. *Proceedings of the Royal Society of London, Series B*, **265**, 913–919.

Girvan, J.R. & Braithwaite, V.A. (2000) Orientation behaviour in sticklebacks: modified by experience or population specific? *Behaviour*, **137**, 833–843.

Goodyear, P.C. (1973) Learned orientation in the predator avoidance of mosquitofish, *Gambusia affinis*. *Behaviour*, **45**, 191–224.

Goodyear, P.C. & Bennet, D.H. (1979) Sun compass orientation of immature bluegill. *Transactions of the American Fisheries Society*, **108**, 555–559.

Goodyear, P.C. & Ferguson, D.E. (1969) Sun-compass orientation in mosquitofish, *Gambusia affinis*. *Animal Behaviour*, **17**, 636–640.

Graff, C., Kaminski, G., Gresty, M. & Ohlmann, T. (2004) Fish perform spatial pattern recognition and abstraction by exclusive use of active electrolocation. *Current Biology* **14**, 181–123.

Green, J.M. (1971) High tide movements and homing behaviour of the tidepool sculpin *Oligocottus maculosus*. *Journal of the Fisheries Research Board of Canada*, **28**, 383–389.

Gross, M.R., Coleman, R.M. & McDowall, R.M. (1988) Aquatic productivity and the evolution of diadromous fish migration. *Science*, **239**, 1291–1293.

Hallacher, L.E. (1984) Relocation of original territories by displaced black-and-yellow rockfish, *Sebastes chrysomelas*, from Carmel Bay, California. *California Fish and Game*, **70**, 158–162.

Hansen, L.P., Døving, K.B. & Jonsson, B. (1987) Migration of farmed adult Atlantic salmon with and without olfactory sense, released on the Norwegian coast. *Journal of Fish Biology*, **30**, 713–721.

Hansen, L.P., Jonsson, N. & Jonsson, B. (1993) Oceanic migration in homing Atlantic salmon. *Animal Behaviour*, **45**, 927–941.

Hard, J.J. & Heard, W.R. (1999) Analysis of straying variation in Alaskan hatchery chinook salmon (*Oncorhynchus tshawytscha*) following transplantation. *Canadian Journal of Fisheries and Aquatic Sciences*, **56**, 578–589.

Hasler, A.D. (1971) Orientation and fish migration. In: W.S. Hoar, & D.J. Randall (eds), *Fish Physiology*, pp. 429–510. Academic Press, London.

Hasler, A.D. & Scholz, A.T. (1983) *Olfactory Imprinting and Homing in Salmon*. Springer-Verlag, Berlin.

Hasler, A.D., Horrall, R.M., Wisby, W.J. & Braemer, W. (1958) Sun-orientation and homing in fishes. *Limnology and Oceanography*, **111**, 353–361.

Hawryshyn, C.W., Arnold, M.G., Bowering, E. & Cole, R.L. (1990) Spatial orientation of rainbow trout to plane-polarised light: the ontogeny of E-vector discrimination and spectral sensitivity characteristics. *Journal of Comparative Physiology A*, **166**, 565–574.

Healy, S.D. (1998) *Spatial Representation in Animals*. Oxford University Press, Oxford.

Heard, W.R. (1996) Sequential imprinting in chinook salmon: is it essential for homing fidelity? *Bulletin of National Research Institute of Aquaculture, Supplement*, **2**, 59–64.

Huber, R., Van Staaden, M.J., Kaufman, L.S. & Liem, K.F. (1997) Microhabitat use, trophic patterns, and the evolution of brain structure in African cichlids. *Brain, behaviour and Evolution*, **50**, 167–182.

Hughes, R.N. & Blight, C.M. (1999) Algorithmic behaviour and spatial memory are used by two intertidal fish species to solve the radial maze. *Animal Behaviour*, **58**, 601–613.

Hughes, R.N. & Blight, C.M. (2000) Two intertidal fish species use visual association learning to track the status of food patches in a radial maze. *Animal Behaviour*, **59**, 613–621.

Hunter, E., Metcalfe, J.D. & Reynolds, J.D. (2003) Migration route and spawning area fidelity by North Sea plaice. *Proceedings of the Royal Society of London, Series B*, **270**, 2097–2103.

Hunter, E., Metcalfe, J.D., Arnold, G.P. & Reynolds, J.D. (2004) Impacts of migratory behaviour on population structure in North Sea plaice. *Journal of Animal Ecology*, **73**, 377–385.

Huntingford, F.A. & Wright, P.J. (1989) How sticklebacks learn to avoid feeding patches. *Behavioural Processes*, **19**, 181–189.

Huntingford, F.A., Braithwaite, V.A., Armstrong, J.D., Aird, D. & Joiner, P. (1998) Homing in juvenile salmon in response to imposed and spontaneous displacement: experiments in an artificial stream. *Journal of Fish Biology*, **53**, 847–852.

Immelmann, K. & Suomi, S.J. (1981) Sensitive phases in development. In: K. Immelmann, G.W. Barlow, L. Petrinovich, & M. Main (eds), *Behavioural Development: The Bielefeld Interdisciplinary Project*, pp. 395–431. Cambridge University Press, Cambridge.

Ingle, D. & Sahagian, D. (1973) Solution of a spatial constancy problem by goldfish. *Physiological Psychology*, **1**, 83–84.

Iwata, M., Tsuboi, H., Yamashita, T., Amemiya, A., Yamada, H., & Chiba, H. (2003) Function and trigger of thyroxine surge in migrating chum salmon *Oncorhynchus keta* fry. *Aquaculture*, **222**, 315–329.

Jones, G.P., Milicich, M.J., Emslie, M.J. & Lunnow, C. (1999) Self-recruitment in a coral reef population. *Nature*, **402**, 802–804.

Jonsson, N., Jonsson, B., Skurdal, J. & Hansen, L.P. (1994) Differential response to water current in offspring of inlet- and outlet-spawning brown trout *Salmo trutta*. *Journal of Fish Biology*, **45**, 356–359.

Jordan, F., Bartolini, M., Nelson, C., Patterson, P.E. & Soulen, H.L. (1997) Risk of predation affects habitat selection by the pinfish *Lagodon rhomboides* (Linnaeus). *Journal of Experimental Marine Biology and Ecology*, **208**, 45–56.

Kalmijn, A.J. (1982) Electric and magnetic field detection in elasmobranch fishes. *Science*, **218**, 916–918.

Kaya, C.M. & Jeanes, E.D. (1995) Retention of adaptive rheotactic behaviour by F_1 fluvial Arctic grayling. *Transactions of the American Fisheries Society*, **124**, 453–457.

Kingsford, M.J., Leis, J.M., Shanks, A., Lindeman, K.C., Morgan, S.G. & Pineda, J. (2002) Sensory environments, larval abilities and local self-recruitment. *Bulletin of Marine Science*, **70**, 309–340.

Kirschvink, J.L., Walker, M.M., Chang, S.-B.R. & Dizon, A.E. (1985) Chains of single domain magnetite particles in chinook salmon, *Oncorhynchus tschawytscha*. *Journal of Comparative Physiology A*, **157**, 375–381.

Kleerokoper, H., Matis, J., Gensler, P. & Maynard, P. (1974) Exploratory behaviour of goldfish, *Carassius auratus*. *Animal Behaviour*, **22**, 124–132.

Kolm, N., Hoffman, E.A., Olsson, J., Berglund, A. & Jones, A.G. (2005) Group stability and homing behaviour but no kin group structures in a coral reef fish. *Behavioral Ecology*, **16**, 521–527.

Kotrschal, K., Van Staaden, M.J. & Huber, R. (1998) Fish brains: evolution and environmental relationships. *Reviews in Fish Biology and Fisheries*, **8**, 373–408.

Kroon, F.J., de Graaf, M. & Liley, N.R. (2000) Social organisation and competition for refuges and nest sites in *Coryphopterus nicholsii* (Gobiidae), a temperate protogynous reef fish. *Environmental Biology of Fishes*, **57**, 401–411.

Leis, J.M., Carson-Ewart, B.M., Hay, A.C. & Cato, D.H. (2003) Coral-reef sounds enable nocturnal navigation by some reef-fish larvae in some places and at some times. *Journal of Fish Biology*, **63**, 724–737.

López, J.C., Broglio, C., Rodríguez, F., Thinus-Blanc, C. & Salas, C. (1999) Multiple spatial learning strategies in goldfish (*Carassius auratus*). *Animal Cognition*, **2**, 109–120.

López, J.C., Bingman, V.P., Rodríguez, F., Gómez, Y. & Salas, C. (2000a) Dissociation of place and cue learning by telencephalic ablation in goldfish. *Behavioural Neuroscience*, **114**, 687–699.

López, J.C., Broglio, C., Rodríguez, F., Thinus-Blanc, C. & Salas, C. (2000b) Reversal learning deficit in a spatial task but not in a cued one after telencephalic ablation in goldfish. *Behavioural Brain Research*, **109**, 91–98.

Markel, R.W. (1994) An adaptive value of spatial learning and memory in the blackeye goby, *Coryphoterus nicholsi*. *Animal Behaviour*, **47**, 1462–1464.

Matthews, K.R. (1990a) An experimental study of the habitat preferences and movement patterns of copper, quillback, and brown rockfishes (*Sebastes* spp.). *Environmental Biology of Fishes*, **29**, 161–178.

Matthews, K.R. (1990b) A telemetric study of the home ranges and homing of copper and quillback rockfishes on shallow rocky reefs. *Canadian Journal of Fisheries and Aquatic Sciences*, **68**, 2243–2250.

Mazeroll, A.I. & Montgomery, W.L. (1995) Structure and organization of local migrations in brown surgeonfish (*Acanthurus nigrofuscus*). *Ethology*, **99**, 89–106.

Mazeroll, A.I. & Montgomery, W.L. (1998) Daily migrations of a coral reef fish in the Red Sea (Gulf of Aqaba, Israel): initiation and orientation. *Copeia*, **4**, 893–905.

McCormick, C.A. (1989) Central lateral line pathways in bony fish. In: S. Coombs, P. Görner & H. Münz (eds), *The Lateral Line: Neurobiology and Evolution*, pp. 341–363. Springer Verlag, New York.

Montgomery, J.C., Tolimieri, N. & Haine, O.S. (2001) Active habitat selection by pre-settlement reef fishes. *Fish and Fisheries*, **2**, 261–277.

Montgomery, J.C., Jeffs, A.G., Simpson, S.D., Meekan, M.G. & Tindle, C. (2006) Sound as an orientation cue for the pelagic larvae of reef fishes and decapod crustaceans. *Advances in Marine Biology*. In press.

Moore, A., Freake, S.M. & Thomas, I.M. (1990) Magnetic particles in the lateral line of the Atlantic salmon (*Salmo salar* L.). *Philosophical Transactions of the Royal Society of London B*, **329**, 11–15.

Moore, A., Bone, Q. & Ryan, K.P. (1991) The anatomy of the lateral line of the Atlantic salmon (*Salmo salar* L.). *Journal of Marine Biology U.K.*, **71**, 949.

Morin, P.-P., Dodson, J.J. & Doré, F.Y. (1989a) Cardiac responses to a natural odorant as evidence of a sensitive period for olfactory imprinting in young Atlantic salmon, *Salmo salar*. *Canadian Journal of Fisheries and Aquatic Sciences*, **46**, 122–130.

Morin, P.-P., Dodson, J.J. & Doré, F.Y. (1989b) Thyroid activity concomitant with olfactory learning and heart rate changes in Atlantic salmon, *Salmo salar*, during smoltification. *Canadian Journal of Fisheries and Aquatic Sciences*, **46**, 131–136.

Morin, P.-P. & Døving, K.B. (1992) Changes in the olfactory function of Atlantic salmon,

Salmo salar, in the course of smoltification. *Canadian Journal of Fisheries and Aquatic Sciences*, **49**, 1704–1713.

Noda, M., Gushima, K. & Kakuda, S. (1994) Local prey search based on spatial memory and expectation in the planktivorous fish, *Chromis chrysurus* (Pomacentridae). *Animal Behaviour*, **47**, 1413–1422.

Odling-Smee, L. & Braithwaite, V.A. (2003) The influence of habitat stability on landmark use during spatial learning in the threespine stickleback. *Animal Behaviour*, **65**, 701–707.

Ogden, J.C. & Buckman, N.S. (1973) Movements, foraging groups, and diurnal migrations of the striped parrotfish *Scarus croicensis* Bloch (Scaridae). *Ecology*, **54**, 598–596.

Ogden, J.C. & Quinn, T.P. (1984) Migration in coral reef fishes: ecological significance and orientation mechanisms. In: J.D. McCleave, G.P. Arnold, J.J. Dodson & W.H. Neill (eds), *Mechanisms of Migration in Fishes*, pp. 293–308. Plenum Press, New York.

Pascual, M.A. & Quinn, T.P. (1994) Geographical patterns of straying of fall chinook salmon, *Oncorhynchus tshawytscha* (Walbaum), from Columbia River (USA) hatcheries. *Aquaculture and Fisheries Management*, **25, (2)**, 17–30.

Pascual, M.A. & Quinn, T.P. (1995) Factors affecting the homing of fall chinook salmon from Columbia River hatcheries. *Transactions of the American Fisheries Society*, **124**, 308–320.

Pitcher, T.L. & Magurran, A.E. (1983) Shoal size, patch profitability and information exchange in foraging goldfish. *Animal Behaviour*, **31**, 546–555.

Planes, S. (2002) Biogeography and larval dispersal inferred from population genetic analysis. In: P.F. Sale (ed.), *Coral Reef Fishes: Dynamics and Diversity in a Complex Ecosystem*, pp. 201–220. Academic Press, San Diego.

Quinn, T.P. (1985) Salmon homing: is the puzzle complete? *Environmental Biology of Fishes*, **12**, 315–317.

Quinn, T.P. & Dittman, A.H. (1990) Pacific salmon migrations and homing: mechanisms and adaptive significance. *Trends in Ecology and Evolution*, **5**, 174–177.

Quinn, T.P. & Ogden, J.C. (1984) Field evidence of compass orientation in migrating juvenile grunts (Haemulidae). *Journal of Experimental Marine Biology and Ecology*, **81**, 181–192.

Reebs, S.G. (1996) Time-place learning in golden shiners (Pisces: *Cyprinidae*). *Behavioural Processes*, **36**, 253–262.

Reebs, S.G. (1999) Time-place learning based on food but not on predation risk in a fish, the inanga (*Galaxias maculatus*). *Ethology*, **105**, 361–371.

Reese, E.S. (1989) Orientation behaviour of butterflyfishes (family Chaetodontidae) on coral reefs: spatial learning of route specific landmarks and cognitive maps. *Environmental Biology of Fishes*, **25**, 79–86.

Rodríguez, F., Durán, E., Vargas, J.P., Torres, B. & Salas, C. (1994) Performance of goldfish trained in allocentric and egocentric maze procedures suggest the presence of a cognitive mapping system in fishes. *Animal Learning and Behaviour*, **22**, 409–420.

Roitblat, H.L., Tham, W. & Golub, L. (1982) Performance of *Betta splendens* in a radial maze. *Animal Learning and Behaviour*, **10**, 108–114.

Salas, C., Broglio, C., Rodríguez, F., López, J.C., Portavella, M. & Torres, B. (1996a) Telencephalic ablation in goldfish impairs performance in a spatial constancy problem but not in a cued one. *Behavioural Brain Research*, **79**, 193–200.

Salas, C., Rodríguez, F., Vargas, J.P., Durán, E. & Torres, B. (1996b) Spatial learning and memory deficits after telencephalic ablation in goldfish trained in place and turn maze procedures. *Behavioural Neuroscience*, **110**, 965–980.

Salas, C., Broglio, C. & Rodriguez, F. (2003) Evolution of forebrain and spatial cognition in vertebrates: conservation across diversity. *Brain Behaviour and Evolution*, **62**, 72–82.

Schluter, D. & McPhail, J.D. (1992) Ecological character displacement and speciation in sticklebacks. *The American Naturalist*, **140**, 85–108.

Scholz, A.T., Horrall, R.M., Cooper, J.C. & Hasler, A.D. (1976) Imprinting to chemical cues: the basis for home stream selection in salmon. *Science*, **192**, 1247–1249.

Schwassmann, H.O. & Braemer, W. (1961) The effect of experimentally changed photoperiod on the sun orientation rhythm of fish. *Physiological Zoology*, **34**, 273–326.

Sherry, D.F. (1998) The ecology and neurobiology of spatial memory. In: R. Dukas (ed.), *Cognitive Ecology*, pp. 261–296. The University of Chicago Press, Chicago.

Shettleworth, S.J. (1993) Where is the comparison in comparative cognition? *American Psychological Society*, **4**, 179–184.

Simpson, S.D., Meekan, M.G., McCauley, R.D. & Jeffs, A.G. (2004) Attraction of settlement-stage coral reef fishes to reef noise. *Marine Ecology Progress Series*, **276**, 263–268.

Simpson, S.D., Meekan, M.G., Montgomery, J.C., McCauley, R.D. & Jeffs, A.G. (2005a) Homeward sound. *Science*, **308**, 221.

Simpson, S.D., Yan, H.Y., Wittenrich, M.L. & Meekan, M.G. (2005b) Response of embryonic coral reef fishes (Pomacentridae: *Amphiprion spp.*) to noise. *Marine Ecology Progress Series*, **287**, 201–208.

Smith, R.J.F. & Smith, M.J. (1998) Rapid acquisition of directional preferences by migratory juveniles of two amphidromous Hawaiin gobies, *Awaous guamensis* and *Sicyopterus stimpsoni*. *Environmental Biology of Fishes*, **53**, 275–282.

Stabell, O.B. (1984) Homing and olfaction in salmonids: a critical review with special reference to the Atlantic salmon. *Biological Reviews*, **59**, 333–388.

Swearer, S.E., Caselle, J.E., Lea, D.W. & Warner, R.R. (1999) Larval retention and recruitment in an island population of a coral-reef fish. *Nature*, **402**, 799–802.

Teyke, T. (1985) Collision with and avoidance of obstacles by blind cave fish *Anoptichthys jordani* (Characidae). *Journal of Comparative Physiology A*, **157**, 837–843.

Teyke, T. (1989) Learning and remembering the environment in the blind cave fish *Anoptichthys jordani*. *Journal of Comparative Physiology A*, **164**, 655–662.

Tolimieri, N., Jeffs, A. & Montgomery, J.C. (2000) Ambient sound as a cue for navigation by the pelagic larvae of reef fishes. *Marine Ecology Progress Series*, **207**, 219–224.

Tsukamoto, K., Aoyama, J. & Miller, M.J. (2003) Migration, speciation, and the evolution of diadromy in anguillid eels. *Canadian Journal of Fisheries and Aquatic Sciences*, **59**, 1989–1998.

Ueda, H., Kaeriyama, M., Urano, A., Kurihara, K. & Yamauchi, K. (1995) Homing mechanism in salmon: roles of vision and olfaction. In: F.W. Goetz (ed.), *International Symposium on Reproductive Physiology of Fish*, pp. 35–37. FishSymp 95, Texas.

Unwin, M.J. & Quinn, T.P. (1993) Homing and straying patterns of chinook salmon (*Oncorhynchus tshawytscha*) from a New Zealand hatchery: spatial distribution of strays and effects of release date. *Canadian Journal of Fisheries and Aquatic Sciences*, **50**, 1168–1175.

Van Staaden, M.J., Huber, R., Kaufman, L.S. & Karel, F.L. (1994) Brain evolution in cichlids of the African Great Lakes: brain and body size, general patterns, and evolutionary trends. *Zoology*, **98**, 165–178.

Walker, M.M. (1984) Learned magnetic field discrimination in yellowfin tuna, *Thunnus albacares*. *Journal of Comparative Physiology A*, **155**, 673–679.

Walker, M.M., Quinn, T.P., Kirschvink, J.L. & Groot, C. (1988) Production of single domain magnetite throughout life by sockeye salmon, *Onchorynchus nerka*. *Journal of Experimental Biology*, **140**, 51–63.

Walker, M.M., Diebel, C.E., Haugh, C.V., Pankhurst, P.M., Montgomery, J.C. & Green, C.R. (1997) Structure and function of the vertebrate magnetic sense. *Nature*, **390**, 371–376.

Warburton, K. (1990) The use of local landmarks by foraging goldfish. *Animal Behaviour*, **40**, 500–505.

Warburton, K. (2003) Learning of foraging skills by fish. *Fish and Fisheries*, **4**, 203–215.

Welker, W.I. & Welker, J. (1958) Reaction of fish (*Eucinostomus gula*) to environmental changes. *Ecology*, **39**, 283–288.

Chapter 8

Learned Recognition of Conspecifics

Siân W. Griffiths and Ashley Ward

8.1 Introduction

One of the key concepts in ecology is the way in which the social behaviour of organisms impacts the structure of their geographical distribution. In fishes, individuals are not equally attracted or aggressive towards all other conspecifics; for example, the preferential association of salmonids with relatives (and reduced aggression towards them) is now well established and, more recently, evidence has been obtained for the remarkable ability of fishes to choose among other fishes on the basis of past experience (preferring familiar individuals). This chapter reviews the evidence for these discriminatory behavioural interactions and shows how learned ability to recognize particular conspecifics plays an important role in influencing patterns of mobility and site fidelity in fishes. It will consider how the nature of behavioural interactions and associations between fishes will have especially important implications for their dispersal and also their ability to recolonize after natural or human disturbances. Information regarding group membership (school structure or territorial assemblages) may also have implications for conservation of declining fish stocks and fisheries management policies, because individuals that choose familiar schoolmates accrue antipredator and foraging benefits. Conferring these benefits on declining fish stocks has important economic and conservation implications. Future work may therefore benefit from an exploration of the ecological contexts in which learned recognition of conspecifics is important in natural streams and rivers.

8.2 Recognition of familiars and condition-dependency

It is well established that fishes are capable of discriminating between individuals with whom they interact; for example, shoaling fishes exert a high degree of choice with regard to their preferred shoalmates, selecting on the basis of species (Keenleyside 1955), colour (McRobert & Bradner 1998) and size (Ward & Krause 2001). In recent years, it has also become clear that fishes are capable of more fine-scale assessment and discrimination; for example, it is now known that fishes are able to recognize individuals on the basis of previous experience. Where this occurs in conjunction with particular phenotypic or behavioural cues, this is referred to as condition-dependent recognition; for example, European minnows (*Phoxinus phoxinus*, Cyprinidae) are able to discriminate between conspecifics according to competitive ability and shoal preferentially with poor competitors to reduce their own competition costs (Metcalfe & Thomson 1995). Similarly, three-spined stickle-backs (*Gasterosteus aculeatus*, Gasterosteidae) select co-operative individuals as predator inspection partners and rapidly acquire the ability to recognize particular individuals in this context (Milinski *et al.* 1990a, b).

Where recognition is founded on previous experience in the absence of any obvious phenotypic or behavioural cues, this is referred to as condition-independent recognition and forms the basis for what is generally referred to as familiarity. Typically, familiarity develops gradually between individuals in a population and subsequently informs their association and dispersal decisions (Griffiths & Magurran 1997a). There is now considerable evidence in the literature demonstrating that the ability to recognize and preferentially associate with familiars is widespread in fishes and has important implications for their social organization.

8.2.1   *Laboratory studies of familiarity*

There is a large and increasing body of work documenting the phenomenon of conspecific familiarity in fishes (Table 8.1). Familiarity impacts a wide range of species, and its importance is a function of the social dynamics of the species in question. In shoaling species, the effect of familiarity tends to enhance the benefits of grouping and mediates association decisions in these species. By contrast, less gregarious fishes interact less frequently and familiarity is often confined to the contexts of territorial interactions, dominance hierarchies and mate recognition.

While the ability to recognize and subsequently associate with familiar conspecifics has been documented in a wide range of social fish species, recent work suggests that this ability may also be expressed in a heterospecific context. Ward *et al.* (2003) reported that chub (*Leuciscus cephalus*, Cyprinidae) preferred to associate with familiar rather than unfamiliar individuals of the closely related and sympatric species, the European minnow. Furthermore, the association preference of chub for conspecifics disappeared when focal fishes were given a choice between familiar heterospecifics and unfamiliar conspecifics.

8.2.2   *Mechanisms of familiarity recognition*

Two differing mechanisms have been proposed to explain the phenomenon of familiarity. A very general form of recognition based on odour cues has been documented in fathead minnows (*Pimephales promelas*, Cyprinidae; Brown & Smith 1994) and in three-spined sticklebacks (Ward *et al.* 2004; 2005). A more specific type of visual recognition, which may allow the recognition of particular individuals, has been reported in guppies (*Poecilia reticulata*, Poeciliidae; Griffiths & Magurran 1997a, 1998). Although further work is required to elucidate the role of each, the likelihood is that the two mechanisms operate in conjunction as each allows a different level of recognition.

Chemical signalling is of great importance in the aquatic environment; water acts both as a solvent and as a medium to disperse cues. Chemical cues may be particularly useful when visual communication is limited; for example during darkness or in deep or turbid water, or where the environment is highly structured. Chemical communication mediates the interactions between fishes in a range of diverse

Table 8.1 Summary of investigations on the role of familiarity in fish schooling decisions. Preference for familiar individuals (Y), no preference (N) or preference for unfamiliar individuals (U) is indicated. The discriminatory ability of individual fishes was tested in 2-choice tests or by looking at association patterns of near neighbours. Familiarity has also been tested in the laboratory in the contexts of shoal cohesion, competition and aggression. Preference for familiars in the wild is inferred from shoal fidelity.

Author (s)	Year	Common name	Scientific name	Method	Preference for familiar
Barber & Ruxton	2000	Three-spined stickleback	*Gasterosteus aculeatus*	Shoal cohesion	Y
Barber & Wright	2001	European minnows	*Phoxinus phoxinus*	2-choice test	Y
Binoy & Thomas	2004	Climbing perch	*Anabas testudineus*	2-choice test	Y
Brown & Colgan	1986	Bluegill sunfish	*Lepomis macrochirus*	2-choice test	Y
		Pumpkinseed sunfish	*Lepomis gibbosus*		N
		Rock bass	*Ambloplites rupestris*		N
Brown *et al.*	1993	Hatchery rainbow trout	*Oncorhynchus mykiss*	2-choice test	N
Brown	2002	Rainbowfish	*Melanotaenia* spp.	2-choice test; near neighbour	Y, Y
Brown & Smith	1994	Fathead minnows	*Pimephales promales*	2-choice test	Y
Chivers *et al.*	1995	Fathead minnows	*Pimephales promales*	Shoal cohesion	Y
Courtenay *et al.*	2001	Coho salmon	*Oncorhynchus kisutch*	2-choice test	Y
Croft *et al.*	2004	Trinidadian guppy	*Poecilia reticulata*	2-choice test	Males, Y
Dugatkin & Wilson	1993	Bluegill sunfish	*Lepomis macrochirus*	2-choice test	Y
Farmer *et al.*	2004	Texas shiner	*Notropis amabilis*	2-choice test	Y
		Blacktail shiner	*Cyprinella venusta*	2-choice test	Y
Giaquinto & Volpato	1997	Nile tilapia	*Oreochromis niloticus*	2-choice test	Y
Godin *et al.*	2003	Trinidadian guppy	*Poecilia reticulata*	2-choice test	N
Griffiths	1997	European minnow	*Phoxinus phoxinus*	Resource competition	Y
Griffiths & Magurran	1997a	Trinidadian guppy	*Poecilia reticulata*	Near neighbour	Y
Griffiths & Magurran	1997b	Trinidadian guppy	*Poecilia reticulata*	2-choice test	Y
				Near neighbour	N

Table 8.1 *(continued)*

Author(s)	Year	Common name	Scientific name	Method	Preference for familiar
Griffiths & Magurran	1998	Trinidadian guppy	*Poecilia reticulata*	2-choice test	Females, Y Males, N
Griffiths & Magurran	1999	Trinidadian guppy	*Poecilia reticulata*	2-choice test	Y
Hay & McKinell	2002	Pacific herring	*Clupea pallasi*	Shoal fidelity	Y
Helfman	1984	Yellow perch	*Perca flavescens*	Shoal fidelity	N
Hilborn	1991	Skipjack tuna	*Katsuwonus pelamis*	Shoal fidelity	N
Hoare *et al.*	2000	Banded killifish	*Fundulus diaphanus*	Shoal fidelity	N
Höjesjö *et al.*	1998	Sea trout	*Salmo trutta*	Resource competition	Y
Johnsson	1997	Hatchery rainbow trout	*Oncorhynchus mykiss*	Resource competition	Y
Johnsson & Åkerman	1998	Hatchery rainbow trout	*Oncorhynchus mykiss*	Resource competition	Y
Kelley *et al.*	1999	Trinidadian guppy	*Poecilia reticulata*	Mating attempts	males, U
Klimley & Holloway	1999	Yellowfin tuna	*Thunnus albacares*	Shoal fidelity	Y
Lachlan *et al.*	1998	Domestic guppy	*Poecilia reticulata*	2-choice test	Y
Magurran *et al.*	1994	Trinidadian guppy	*Poecilia reticulata*	2-choice test	Y
Metcalfe & Thomson	1995	European minnow	*Phoxinus phoxinus*	Resource competition	Y
Miklosi *et al.*	1992	Paradise fish	*Macropodus opercularis*	Resource competition	Y
O'Connor *et al.*	2000	Atlantic salmon	*Salmo salar*	Resource competition	Y
Seppä *et al.*	2001	Hatchery Arctic charr	*Salvelinus alpinus*	Growth and mortality	Y
Swaney *et al.*	2001	Domestic guppy	*Poecilia reticulata*	Shoal cohesion	Y
Utne-Palm & Hart	2000	Three-spined stickleback	*Gasterosteus aculeatus*	Resource competition	Y
Van Havre & FitzGerald	1988	Three-spined stickleback	*Gasterosteus aculeatus*	2-choice test	Y
Warburton & Lees	1996	Domestic guppy	*Poecilia reticulata*	2-choice test	Y
Ward *et al.*	2002	Three-spined stickleback	*Gasterosteus aculeatus*	Shoal fidelity	Y
Wisenden & Smith	1998	Fathead minnow	*Pimephales promelas*	Alarm cell proliferation	Y

contexts, including mate choice (Milinski *et al.* 2005), homing and migration (Hasler & Scholz 1983; Dittman & Quinn 1996), predator recognition (Wisenden 2000; and see Chapter 4 for a review) and dominance relationships (Bryant & Atema 1987).

The recognition of familiar individuals on the basis of chemical cues potentially allows a simple means of recognition. Brown & Smith (1994) investigated the sensory modalities involved in the recognition of familiar individuals in fathead minnows, finding that there was no preference for familiars when visual cues alone were available but that a preference for familiars was expressed when olfactory cues were available. The same pattern was also observed in three-spined stickle-backs (Ward *et al.* 2004), while in Nile tilapia (*Oreochromis niloticus*, Cichlidae) olfactory cues are also required for complete recognition of conspecifics (Giaquinto & Volpato 1997). The olfactory cues expressed by individuals are known to be influenced by both recent habitat use and recent diet (Bryant & Atema 1987; Olsén *et al.* 2003; Ward *et al.* 2004) and these cues, in turn, mediate association preferences; fish prefer individuals that smell similar to themselves (Ward *et al.* 2004, 2005), suggesting chemical self-referencing. Ward *et al.* (2005) reported that three-spined sticklebacks recognized and subsequently expressed an association preference for conspecifics and heterospecifics that had experienced similar habitat and diet condi-tions as themselves over the previous 24 h. Perhaps most interestingly, when direct experience was titrated against these general odour cues, fish preferred individuals who expressed the same environmental cues as themselves to individuals with whom they had previously interacted but who expressed different environmental cues (Ward *et al.* 2004, 2005).

A more specific form of recognition appears to underpin familiarity preferences in guppies. A study by Griffiths & Magurran (1997a) showed that familiarity develops gradually as individuals interact repeatedly over a period of 12 days (Fig. 8.1). Subsequent work by Griffiths & Magurran (1997b) showed that there is an upper

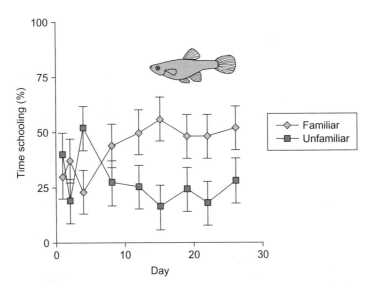

Fig. 8.1   The schooling preference for familiar individuals develops gradually in guppies (*Poecilia reticulata*, Poeciliidae), becoming significant after 12 days.

limit of around 40 different individuals that a single fish can identify as familiar. Furthermore, in contrast to other species, which require the presence of olfactory cues in order to demonstrate an association preference for familiars, guppies have been shown to recognize familiars on the basis of visual cues alone (Griffiths & Magurran 1998).

The gradual development of familiarity is a consistent feature among similar studies across different species and is suggestive of a learning process. A corollary of this is the gradual decrease in familiarity preferences that occurs if familiar individuals are separated; for example, Utne-Palm & Hart's (2000) study of familiarity in three-spined sticklebacks revealed a gradual increase in familiarity over a 4-week period of association and a subsequent decrease in familiarity over a 4-week period of isolation. In a study of rainbow trout (*Oncorhynchus mykiss*, Salmonidae), Johnsson (1997) found that the separation of pairs of territorial combatants for a period of 3 days was sufficient for the familiarity-biased aggression levels to decay. It may be that, under certain circumstances, memory decay may be adaptive (Miklósi *et al.* 1992; Warburton 2003, and see Chapter 2).

In contrast to the gradual development of individual recognition over several days in a context-independent situation, fishes are capable of extremely rapid learning of identities in context-dependent circumstances; for example, Milinski *et al.* (1990a) observed that three-spined sticklebacks accurately associated with individuals who were less likely to defect during predator inspection interactions after observing them perform predator inspection on just four previous occasions. Similarly, Dugatkin & Alfieri (1991) observed a preference among guppies for the better of two inspectors after a period of less than 4 min. However, it is difficult to argue that this represents familiarity in its strictest sense, given that the extent to which the documented recognition persists outside of the immediate context is not known. This limited level of recognition is probably a precursor to familiarity.

The variety of conspecific cues available to fishes at any given time raises the possibility that more than one recognition mechanism may be used. It is plausible that fishes may use chemical cues relating to recent habitat and diet experience, which are rapidly acquired, to provide general recognition of the individuals encountered, while also using learned individual recognition, which takes longer to develop and may be associated with higher memory costs, for more fine-scale recognition in specific contexts; for example, it is known that three-spined sticklebacks are capable of general recognition by environmentally mediated chemical cues (Ward *et al.* 2005), as well as highly specific individual recognition in the context of predator inspection (Milinski *et al.* 1990a).

8.2.3 *Functions of associating with familiar fishes*

By whichever means familiarity is achieved, substantial benefits may be obtained by associating with familiar individuals. Indeed, familiar shoals may be preferred over large shoals under certain conditions (Hager & Helfman 1991; Barber & Wright 2001). In a study on European minnows which titrated familiarity against the general preference for larger shoal sizes, focal fishes were given a choice of associating with a shoal of familiar conspecifics or a larger shoal of unfamiliar fishes. The results clearly indicate the importance of shoalmate familiarity as the number of fishes in the alternative, unfamiliar, shoal needed to be almost double that of the familiar shoal before the preference of the focal minnow for the familiar shoal finally disappeared (Barber & Wright 2001).

In essence, familiarity appears to enhance the benefits of group living by formalizing social relationships. One of the first tests of the functions of familiarity revealed that shoals of familiars were more cohesive than shoals of unfamiliar fishes (Chivers *et al.* 1995), providing a significant advantage to all group members when threatened with predation. Fishes associating with familiars also display a reduced investment in epithelial alarm cells, potentially reflecting lower levels of perceived risk (Wisenden & Smith 1998). Greater shoal cohesion enhances the predator confusion effect, effectively reducing the probability that a predator will be able to make a successful attack (Landeau & Terborgh 1986). Given the clear anti-predator benefits attached to familiarity, it has been predicted that the preference for familiars should increase under a predation threat. Despite this, Griffiths (1997) found no significant increase in the preference of minnows for familiar individuals on the appearance of a model predator. Brown (2002) reported similar results using rainbowfish and suggested that in habitats where fishes often experience a predation risk, it is adaptive to maintain a consistent preference for familiars. A uniform association preference is also suggestive of familiarity benefits beyond the scope of an anti-predator context (e.g. foraging). Nevertheless, the tendency to associate with familiars may vary between populations according to predator densities (Magurran *et al.* 1994). The adaptive benefits of familiarity should theoretically increase in line with predation threat, where predation pressure is low, shoaling tendencies and the preference for familiar individuals may be correspondingly low (Godin *et al.* 2003).

Familiarity also has the very significant effect of reducing competition and aggression (Jakobsson 1987; Brick 1998; and see Chapter 6). The ability to recognize and recall the outcome of previous exchanges with competitors should theoretically affect the strategy that an individual employs in subsequent interactions with the same competitor. Specifically, the expectation of future interactions should predispose animals to cooperative behaviour (Dugatkin 1997; and see Chapter 11). Utne-Palm & Hart's (2000) study of competition between sticklebacks revealed that familiar pairs were less likely to contest each prey item and, overall, were more likely to gain an equal share of the available prey items than pairs of unfamiliar fishes.

Some of the earliest studies of familiarity were concerned with territoriality and what became known as the 'dear enemy' effect (Fisher 1954; Getty 1987), where the development of familiarity between holders of adjacent territories is usually accompanied by a reduction in aggression between them. This type of reciprocal arrangement has been reported across a number of different territorial fish species including cichlids (*Neolamprologus pulcher*, Cichlidae; Frostman & Sherman 2004) and variegated pupfish (*Cyprinodon variegates*, Cyprinodontidae; Leiser 2003). Such 'dear enemy' arrangements increase the efficiency of territory defence by allowing territory holders to concentrate their efforts on repelling unfamiliar intruders (Leiser & Itzkowitz, 1999). Familiarity is known to stabilize dominance hierarchies, presumably because it allows individuals to remember the outcome of previous contests and to avoid further costly fighting. In a study on sea trout (*Salmo trutta trutta*, Salmonidae) Höjesjö *et al.* (1998) reported that familiarity stabilized dominance hierarchies and generally reduced the number of aggressive interactions between fishes. Juvenile Atlantic salmon (*Salmo salar*, Salmonidae) signal their submissiveness to familiar dominant individuals by darkening their body colouration, thereby incurring less direct aggression (O'Connor *et al.* 2000). Similarly, the ability of rainbow trout to recognize individuals allows third-party observers to gauge the competitive ability of a pair of combatants by watching contests (eavesdropping; Peake & McGregor 2004). The observer may then use this information to assess its own chances of prevailing against either of the two fighters in any subsequent contest (Johnsson & Akerman 1998). Perhaps as a consequence, Arctic charr (*Salvelinus alpinus*, Salmonidae) that were maintained in familiar conspecific groups showed increased survivorship and better overall body condition than those maintained in non-familiar groups over a 21-day period (Seppä *et al.* 2001).

Recently, Griffiths *et al.* (2004) proposed a more general benefit of familiarity that is likely to apply across different contexts. They proposed that familiarity acts to release some of the constraints on an individual's time budget, a benefit which is akin to that enjoyed by territorial individuals engaging in 'dear enemy' reciprocity (Fisher 1954; Getty 1987). The stable social conditions and the attendant reduction in aggression found in familiar groups allow individuals to spend more time foraging or being vigilant. Empirical findings supported this hypothesis; juvenile brown trout (*Salmo trutta*, Salmonidae) reacted more quickly to a simulated predator attack and gathered more food items when in a group of familiars than when in a group of unfamiliar fishes (Griffiths *et al.* 2004). Familiarity may also lead to greater foraging efficiency in groups by enabling more effective information transfer; for example, Swaney *et al.* (2001) reported that familiarity facilitated social learning of foraging information in guppies. This finding helps to explain the results of Ward *et al.* (2005) on sticklebacks where groups of familiar individuals located foraging patches more quickly than groups of unfamiliar fishes and, consequently, had a higher *per capita* feeding rate.

Familiarity is documented as an important component in the mate-choice decisions of some species, particularly where a promiscuous breeding system operates. Male guppies direct more of their courtship and mating effort towards unfamiliar females than towards familiar females, thereby reducing duplication of mating attempts towards the same individual females and maximizing reproductive success (Kelley *et al.* 1999). This same preference for unfamiliar females has also been observed in another livebearer, the Panamanian bishop (*Brachyraphis episcopi*, Poeciliidae), in both laboratory and field trials (Farr 1977; Simcox *et al.* 2005).

8.2.4   *Familiarity in free-ranging fishes*

Numerous demonstrations of familiarity-biased association have been reported from laboratory experiments. However, in common with other observed behavioural phenomena it is important to establish that these behaviour patterns also occur in the field, where they may be subject to natural selection. This is especially important in familiarity studies as it may be argued that the space constraints of laboratory aquariums produce environments where the number and frequency of repeated interactions between any two fishes are artificially high. Barber & Ruxton (2000) attempted to address this by designing a laboratory experiment where fishes could move freely between groups. Their results showed non-random patterns of association between fishes, indicating a significant preference for shoaling with familiar individuals. The more naturalistic conditions of Barber & Ruxton's experiment clearly demonstrated the potential of familiarity to play a significant role in the social organization of free-ranging fishes.

There have been comparatively few investigations into the social dynamics of free-ranging fishes, which probably reflects the logistical difficulties of such studies. The results of those studies that have been conducted in the field are somewhat equivocal. Hoare *et al.* (2000) and Hilborn (1991) report low levels of shoal fidelity in banded killifish (*Fundulus diaphanus*, Cyprinodontidae) and skipjack tuna (*Katsuwonus pelamis*, Scombridae), and Helfman (1984) found no significant association patterns between individually marked yellow perch (*Perca flavescens*, Percidae). Others have provided at least tentative evidence of familiarity-biased association in the field; for example, Klimley & Holloway (1999) report repeated co-occurrences of electronically tagged yellowfin tuna (*Thunnus albacares*, Scombridae) at particular times and places. Ward *et al.* (2002) observed non-random patterns of association between pairs of free-ranging three-spined sticklebacks over consecutive days that could not be explained by site fidelity or phenotype matching.

Large-scale studies of migrating populations also indicate a tendency for large groups of individuals to associate over considerable periods of time and distances (McKinnell *et al.* 1997; Hay & McKinnell 2002). McKinnell *et al.*'s study (1997) of migrating steelhead trout (*Oncorhynchus mykiss*, Salmonidae) demonstrated that individuals form long-term associations in the North Pacific. Similarly, Hay & McKinnell (2002) documented the movement patterns of over half a million Pacific herring (*Clupea pallasi*, Clupeidae) over a period of 14 years and concluded that individuals formed stable temporal and spatial associations. It seems unlikely, however, that familiarity is entirely responsible for these patterns. Shoal fidelity in mass migrations of fishes may be promoted by one or more of several different mechanisms, including: kin-specific or population-specific recognition (Quinn & Tolson 1986); activity synchronization (cf. Conradt & Roper 2000); population-specific migration traditions (Warner 1988); or pheromonal attraction (Baker & Montgomery 2001). Alternatively, because migrating fishes tend to remain in large, high density shoals rather than forming a number of small, loose aggregations, patterns of association may be explained simply by long-term shoal cohesion.

The recent application of network theory and analysis to free-ranging fish populations provides us with an excellent opportunity to gain a greater understanding of their social dynamics (Watts & Strogatz 1998; Croft *et al.* 2005). Using network analysis we can determine the extent of the social connections between individuals in a population. This allows us to make predictions about a range of things, including the transmission of information through a network and the alliances between individuals (Connor *et al.* 1999; and see Chapter 9). In the context of familiarity, network analysis enables us to establish the conditions under which familiarity develops and potentially to observe the relevance of familiarity to the social organization of free-ranging fish populations. To date, studies of populations of two species have been re-analysed using these methods, the Trinidadian guppy (Croft *et al.* 2003) and the three-spined stickleback (Ward *et al.* 2002). In both cases, the populations exhibited highly structured and connected network patterns, consistent with 'small world' social networks. Such networks tend to be characterized by a high number of cliques, meaning that pairs and small groups of fishes repeatedly co-occur. Such non-random patterns of association in the absence of phenotypic variability suggest that familiarity plays a significant role in structuring the interactions of fishes in the studied populations (Croft *et al.* 2005).

8.2.5 *Determinants of familiarity*

A number of factors affect the adaptive value of familiarity to wild fishes, including the ecological conditions that are experienced in the wild and the social dynamics of the species under consideration; for example, the social organization of female guppies differs from that of males (Croft *et al.* 2003). Females spend a greater proportion of their time shoaling and repeatedly interacting with the same individuals, whereas males devote more of their time to seeking mating opportunities. Females have more chance of developing familiarity in the first place, therefore, and may also enjoy greater benefits from such associations than males. This may explain the finding of Griffiths & Magurran (1998) that females demonstrated a preference for same-sex familiars, whereas males did not (although see Croft *et al.* 2004). This is unlikely to be a result of limited cognitive ability of males. Rather, males are likely to recognize familiars, but only behave differentially towards them in the context of mate choice. Male guppies are capable of recognizing and avoiding familiar females (Kelley *et al.* 1999). Similarly, male Panamanian bishop fish use familiarity to avoid mating repeatedly with the same subset of females, although the tendency to be choosy decreases with increased predation risk (Simcox *et al.* 2005).

Familiarity is most likely to develop under stable social conditions, which are dependent, in many instances, on ecological conditions; for example, during the dry season guppies are restricted to small pools with relatively small populations, which create ideal conditions for the development of familiarity, whereas during the rainy season the pools are joined into a more continuous river environment. This expands the potential home range of the fishes and increases the numbers of different individuals that each might encounter. As habitat complexity and the number of conspecifics encountered increases, the development of familiarity is constrained and the difficulties faced by individual fishes in discriminating familiar from unfamiliar may also increase. Long-term shoal fidelity is typically rare among fishes of shallow freshwater habitats (Hoare *et al.* 2000). While such conditions tend to militate against the development of learned familiarity, it remains possible that small subsets of shoals may remain together for extended periods (Ward *et al.* 2002). It has been suggested that group membership may be more stable among many coral reef fishes, which typically show a high degree of site fidelity, promoting high levels of local familiarity (Mapstone & Fowler 1988).

8.3   Familiarity or kin recognition?

Studies showing strong partner-choice preferences by fishes lead to the prediction that aggregations of fishes in the wild may be composed of non-random assortments of individuals. However, the possibility that fishes captured together are related to one another and that recognition is achieved on the basis of kinship, cannot be discounted. Indeed many fishes are now understood to have the cognitive ability to distinguish kin from non-kin (Table 8.2). This section reviews kin recognition abilities among fishes and describes the putative benefits of kin-biased behaviour.

8.3.1 *Kin recognition theory*

In the 1960s Hamilton published his revolutionary model of the evolution of social behaviour (Hamilton 1964), solving the paradox of why animals behave in ways that benefit others but are costly to themselves. Hamilton explained that individuals only assist (or avoid impeding) close relatives, because by doing so they gain inclusive fitness (kin selection) advantages: increasing the fitness of both the individual and kin. By behaving altruistically, genes shared by close relatives that are identical by descent are preferentially propagated to the next generation.

A prerequisite to individuals accruing kin selection advantages is interaction among relatives, which usually involves kin aggregation (Fletcher & Mitchener 1987; Hepper 1991; Pfennig & Sherman 1995). Initial studies of kin recognition and kin-biased association therefore focused on species with complex social structures, where individuals live in close proximity to family members. In birds, most work has been undertaken with species that breed cooperatively: some individuals relinquishing their own breeding attempts to help rear the offspring of kin (Brown 1970; Skutch 1987). More recently, fishes have also proved to be worthwhile study organisms. They clearly have the cognitive skills to distinguish kin from non-kin (Griffiths 2003; Ward & Hart 2003), and in many cases the acquisition of inclusive fitness advantages has important implications for fisheries managers and conservation biologists, as well as for individual fish (Parrish *et al.* 1998; Fréon & Misund 1999).

8.3.2 *Evidence for kin recognition from laboratory studies*

The main tranche of evidence for kin discrimination in fishes comes from juvenile salmonids, which prefer the water-borne odour of close relatives to that of unrelated conspecifics (Table 8.2; and reviewed by Brown & Brown 1996a; Olsén 1999; Griffiths 2003; Ward & Hart 2003). Juvenile salmonids are ideal candidates for these investigations because they are territorial, and the effect of group composition on patterns of association is likely to be particularly strong where dispersal from nests is limited (Elliott 1987) and where dominance hierarchies are stable (Huntingford & Turner 1987). Kinship is also likely to influence patterns of association in other nest-building species such as stickleback (VanHavre & FitzGerald 1988; FitzGerald & Morrissette 1992). Stickleback offspring are reared together in sibling groups and thus have the opportunity to learn the identity of kin (Wootton 1985).

This raises the interesting question of whether learned recognition of related conspecifics can more parsimoniously be attributed to prior experience than to kinship (Grafen 1990). In some studies, test fishes are not only related to, but are familiar with the stimulus fishes (having previously been housed together). Many studies have been unable to discount this straightforward explanation (see discussions in Quinn & Hara 1986; Winberg & Olsén 1992; Brown & Brown 1996a; Griffiths & Magurran 1999; Steck *et al.* 1999; Krause *et al.* 2000). Indeed Courtenay *et al.* (1997) found that common rearing increased the preference of juvenile coho

Table 8.2 Summary of investigations of the role of kinship in fish schooling decisions. Kin discrimination was absent (N), or resulted in kin-association (Y) or kin-avoidance (A). The discriminatory ability of individual fishes was tested in 2-choice tests or 2-choice flow tanks (where water was through-flow). The social behaviour of groups of fishes was observed in aquaria and stream tanks. The possible confounding effects of familiarity on these results are indicated as follows: test and stimulus fishes housed together before trial and therefore possibly familiar with one another (F); naturally occurring groups of fishes taken from the wild (W); test and stimulus fishes were parent and offspring (P/O); and levels of familiarity between stimulus fishes controlled (C) and unknown (?).

Author (s)	Year	Common name	Scientific name	Method	Preference for kin	Level of familiarity
Arnold	2000	Rainbowfish	*Melanotaenia eachamensis*	2-choice test	Y	C
Avise & Shapiro	1986	Serranid reef fish	*Anthias squamipinnis*	Molecular analysis	N	W
Baras & d'Almeida	2001	Sharptooth catfish	*Clarias gariepinus*	Aquarium tanks	N	F
Barnett	1982	Midas cichlid	*Cichlasoma citrinellum*	2-choice flow tank	Y	P/O
Beacham	1989	Coho salmon	*Oncorhynchus kisutch*	Aquarium tanks	N	C
Brown & Brown	1992	Hatchery rainbow trout	*Oncorhynchus mykiss*	2-choice flow tank	Y	C
		Atlantic salmon	*Salmo salar*		Y	C?
Brown & Brown	1993a	Rainbow trout	*Oncorhynchus mykiss*	Stream tank	Y	F
		Atlantic salmon	*Salmo salar*		Y	F
Brown & Brown	1993b	Hatchery rainbow trout	*Oncorhynchus mykiss*	Stream tank	Y	C
		Atlantic salmon	*Salmo salar*		Y	C
Brown & Brown	1996b	Rainbow trout	*Oncorhynchus mykiss*	Stream tank	Y	C
		Atlantic salmon	*Salmo salar*		Y	C
Brown *et al.*	1993	Hatchery rainbow trout	*Oncorhynchus mykiss*	2-choice flow tank	Y	C
Brown *et al.*	1996	Arctic charr	*Salvelinus alpinus*	Aquarium tanks	Y	C
Carlsson & Carlsson	2002	Brown trout	*Salmo trutta*	Molecular analysis	A	W
Carlsson *et al.*	2004	Brown trout	*Salmo trutta*	Molecular analysis	Y	W
Courtenay *et al.*	2001	Coho salmon	*Oncorhyncus kisutch*	2-choice flow tank	Y	F
Dowling & Moore	1986	Common shiner	*Notropis cornutus*	Molecular analysis	N	W
Ferguson & Noakes	1980	Common shiner	*Notropis cornutus*	Molecular analysis	Y	W
FitzGerald & Morrissette	1992	Three-spined stickleback	*Gasterosteus aculeatus*	2-choice test	?	C
Fontaine & Dodson	1999	Atlantic salmon	*Salmo salar*	Molecular analysis	N	W
Frommen & Bakker	2004	Three-spined stickleback	*Gasterosteus aculeatus*	2-choice test	Y	F
Garant *et al.*	2000	Atlantic salmon	*Salmo salar*	Molecular analysis	N	W
Gerlach *et al.*	2001	Eurasian perch	*Perca fluviatilis*	Molecular analysis	Y	W
Griffiths & Armstrong	2002	Atlantic salmon	*Salmo salar*	Stream tank	Y	W
Griffiths & Armstrong	2001	Atlantic salmon	*Salmo salar*	Field test	Y	C

Author	Year	Common name	Scientific name	Method		
Griffiths & Armstrong	2000	Atlantic salmon	*Salmo salar*	Aquarium	Y	C
Griffiths & Magurran	1999	Trinidadian guppies	*Poecilia reticulata*	2-choice test	N	C
Griffiths et al.	2003	Atlantic salmon	*Salmo salar*	Stream tank	A	C
Hauser et al.	1998	Tanganyika sardine	*Limnonthrissa miodon*	Molecular analysis	N	W
Herbinger et al.	1997	Cod	*Gadus morhua*	Molecular analysis	Y	W
Herbinger et al.	1999	Hatchery Atlantic salmon	*Salmo salar*	Aquarium tanks	Y	C
Hiscock & Brown	2000	Brook trout	*Salvelinus fontalis*	2-choice flow tank	Y	F
Kolm et al.	2005	Banggai cardinalfish	*Pterapogon kauderni*	Molecular analysis	N	W
Lima & Vrijenhoek	1996		*Poecilia monacha*	Aquarium tank	Y	F
Loekle et al.	1982	Guppy	*Poecilia reticulata*	2-choice test	Y	P/O
		Black molly	*Poecilia shenops*		Y	P/O
Magurran et al.	1994		*Poecilia reticulata*	Molecular analysis	Y	W
Mann et al.	2003	Zebrafish	*Danio rerio*	2-choice test	Y	F
McKaye & Barlow	1976	Midas cichlid	*Cichlasoma citrinellum*	2-choice test	Y	P/O
Mjølnerød et al.	1999	Atlantic salmon	*Salmo salar*	Molecular analysis	Y	W
Moore et al.	1994	Atlantic salmon	*Salmo salar*	Flow tank	Y	F
Naish et al.	1993	European minnow	*Phoxinus phoxinus*	Molecular analysis	N	W
Neff	2003	Bluegill sunfish	*Lepomis macrochirus*	Field test	A	W
Neff & Sherman	2005	Bluegill sunfish	*Lepomis macrochirus*	2-choice test	Y	Y
Ojanguren & Braña	1999	Brown trout	*Salmo trutta*	2-choice flow tank	A & N	C
Olsén	1989	Arctic charr	*Salvelinus alpinus*	2-choice flow tank	Y	F?
Olsén & Järvi	1997	Arctic charr	*Salvelinus alpinus*	Aquarium tanks	Y	C
Olsén & Winberg	1996	Arctic charr	*Salvelinus alpinus*	2-choice flow tank	Y	C
Olsén et al.	1996	Brown trout	*Salmo salar*	Aquarium tanks	Y	C
Olsén et al.	1998	Arctic charr	*Salvelinus alpinus*	2-choice flow tank	Y	C
Olsén et al.	2002	Arctic charr	*Salvelinus alpinus*	2-choice flow tank	Y	C
Olsén et al.	2004	Atlantic salmon	*Salmo salar*	Molecular analysis	Y	C
Peuhkuri & Seppä	1998	Three-spined stickleback	*Gasterosteus aculeatus*	Molecular analysis	N	W
Pouyaud et al.	1999	Mouthbrooding tilapia	*Sarotherodon melanotheron*	Molecular analysis	Y	F
Quinn & Busack	1985	Coho salmon	*Oncorhynchus kisutch*	2-choice flow tank	Y	C
Quinn & Hara	1986	Coho salmon	*Oncorhynchus kisutch*	2-choice flow tank	N	F
Russel et al.	2004	Trinidadian guppy	*Poecilia reticulata*	Molecular analysis	N	W
Steck et al.	1999	Three-spined stickleback	*Gasterosteus aculeatus*	2-choice flow tank	N	C
Svensson et al.	1998	Common goby	*Pomatoschistus microps*	Aquarium tank	N	N
VanHavre & FitzGerald	1988	Three-spined stickleback	*Gasterosteus aculeatus*	2-choice test	Y	C
Warburton & Lees	1996	Domestic guppy	*Poecilia reticulata*	2-choice test	Y	F
Winberg & Olsén	1992	Arctic charr	*Salvelinus alpinus*	2-choice flow tank	Y	F

salmon (*Oncorhynchus kisutch*, Salmonidae) for their siblings. Furthermore, Arctic charr reared in isolation from the egg stage do not discriminate kin from non-kin (Winberg & Olsén 1992; Olsén & Winberg 1996), suggesting that the recognition template by which fishes discriminate siblings from unrelated individuals is probably learned. Nevertheless, studies that control for familiarity demonstrate conclusively that relatedness has an important influence on patterns of association in many species of fish. In two-choice tests, preference for unfamiliar kin has been demonstrated for species ranging from Atlantic salmon, coho salmon, Arctic charr, rainbow trout and brown trout to Trinidadian guppies, three-spined stickleback and rainbow fish (Table 8.2).

8.3.3   *Advantages of kin recognition*

The advantages of such discriminatory behaviour are twofold. Direct fitness advantages are conferred by reduced levels of aggression among groups of kin compared to non-kin (Brown & Brown 1993a,b; Brown *et al.* 1996; Olsén *et al.* 1996), which lessens the risk of injury (Huntingford & Turner 1987) and reduces the risk of being preyed on (Jakobsson 1987; Brick 1998). Furthermore, when aggression is low, animals have more time to forage and attend to other activities (Roitblat 1987; Krause & Godin 1996). The formation and maintenance of stable kin groups therefore provides many potential advantages. Indeed, siblings housed together in aquaria or enclosures have reduced variation in size among individuals (Brown *et al.* 1996), smaller territories and faster growth (Brown & Brown 1993a, 1996b; Greenberg *et al.* 2002).

Clearly, there are significant incentives for individuals joining a group of related rather than unrelated conspecifics. In nature, however, resources may be limited, and fishes compete with one another for food and suitable habitat. Under these conditions individuals face a dilemma: to behave selfishly by monopolizing resources, or to share a proportion of the available resources with close relatives. Future inclusive fitness benefits are weighed against the cost of immediate reduction in resources. Moreover, because the cost of reduced resource domination is likely to be disproportionately visited upon high-ranking individuals that are resource-rich, the decision to behave nepotistically is predicted to be mediated by dominance rank. Griffiths & Armstrong (2002) investigated this possibility in a study of juvenile Atlantic salmon parr in a large indoor stream. It was found that dominant individuals monopolized feeding territories and food until they became satiated. After achieving satiation, however, dominant fishes shared excess food with subordinate neighbouring territory holders if the neighbours were close relatives (Griffiths & Armstrong 2002). It seems that while the food intake of dominant fishes is not reduced by either kin or non-kin neighbouring fishes, subordinate individuals are able to increase their food intake by feeding within the territories of dominant kin (Griffiths & Armstrong 2002). Faster growth among subordinate kin, but not subordinate non-kin has also been documented for hatchery-reared Atlantic salmon and rainbow trout (Brown & Brown 1996b). It seems therefore, that kin may accrue advantages from occasional association within a more fluid territorial structure rather than forming fixed associations. Indeed, recent work has shown that parr have large overlapping home ranges rather than fixed territories (Armstrong *et al.* 1999).

8.3.4   *Kin association in the wild*

The strong laboratory evidence that fishes prefer to associate with siblings, gaining direct fitness advantages through increased feeding and growth (reviewed by Brown & Brown 1996a; Olsén 1999; Griffiths 2003; Ward & Hart 2003) and inclusive fitness advantages by kin selection theory (Hamilton 1964), leads to the prediction that aggregations of fishes in the wild should be composed of non-random assortments of related individuals (although see below for exception). In territorial species, individuals should preferentially occupy feeding territories adjacent to relatives in order to gain benefits from reduced aggression, while in shoaling species, genetic relatedness should be higher within than between shoals if fishes are to benefit from increased growth and survival.

Paradoxically, there is little evidence for kin-biased association patterns in the wild among territorial fishes (Fontaine & Dodson 1999; Garant *et al.* 2000; Carlsson & Carlsson 2002) but see Mjølnerød *et al.* (1999) and Carlsson *et al.* (2004). However, most genetic analyses of fish populations have calculated the relatedness of fishes caught together in relatively long stream stretches varying from 50 to 300 m (Carlsson *et al.*1999; 2004; Hansen *et al.* 1997; Mjølnerød *et al.* 1999; Garant *et al.* 2000). Only one previous study has investigated how relatedness influences patterns of distribution of territorial fishes at high resolution. Fontaine & Dodson (1999) tested whether juvenile Atlantic salmon occupied territories adjacent to kin. They measured the relatedness of pairs of fish within five 20-m-long stream stretches but, contrary to their predictions, found little evidence of siblings defending adjacent territories. It seems that salmon tend not to cluster in kin groups, except as they disperse from nests.

Similarly, there is little evidence for kin-biased association patterns among shoals (Table 8.2), although there is some weak evidence of kin structuring in shoals of Eurasian perch (*Perca fluviatilis*, Percidae; Gerlach *et al.* 2001), tilapia (*Sarotherodon melanotheron*, Cichlidae; Pouyaud *et al.* 1999) and Atlantic salmon smolts (Olsén *et al.* 2004). Olsén's recent work is interesting because it suggests that juvenile Atlantic salmon smolts migrate downstream in shoals that are largely composed of relatives (Olsén *et al.* 2004). Individually tagged (and genetically typed) fishes were identified as they swam past a detector located at the downstream end of the stream, and the time between fish was used as an index of shoal cohesion (Olsén *et al.* 2004). Related shoalmates were separated by approximately 35 s, while unrelated shoalmates were approximately 42 s apart, corresponding to spatial separation of 7.7–15.0 body lengths [based on fishes' migration speeds ranging from 0.044 m s^{-1} (drift speed) to 0.089 m s^{-1} (fastest fishes)]. These values exceed the customary definition of shoal membership (3–5 body lengths; Pitcher & Parrish 1993), and suggest that the smolts did not form cohesive schools. Nevertheless there is a clear difference in spacing between related and unrelated fishes. Perhaps this can be explained by work that suggests smolt migration occurs in two phases, beginning with solitary movement, followed by schooling as singletons join together (Bakshtanskiy *et al.* 1980, 1988; Riley *et al.* 2002). The smolts in Olsén's study may have been in the process of forming shoals and the intriguing possibility remains that the effect of kinship on shoal composition strengthens as migration proceeds. Additionally, related fishes that are spawned and reared together may have a high probability of maturing and migrating synchronously due to 'common-garden' conditions. In the future it may be informative to investigate temporal variation in shoal structure, as fishes make moment-by-moment decisions about whether to join or leave a group.

8.3.5   *Explaining the discrepancies between laboratory and field*

Discrepancies between behavioural observations of kin recognition among fishes under laboratory conditions and genetic studies of wild fish populations are perplexing. Even within species, laboratory observations of preferential association with related individuals (VanHavre & FitzGerald 1988; FitzGerald & Morrissette 1992) have not been confirmed by field experiments (Peuhkuri & Seppä 1998). Part of the reason is that unnatural conditions are prevalent in many laboratory environments. The confinement of fishes to small and simple habitats for prolonged periods may allow tank-mate identity to be learned and stronger associations to be formed than would occur naturally. Furthermore, recirculation of water and the increased concentration of odour cues in laboratory aquaria may allow kin recognition to be achieved relatively easily, although also instigating heightened aggression in dominant fishes, but only towards non-kin subordinates (Griffiths & Armstrong 2000). When water is not recirculated there is no difference in aggression between kin and non-kin, which suggests that associating with kin may be advantageous, but only in areas of streams where water recirculates, for example in eddies behind rocks or in streambed refugia. In natural streams and rivers kin selection benefits may be modulated by local patterns of water flow (Griffiths & Armstrong 2000).

Further evidence that concentration of water-borne odour cues is important in discriminating conspecifics is given by Steck *et al.* (1999), Hiscock & Brown (2000) and Courtenay *et al.* (2001). Three-spined sticklebacks, brook trout (*Salvelinus fontinalis*, Salmonidae) and Juvenile coho salmon given the choice of water containing a high or low concentration of chemical cues from unrelated conspecifics prefer highly concentrated odour (Steck *et al.* 1999; Hiscock & Brown 2000; Courtenay *et al.* 2001). Interestingly the preference by brook trout for strong odour of non-kin over weak odour of kin suggests that under certain conditions, signal strength (odour concentration) overrides information describing relatedness of the putative shoalmates (Hiscock & Brown 2000). Recent work on the relative importance of chemical and visual signals during anti-predator behaviour shows that environmental conditions (e.g. water clarity) affect the response of fishes to chemical alarm cues (Hartman & Abrahams 2000; and see Chapter 4). It is possible, therefore, that the clarity of kin odour cues relative to background levels of other chemical information may also change temporally and/or spatially, and may explain why some studies have failed to find any influence of kinship on patterns of growth and association (Beacham 1989; Ojanguren & Braña 1999).

The exact nature of the kinship signal is not fully understood. Olsén (1987) and Moore *et al.* (1994) showed that odours in the urine are important, at least for Arctic charr and juvenile Atlantic salmon. Recently, the role of the major histocompatibility complex (MHC) has been highlighted. The MHC genes are highly polymorphic (Rammensee *et al.* 1997), and allow detection and recognition of individuals (reviewed by Penn 2002). Information about kinship may be conveyed by MHC breakdown products as MHC molecules are replaced and degraded into small proteins and peptides that are released through the urine (Brown & Eklund 1994;

Milinski *et al.* 2005). Olsén *et al.* 2002 reported that juvenile salmonids showed no preference in a choice between the water-borne odour of a sibling with a different MHC genotype and a non-sibling with an identical MHC genotype to the choosing fish. When the choice was between two sibling odours, focal fishes demonstrated a preference for siblings with matching MHC genotypes. In the context of mate choice, MHC is used by female three-spined sticklebacks to optimize the degree of MHC polymorphism and disease resistance in their offspring (Reusch *et al.* 2001; Aeschlimann *et al.* 2003; Wegner *et al.* 2003a,b). It seems that while reduced aggression among kin may promote closely related juveniles to associate with one another, adults may strike an optimal balance between inbreeding and outbreeding by choosing a mate slightly different from immediate kin (Bateson 1983).

8.3.6   *Kin avoidance*

Most studies of kin discrimination have focused on patterns of kin association, because aggregation of relatives is usually a prerequisite to performing altruistic behaviour and accruing kin selection advantages (Fletcher & Mitchener 1987; Hepper 1991; Emlen 1995; Pfennig & Sherman 1995). However, patterns of distribution other than kin association may also provide opportunities for kin-directed behaviour. Under specific circumstances (such as mate choice, see above), kin selection advantages may be afforded to individuals that avoid rather than associate with kin. Apart from mate-choice decisions, patterns of kin avoidance are also predicted in species where cannibalism is prevalent, or where sibling competition is intense.

Perhaps the best examples of kin avoidance in fishes come from cannibalistic species (reviewed by Manica 2002); for example, Loekle *et al.* (1982) observed that female Trinidadian guppies and black mollies (*Poecilia sphenops*, Poeciliidae) cannibalized the offspring of other females in preference to their own. One explanation is that adult females gain kin selection advantages by avoiding close relatives, and eating only unrelated juveniles (Loekle *et al.* 1982). However, the possibility also exists that kin-biased dispersal is achieved as a result of the kin-recognition abilities of offspring, as is the case for *Poecilia monach* juveniles, who actively disperse from their mother (Lima & Vrijenhoek 1996). In male fishes, filial cannibalism is expected to be most strongly influenced by kinship in species where cuckoldry is commonplace. Indeed in Bluegill sunfish (*Lepomis macrochirus*, Centrarchidae), males distinguish between the water-borne odour of their own offspring and unrelated fry fathered by sneaker males (Neff & Sherman 2005). In contrast, however, Svensson *et al.* (1998) were unable to show any increases in filial cannibalism in male common gobies (*Pomatoschistus microps*, Gobiidae) whose offspring were likely to have been sired by sneaker males.

Other examples of kin-biased cannibalism come from work on three-spined sticklebacks, where females raid male nests to eat the eggs (Wootton 1985). Males only eat eggs from their own nests if other fishes have begun an attack and it becomes impossible to defend the nest (FitzGerald & VanHavre 1987). Females, on the other hand, avoid raiding nests in which they have spawned (FitzGerald & VanHavre 1987), and after spawning, switch their association preferences from males that provide offspring with optimal allelic diversity, to suboptimal males (Milinski *et al.* 2005). Presumably this behaviour restricts egg raiding to unrelated eggs (Milinski *et al.* 2005).

Could sibling competition lead to kin avoidance? At low water temperatures, Atlantic salmon become nocturnal, hiding in streambed crevices during the day and emerging at night to forage in the water column. A study of winter sheltering behaviour in juvenile Atlantic salmon found that at this time salmon preferentially associate with non-kin and avoid sharing shelters with kin (Griffiths *et al.* 2003), perhaps to avoid imposing the costs of aggregation on close relatives (Waldman 1988). It is possible that kin avoidance may also reduce the degree of competition among siblings because different species of salmonid fishes specialize in different microhabitats (McLaughlin *et al.* 1999). In agreement with this prediction Carlsson & Carlsson (2002) found that for juvenile brown trout the distance between related territory holders was greater than predicted by chance. It seems that the decision of juvenile salmon to preferentially associate or avoid related conspecifics is not a simple one, but varies seasonally and with local environmental conditions.

Indeed, in two recent field studies, of Atlantic salmon (Griffiths & Armstrong 2001) and brown trout (Greenberg *et al.* 2002), the prevailing conditions favoured greatest advantages being afforded to fishes that associated with non-kin. When Atlantic salmon were released into sections of a stream in groups of full siblings or groups from a mixture of eight separate families the density of fishes in genetically diverse groups was almost twice as high as in groups where genetic diversity was low (Griffiths & Armstrong 2001). Similarly Greenberg *et al.* (2002) found that juvenile brown trout had higher growth rates when reared in large enclosures with a mixture of siblings and non-siblings compared to single-family groups. It seems that under these conditions the conflicting outcomes of kin selection and heterogeneous advantage (where genetically dissimilar individuals compete less intensely) balance so as to encourage dispersal rather than preferential association of kin. These observations may explain why genetic studies have failed to find evidence of kin aggregating in the wild despite the apparent advantages of such behaviour implicit from the results of laboratory studies. Indeed recent work found that for juvenile brown trout the distance between related territory holders was greater than predicted by chance (Carlsson & Carlsson 2002). Importantly, the results suggest not only an ecological advantage to individual juvenile salmon of avoiding relatives, but also an advantage to parents of producing genetically diverse progeny, as these offspring may realize higher densities than juveniles of low genetic diversity (Griffiths & Armstrong 2001).

8.4   Conclusions

Clearly, the ability to recognize and discriminate among conspecifics on the basis of relatedness and familiarity is widespread among fishes. The challenge for researchers in the future will be to examine these patterns in a broader context. First, more data need to be obtained from field studies to provide us with a greater understanding of the importance of relatedness and familiarity for free-ranging fishes. Second, the studies of relatedness and familiarity have focused on a relatively small number of species. This has produced a skew in the literature in favour of freshwater over marine species and for those species that inhabit the shallow margins of aquatic habitats, as these are easier to obtain and usually to maintain. In studies of relatedness in fishes, most work has similarly focused on species where there are good *a priori* reasons for predicting individuals will have the opportunity to learn the identity of kin; for example, species whose life history strategies result in offspring being reared together (e.g. mouth brooders, nest-builders or species where juveniles shoal immediately after birth) are far more likely to have kin-biased behaviour patterns than species with broadcast spawning. Clearly, it is important to examine the importance of relatedness and familiarity in less structured environments and among limnetic and pelagic species. Third, as a corollary of this, the influence of the ecological conditions on the recognition and preference for related and familiar individuals requires elucidation; for example, habitat complexity and population density are both likely to affect encounter probabilities between related and/or familiar fishes (Pollock & Chivers 2003). This may be especially important in allowing us to predict the most effective conditions for the generation of familiarity preferences in particular. Fisheries managers can modify the physical structures of habitats and social structures of fish communities that will affect the cost–benefit ratios that may influence shoaling behaviour. This may then enable the fisheries industry to harness further the growth and survivorship benefits associated with familiarity.

Recent technological advances may allow us to build up a clearer picture of patterns of relatedness and familiarity among free-ranging fishes; for example, field studies employing passive integrated transponder tags may become more frequent as they are predicted to reduce in both size and cost and improve in terms of their spatial and temporal resolution. Similarly, the use of microsatellite markers permits us to establish genetic relatedness with much greater accuracy. Furthermore, closer scrutiny of highly polymorphic areas of the genome, notably the MHC, has allowed us to gain a much greater understanding of the role of the genome in mediating association and mating decisions. Given that chemical cues mediated by the MHC are highly individualistic, it is a distinct possibility that this plays a major role in the recognition of both kin and familiar individuals (Olsén *et al.* 2002; Penn 2002).

8.5   Acknowledgements

We would like to thank all those who have contributed to the early drafts of this chapter and to the editors for the invitation to make this contribution.

8.6   References

Aeschlimann, P.B., Häberli, M.A., Reusch, T.B.H., Boehm, T. & Milinski, M. (2003) Female sticklebacks *Gasterosteus aculeatus* use self-reference to optimize MHC allele number during mate selection. *Behavioral Ecology and Sociobiology*, **54**, 119–126.

Armstrong, J.D., Huntingford, F.A. & Herbert, N.A. (1999) Individual space use strategies of wild juvenile Atlantic salmon. *Journal of Fish Biology*, **55**, 1201–1212.

Arnold, K. (2000) Kin recognition in rainbowfish (*Melanotaenia eachamensis*): sex, sibs and shoaling. *Behavioral Ecology and Sociobiology*, **48**, 385–391.

Avise, J.C. & Shapiro, D.Y. (1986) Evaluating kinship of newly settled juveniles within social groups of the coral reef fish *Anthia squamippinis*. *Evolution*, **40**, 1051–1059.

Baker, C.F. & Montgomery, J.C. (2001) Species-specific attraction of migratory banded kokopu juveniles to adult pheromones. *Journal of Fish Biology*, **58**, 1221–1229.

Bakshtanskiy, E.L., Nesterov, V.S. & Neklyudov, M.N. (1980) The behaviour of young Atlantic salmon, *Salmo salar*, during downstream migration. *Journal of Ichthyology*, **20**, 93–100.

Bakshtanskiy, E.L., Nesterov, V.S. & Neklyudov, M.N. (1988) Development of schooling behavior in Atlantic salmon, *Salmo salar*, during seaward migration. *Journal of Ichthyology*, **28**, 91–101.

Baras, E. & d'Almeida, A. (2001) Size heterogeneity prevails over kinship in shaping cannibalism among larvae of sharptooth catfish *Clarias gariepinus*. *Aquatic Living Resources*, **14**, 251–256.

Barber, I. & Ruxton, G. (2000) The importance of stable schooling: do familiar sticklebacks stick together? *Proceedings of the Royal Society of London, Series B*, **267**, 151–155.

Barber, I. & Wright, H. (2001) How strong are familiarity preferences in shoaling fish? *Animal Behaviour*, **61**, 975–979.

Barnett, C. (1982) The chemosensory responses of young cichlid fish to parents and predators. *Animal Behaviour*, **30**, 35–42.

Bateson, P. (1983) Optimal outbreeding. In: P. Bateson (ed.), *Mate Choice*, pp. 257–277. Cambridge University Press, Cambridge.

Beacham, T.D. (1989) Effect of siblings on growth of juvenile coho salmon (*Oncorhynchus kisutch*). *Canadian Journal of Zoology*, **67**, 601–605.

Binoy, V.V. & Thomas, K.J. (2004) The climbing perch (*Anabas testudineus* Bloch), a freshwater fish, prefers larger unfamiliar shoals to smaller familiar shoals. *Current Science*, **86**, 207–211.

Brick, O. (1998) Fighting behaviour, vigilance and predation risk in the cichlid fish, *Nannacara anomala*. *Animal Behaviour*, **56**, 309–317.

Brown, C. (2002) Do female rainbow fish (Melanotaenia spp.) prefer to shoal with individuals under predation pressure? *Journal of Ethology*, **20**, 89–94.

Brown, G.E. & Brown, J.A. (1992) Do rainbow trout and Atlantic salmon discriminate kin? *Canadian Journal of Zoology*, **70**, 1636–1640.

Brown, G.E. & Brown, J.A. (1993a) Do kin always make better neighbours?: the effects of territory quality. *Behavioral Ecology and Sociobiology*, **33**, 225–231.

Brown, G.E. & Brown, J.A. (1993b) Social dynamics in salmonid fishes: do kin make better neighbours? *Animal Behaviour*, **45**, 863–871.

Brown, G.E. & Brown, J.A. (1996a) Kin discrimination in salmonids. *Reviews in Fish Biology and Fisheries*, **6**, 201–219.

Brown, G.E. & Brown, J.A. (1996b) Does kin-biased territorial behavior increase kin-biased foraging in juvenile salmonids? *Behavioral Ecology*, **7**, 24–29.

Brown, G.E. & Smith, R.J.F. (1994) Fathead minnows use chemical cues to discriminate shoalmates from unfamiliar conspecifics. *Journal of Chemical Ecology*, **20**, 3051–3061.

Brown, G.E., Brown, J.A. & Crosbie, A.M. (1993) Phenotype matching in juvenile rainbow trout. *Animal Behaviour*, **46**, 1223–1225.

Brown, G.E., Brown, J.A. & Wilson, W.R. (1996) The effects of kinship on the growth of juvenile Arctic charr. *Journal of Fish Biology*, **48**, 313–320.

Brown, J.A. & Colgan, P.W. (1986) Individual and species recognition in centrarchid fishes: evidence and hypotheses. *Behavioral Ecology and Sociobiology*, **19**, 373–379.

Brown, J.L. (1970) Cooperative breeding and altruistic behavior in the Mexican jay, *Apelocoma ultramarina*. *Animal Behaviour*, **18**, 366–378.

Brown, J.L. & Eklund, A. (1994) Kin recognition and the major histocompatibility complex – an integrative review. *American Naturalist*, **143**, 435–461.

Bryant, B.P. & Atema, J. (1987) Diet manipulation affects social-behavior of catfish – importance of body odor. *Journal of Chemical Ecology*, **13**, 1645–1661.

Carlsson, J. & Carlsson, J.E.L. (2002) Micro-scale distribution of brown trout: an opportunity for kin selection? *Ecology of Freshwater Fish*, **11**, 234–239.

Carlsson, J., Olsén, K.H., Nilsson, J., Øverli, Ø. & Stabell, O.B. (1999) Microsatellites reveal fine-scale genetic structure in stream-living brown trout. *Journal of Fish Biology*, **55**, 1290–1303.

Carlsson, J., Carlsson, J.E.L., Olsén K.H., Hansen, M.M., Eriksson, T. & Nilsson, J. (2004) Kin-biased distribution in brown trout: an effect of redd location of kin recognition? *Heredity*, **92**, 53–60.

Chivers, D., Brown, G. & Smith, R. (1995) Familiarity and shoal cohesion in fathead minnows (*Pimephales promelas*): implications for antipredator behaviour. *Canadian Journal of Zoology*, **73**, 955–960.

Connor, R.C., Heithaus, M.R. & Barre, L.M. (1999) Superalliance of bottlenose dolphins. *Nature*, **397**, 571–572.

Conradt, L. & Roper, T.J. (2000) Activity synchrony and social cohesion: a fission–fusion model. *Proceedings of the Royal Society of London, Series B*, **267**, 2213–2218.

Courtenay, S.C., Quinn, T.P., Dupuis, H.M.C., Groot, C. & Larkin, P.A. (2000) Factors affecting the recognition of population-specific odours by juvenile coho salmon. *Journal of Fish Biology*, **50**, 1042–1060.

Courtenay, S., Quinn, T., Dupuis, H., Groot, C. & Larkin, P. (2001) Discrimination of family-specific odours by juvenile coho salmon: roles of learning and odour concentration. *Journal of Fish Biology*, **58**, 107–125.

Croft, D.P., Arrowsmith, B.J., Bielby, J., Skinner, K., White, E., Couzin, I.D., Magurran, A.E., Ramnarine, I. & Krause, J. (2003) Mechanisms underlying shoal composition in the Trinidadian guppy, *Poecilia reticulata*. *Oikos*, **100**, 429–438.

Croft, D.P., Arrowsmith, B.J., Webster, M. & Krause, J. (2004) Intra-sexual preferences for familiar fish in male guppies. *Journal of Fish Biology*, **64**, 279–283.

Croft, D.P., James, R., Ward, A.J.W., Botham, M.S., Mawdsley, D. & Krause, J. (2005) Assortative interactions and social networks in fish. *Oecologia*, **143**, 211–219.

Dittman, A.H. & Quinn, T.P. (1996) Homing in Pacific salmon: mechanisms and ecological basis. *Journal of Experimental Biology*, **199**, 83–91.

Dowling, T.E. & Moore, W.S. (1986) Absence of population subdivision in the common shiner, *Notropis cornutus* (Cyprinidae). *Environmental Biology of Fishes*, **15**, 151–155.

Dugatkin, L. (1997) *Cooperation Among Animals: An Evolutionary Perspective*. Oxford University Press, Oxford.

Dugatkin, L. & Alfieri, M. (1991) Tit-for-Tat in guppies (*Poecilia reticulata*): the relative nature of cooperation and defection during predator inspection. *Evolutionary Ecology*, **5**, 300–309.

Dugatkin, L.A. & Wilson, D.S. (1993) Fish behaviour, partner choice experiments and cognitive ethology. *Reviews in Fish Biology and Fisheries*, **33**, 368–372.

Elliott, J.M. (1987) The distances travelled by downstream moving trout fry, *Salmo trutta*, in a lake district stream. *Freshwater Biology*, **17**, 491–499.

Emlen, S.T. (1995) An evolutionary theory of the family. *Proceedings of the National Academy of Sciences, USA*, **92**, 8092–8099.

Farmer, N.A., Ribble, D.O. & Miller, D.G. (2004) Influence of familiarity on shoaling behaviour in Texas and blacktail shiners. *Journal of Fish Biology*, **64**, 776–782.

Farr, J.A. (1977) Male rarity or novelty, female choice behavior, and sexual selection in guppy, *Poecilia reticulata* Peters (Pisces, Poeciliidae). *Evolution*, **31 (1)**, 162–168.

Ferguson, M.M. & Noakes, D.L.G. (1980) Social grouping and genetic variation in common shiners, *Notropis cornutus* (Pisces, Cyprinidae). *Environmental Biology of Fishes*, **6**, 357–360.

Fisher, J. (1954) Evolution and bird sociality. In: J. Huxley, A. Hardy & E. Ford (eds), *Evolution as a Process*, pp. 71–83. Allen & Unwin, London.

FitzGerald, G.J. & Morrissette, J. (1992) Kin recognition and choice of shoal mates by three-spine sticklebacks. *Ethology Ecology and Evolution*, **4**, 273–283.

Fletcher, D. & Michener, C. (eds) (1987) *Kin Recognition in Animals*. Wiley, New York.

Fontaine, P.M. & Dodson, J.J. (1999) An analysis of the distribution of juvenile Atlantic salmon (*Salmo salar*) in nature as a function of relatedness using microsatellites. *Molecular Ecology*, **8**, 189–198.

Fréon, P. & Misund, O. (1999) *Dynamics of Pelagic Fish Distribution and Behaviour: Effects on Fisheries and Stock Assessment*. Blackwell Science Ltd, Oxford.

Frommen, J.G. & Bakker, T.C.M. (2004) Adult three-spined sticklebacks prefer to shoal with familiar kin. *Behaviour*, **141**, 1401–1409.

Frostman, P. & Sherman, P.T. (2004) Behavioral response to familiar and unfamiliar neighbors in a territorial cichlid, *Neolamprologus pulcher*. *Ichthyological Research*, **51**, 283–285.

Garant, D., Dodson, J.J. & Bernatchez, L. (2000) Ecological determinants and temporal stability of the within-river population structure in Atlantic salmon (*Salmo salar* L.). *Molecular Ecology*, **9**, 615–628.

Gerlach, G., Schardt, U., Eckmann, R. & Meyer, A. (2001) Kin-structured subpopulations in Eurasian Perch (*Perca fluviatilis* L.). *Heredity*, **86**, 213–221.

Getty, T. (1987) Dear enemies and the prisoners dilemma – why should territorial neighbors form defensive coalitions? *American Zoologist*, **27**, 327–336.

Giaquinto, P. & Volpato, G. (1997) Chemical communication, aggression, and conspecific recognition in the fish Nile tilapia. *Physiology and Behavior*, **62**, 1333–1338.

Godin, J.-G.J., Alfieri, M.S., Hoare, D.J. & Sadowski, J.A. (2003) Conspecific familiarity and shoaling preferences in a wild guppy population. *Canadian Journal of Zoology*, **81**, 1899–1904.

Grafen, A. (1990) Do animals really recognize kin? *Animal Behaviour*, **39**, 42–54.

Greenberg, L.A., Hernnäs, B., Brönmark, C., Dahl, J., Eklöv, A. & Olsén, K.H. (2002) Effects of kinship on growth and movements of brown trout in field enclosures. *Ecology of Freshwater Fish*, **11**, 251–259.

Griffiths, S. (1997) Preferences for familiar fish do not vary with predation. *Journal of Fish Biology*, **51**, 489–495.

Griffiths, S.W. (2003) Learned recognition of conspecifics by fishes. *Fish and Fisheries*, **4**, 256–268.

Griffiths, S. & Armstrong, J. (2000) Differential responses of kin and non-kin salmon to patterns of water flow: does recirculation influence aggression? *Animal Behaviour*, **59**, 1019–1023.

Griffiths, S.W. & Armstrong, J.D. (2001) The benefits of genetic diversity outweigh those of kin association in a territorial animal. *Proceedings of the Royal Society of London, Series B*, **268**, 1293–1296.

Griffiths, S. & Armstrong, J. (2002) Kin-biased territory overlap and food sharing among Atlantic salmon juveniles. *Journal of Animal Ecology*, **71**, 480–486.

Griffiths, S. & Magurran, A. (1997a) Familiarity in schooling fish: how long does it take to acquire? *Animal Behaviour*, **53**, 945–949.

Griffiths, S. & Magurran, A. (1997b) Schooling preferences for familiar fish vary with group size in a wild guppy population. *Proceedings of the Royal Society of London, Series B*, **264**, 547–551.

Griffiths, S. & Magurran, A. (1998) Sex and schooling behaviour in the Trinidadian guppy. *Animal Behaviour*, **56**, 689–693.

Griffiths, S. & Magurran, A. (1999) Schooling decisions in guppies (*Poecilia reticulata*) are based on familiarity rather than kin recognition by phenotype matching. *Behavioral Ecology and Sociobiology*, **45**, 437–443.

Griffiths, S., Armstrong, J. & Metcalfe, N. (2003) The cost of aggregation: juvenile salmon avoid sharing winter refuges with siblings. *Behavioral Ecology*, **14**, 602–606.

Griffiths, S.W., Brockmark, S., Hojesjo, J. & Johnsson, J.I. (2004) Coping with divided attention: the advantage of familiarity. *Proceedings of the Royal Society of London, Series B*, **271**, 695–699.

Hager, M.C. & Helfman, G.S. (1991) Safety in numbers – shoal size choice by minnows under predatory threat. *Behavioral Ecology and Sociobiology*, **29**, 271–276.

Hamilton, W.D. (1964) The genetical evolution of social behaviour. *Journal of Theoretical Biology*, **7**, 1–52.

Hansen, M.M., Nielsen, E.E. & Mensberg, K.-L.D. (1997) The problem of sampling families rather than populations: relatedness among individuals in samples of juvenile brown trout *Salmo trutta* L. *Molecular Ecology*, **6**, 469–474.

Hartman, E.J. & Abrahams, M.V. (2000) Sensory compensation and the detection of predators: the interaction between chemical and visual information. *Proceedings of the Royal Society of London, Series B*, **267**, 571–575.

Hasler, A.D. & Scholz, A.T. (1983) *Olfactory Imprinting and Homing in Salmon*. Springer-Verlag, Berlin.

Hauser, L., Carvalho, G.R. & Pitcher, T.J. (1998) Genetic population structure in the Lake Tanganyika sardine *Limnothrissa miodon*. *Journal of Fish Biology*, **53**, 413–429.

Hay, D.E. & McKinnell, S.M. (2002) Tagging along: association among individual Pacific herring (*Clupea pallasi*) revealed by tagging. *Canadian Journal of Fisheries and Aquatic Sciences*, **59**, 1960–1968.

Helfman, G.S. (1984) School fidelity in fishes: the yellow perch pattern. *Animal Behaviour*, **32**, 663–672.

Hepper, P.G. (ed.) (1991) *Kin Recognition*. Cambridge University Press, Cambridge.

Herbinger, C., Doyle, R., Taggart, C., Lochmann, S., Brooker, A., Wright, J. & Cook, D. (1997) Family relationships and effective population size in a natural cohort of Atlantic cod (*Gadus morhua*) larvae. *Canadian Journal of Fisheries and Aquatic Sciences*, **54**, 11–18.

Herbinger, C.M., O'Reilly, P.T., Doyle, R.W., Wright, J.M. & O'Flynn, F. (1999) Early growth performance of Atlantic salmon full-sib families reared in single family tanks versus in mixed family tanks. *Aquaculture*, **173**, 105–116.

Hilborn, R. (1991) Modelling the stability of fish schools: exchange of individual fish between schools of skipjack tuna (*Katsuwonus pelamis*). *Canadian Journal of Fisheries and Aquatic Sciences*, **48**, 1081–1091.

Hiscock, M.J. & Brown, J.A. (2000) Kin discrimination in juvenile brook trout (*Salvelinus fontinalis*) and the effect of odour concentration on kin preferences. *Canadian Journal of Zoology*, **78**, 278–282.

Hoare, D.J., Ruxton, G.D., Godin, J.-G.J. & Krause, J. (2000) The social organization of free-ranging fish shoals. *Oikos*, **89**, 546–554.

Höjesjö, J., Johnsson, J.I., Petersson, E. & Järvi, T. (1998) The importance of being familiar: individual recognition and social behavior in sea trout (*Salmo trutta*). *Behavioral Ecology*, **9**, 445–451.

Huntingford, F.A. & Turner, A. (1987) *Animal Conflict*. Chapman and Hall, London.

Jakobsson, S. (1987) Male behaviour in conflicts over mates and territories. PhD thesis. University of Stockholm.

Johnsson, J.J. (1997) Individual recognition affects aggression and dominance relations in rainbow trout, *Oncorhynchus mykiss*. *Ethology*, **103**, 267–282.

Johnsson, J.J. & Åkerman, A. (1998) Watch and learn: preview of the fighting ability of opponents alters contest behaviour in rainbow trout. *Animal Behaviour*, **56**, 771–776.

Keenleyside, M.H.A. (1955) Some aspects of the schooling behaviour of fish. *Behavior*, **8**, 83–248.

Kelley, J., Graves, J. & Magurran, A.E. (1999) Familiarity breeds contempt in guppies. *Nature*, **401**, 661–662.

Klimley, A.P. & Holloway, C.F. (1999) School fidelity and homing synchronicity of yellowfin tuna, *Thunnus albacares*. *Marine Biology*, **133**, 307–317.

Kolm, N., Hoffman, E.A., Olsson, J., Berglund, A. & Jones, A.G. (2005) Group stability and homing behavior but no kin group structures in a coral reef fish. *Behavioral Ecology*, **16**, 521–527.

Krause, J. & Godin, J.-G.J. (1996) Influence of prey foraging posture on flight behavior and predation risk: predators take advantage of unwary prey. *Behavioral Ecology*, **7**, 264–271.

Krause, J., Butlin, R., Peuhkuri, N. & Pritchard, V. (2000) The social organization of fish shoals: a test of the predictive power of laboratory experiments for the field. *Biological Reviews*, **75**, 477–501.

Lachlan, R., Crooks, L. & Laland, K. (1998) Who follows whom? Shoaling preferences and social learning of foraging information in guppies. *Animal Behaviour*, **56**, 181–190.

Landeau, L. & Terborgh, J. (1986) Oddity and the 'confusion effect' in predation. *Animal Behaviour*, **34**, 1372–1380.

Leiser, J.K. (2003) When are neighbours 'dear enemies' and when are they not? The responses of territorial male variegated pupfish, *Cyprinodon variegatus*, to neighbours, strangers and heterospecifics. *Animal Behaviour*, **65**, 453–462.

Leiser, J.K. & Itzkowitz, M. (1999) The benefits of dear enemy recognition in three-contender convict cichlid (*Cichlasoma nigrofasciatum*) contests. *Behaviour*, **136**, 983–1003.

Lima, N.R.W. & Vrijenhoek, R.C. (1996) Avoidance of filial cannibalism by sexual and clonal forms of Poeciliopsis (Pisces: Peociliidae). *Animal Behaviour*, **51**, 293–301.

Loekle, D.M., Madison, D.M. & Christian, J.J. (1982) Time dependency and kin recognition of cannibalistic behaviour among Poeciliid fishes. *Behavioural and Neural Biology*, **35**, 315–318.

Magurran, A., Seghers, B., Shaw, P. & Carvalho, G. (1994) Schooling preferences for familiar fish in the guppy, *Poecilia reticulata*. *Journal of Fish Biology*, **45**, 401–406.

Manica, A. (2002) Filial cannibalism in teleost fish. *Biological Reviews*, **77**, 261–277.

Mann, K.D., Turnell, E.R., Atema, J. & Gerlach, G. (2003) Kin recognition in juvenile zebrafish (*Danio rerio*) based on olfactory cues. *Biological Bulletin*, **205**, 224–225.

Mapstone, B.D. & Fowler, A.J. (1988) Recruitment and structure of assemblages of fish on coral reefs. *Trends in Ecology and Evolution*, **3**, 72–77.

McKaye, K. & Barlow, G. (1976) Chemical recognition of young by the midas cichlid, *Cichlasoma citrinellum*. *Copeia*, **1965**, 276–282.

McKinnell, S., Pella, J.J. & Dahlberg, M.L. (1997) Population-specific aggregations of steelhead trout (*Oncorhynchus mykiss*) in the North Pacific Ocean. *Canadian Journal of Fisheries and Aquatic Sciences*, **54**, 2368–2376.

McLaughlin, R.L., Ferguson, M.M. & Noakes, D.L.G. (1999) Adaptive peaks and alternative foraging tactics in brook charr: evidence of short-term divergent selection for sitting-and-waiting and actively searching. *Behavioral Ecology and Sociobiology*, **45**, 386–395.

McRobert, S.P. & Bradner, J. (1998) The influence of body coloration on shoaling preferences in fish. *Animal Behaviour*, **56**, 611–615.

Metcalfe, N.B. & Thomson, B.C. (1995) Fish recognize and prefer to shoal with poor competitors. *Proceedings of the Royal Society of London, Series B*, **259**, 207–210.

Miklósi, Á., Haller, J. & Csányi, V. (1992) Different duration of memory for conspecific and heterospecific fish in the paradise fish (*Macropodus opercularis* L.). *Ethology*, **90**, 29–36.

Milinski, M., Pfluger, D., Kulling, D. & Kettler, R. (1990a) Do sticklebacks cooperate repeatedly in reciprocal pairs? *Behavioral Ecology and Sociobiology*, **27**, 17–21.

Milinski, M., Kulling, D. & Kettler, R. (1990b) Tit for tat: sticklebacks 'trusting' a cooperating partner. *Behavioral Ecology*, **1**, 7–12.

Milinski, M., Griffiths, S., Wegner, K.M., Reusch, T.B.H., Haas-Assenbaum, A. & Boehm, T. (2005) Mate choice decisions of stickleback females predictably modified by MHC peptide ligands. *Proceedings of the National Academy of Sciences of the USA*, **102**, 4414–4418.

Mjølnerød, I.B., Refseth, U.H. & Hindar, K. (1999) Spatial association of genetically similar Atlantic salmon juveniles and sex bias in spatial patterns in a river. *Journal of Fish Biology*, **55**, 1–8.

Moore, A., Ives, M.J. & Kell, L.T. (1994) The role of urine in sibling recognition in Atlantic salmon *Salmo salar* (L.) parr. *Proceedings of the Royal Society of London, Series B*, **255**, 173–180.

Naish, K.A., Carvalho, G.R. & Pitcher, T.J. (1993) The genetic structure and microdistribution of shoals of *Phoxinus phoxinus*, the European minnow. *Journal of Fish Biology*, **43**(**A**), 75–89.

Neff, B.D. (2003) Paternity and condition affect cannibalistic behavior in nest-tending bluegill sunfish. *Behavioral Ecology and Sociobiology*, **54**, 377–384.

Neff, B.D. & Sherman, P.W. (2005) In vitro fertilization reveals offspring recognition via self-referencing in a fish with paternal care and cuckoldry. *Ethology*, **111**, 425–438.

O'Connor, K.I., Metcalfe, N.B. & Taylor, A.C. (2000) Familiarity influences body darkening in territorial disputes between juvenile salmon. *Animal Behaviour*, **59**, 1095–1101.

Ojanguren, A. & Braña, F. (1999) Discrimination against water containing unrelated conspecifics and a marginal effect of relatedness on spacing behaviour and growth in juvenile brown trout, *Salmo trutta* L. *Ethology*, **105**, 937–948.

Olsén, K.H. (1987) Chemoattraction of juvenile Arctic charr (*Salvelinus alpinus*, L.) to water scented by conspecific intestinal content and urine. *Comparative Biochemistry and Physiology A*, **87**, 641–643.

Olsén, K.H. (1989) Sibling recognition in juvenile Arctic charr, *Salvelinus alpinus* (L.). *Journal of Fish Biology*, **34**, 571–581.

Olsén, K.H. (1999) Present knowledge of kin discrimination in salmonids. *Genetica*, **104**, 295–299.

Olsén, K.H. & Järvi, T. (1997) Effects of kinship on aggression and RNA content in juvenile Arctic charr. *Journal of Fish Biology*, **51**, 422–435.

Olsén, K. & Winberg, S. (1996) Learning and sibling odor preference in juvenile Arctic charr, *Salvelinus alpinus* (L.). *Journal of Chemical Ecology*, **22**, 773–786.

Olsén, K.H., Järvi, T. & Löf, A.-C. (1996) Aggressiveness and kinship in brown trout. *Behavioral Ecology*, **7**, 445–450.

Olsén, K.H., Grahn, M., Lohm, J. & Langefors, A. (1998) MHC and kin discrimination in juvenile Arctic charr, *Salvelinus alpinus* (L.). *Animal Behaviour*, **56**, 319–327.

Olsén, K., Grahn, M. & Lohm, J. (2002) Influence of MHC on sibling discrimination in Arctic charr, *Salvelinus alpinus* (L.). *Journal of Chemical Ecology*, **28**, 783–795.

Olsén, K.H., Grahn, M. & Lohm, J. (2003) The influence of dominance and diet on individual odours in MHC identical juvenile Arctic charr siblings. *Journal of Fish Biology*, **63**, 855–862.

Olsén, K.H., Petersson, E., Ragnarsson, B., Lundqvist, H. & Jarvi, T. (2004) Downstream migration in Atlantic salmon (*Salmo salar*) smolt sibling groups. *Canadian Journal of Fisheries and Aquatic Sciences*, **61**, 328–331.

Parrish, D., Behnke, R., Gephard, S., McCormick, S. & Reeves, G. (1998) Why aren't there more Atlantic salmon (*Salmo salar*)? *Canadian Journal of Fisheries and Aquatic Sciences*, **55**, 281–287.

Peake, T.M. & McGregor, P.K. (2004) Information and aggression in fishes. *Learning & Behavior*, **32**, 114–121.

Penn, D.J. (2002) The scent of genetic compatibility: sexual selection and the major histocompatibility complex. *Ethology*, **108**, 1–21.

Peuhkuri, N. & Seppä, P. (1998) Do three-spined sticklebacks group with kin? *Annales Zoologici Fennici*, **35**, 21–27.

Pfennig, D.W. & Sherman, P.W. (1995) Kin recognition. *Scientific American*. June, 98–103.

Pitcher, T.J. & Parrish, J.K. (1993) Functions of shoaling behaviour in teleosts. In: T.J. Pitcher (ed.), *Behaviour of Teleost Fishes*, pp. 363–439. Chapman & Hall, London.

Pollock, M.S. & Chivers, D.P. (2003) Does habitat complexity influence the ability of fathead minnows to learn heterospecific chemical alarm cues? *Canadian Journal of Zoology*, **81**, 923–927.

Pouyaud, L., Desmarais, E., Chenuil, A., Agnese, J. & Bonhomme, F. (1999) Kin cohesiveness and possible inbreeding in the mouthbrooding tilapia *Sarotherodon melanotheron* (Pisces Cichlidae). *Molecular Ecology*, **8**, 803–812.

Quinn, T. & Hara, T. (1986) Sibling recognition and olfactory sensitivity in juvenile coho salmon (*Oncorhynchus kisutch*). *Canadian Journal of Zoology*, **64**, 921–925.

Quinn, T.P. & Busack, C.A. (1985) Chemosensory recognition of siblings in juvenile coho salmon (*Oncorhynchus kisutch*). *Animal Behaviour*, **33**, 51–56.

Quinn, T.P. & Tolson, G.M. (1986) Evidence of chemically mediated population recognition in coho salmon (*Oncorhynchus kisutch*). *Canadian Journal of Zoology*, **64**, 84–87.

Rammensee, H.G., Bachmann, J. & Stefanovic, S. (1997) *MHC Ligands and Peptide Motifs*. Landes Bioscience, Georgetown.

Reusch, T.B.H., Häberli, M.A., Aeschlimann, P.B. & Milinski, M. (2001) Female sticklebacks count alleles in a strategy of sexual selection explaining MHC polymorphism. *Nature*, **414**, 300–302.

Riley, W.D., Eagle, M.O. & Ives, S.J. (2002) The onset of downstream movement of juvenile Atlantic salmon, *Salmo salar* L., in a chalk stream. *Fisheries Management & Ecology*, **9**, 87–94.

Roitblat, H.L. (1987) *Introduction to Comparative Cognition*. W.H. Freeman & Co., New York.

Russel, S.T., Kelley, J.L., Graves, J.A. & Magurran, A.E. (2004) Kin structure and shoal composition dynamics in the guppy, *Poecilia reticulata*. *Oikos*, **106**, 520–526.

Seppä, T., Laurila, A., Peuhkuri, N., Piironen, J. & Lower, N. (2001) Early familiarity has fitness consequences for Arctic charr (*Salvelinus alpinus*) juveniles. *Canadian Journal of Fisheries and Aquatic Science*, **58**, 1380–1385.

Simcox, H., Colegrave, N., Heenan, A., Howard, C. & Braithwaite, V.A. (2005) Context-dependent male mating preferences for unfamiliar females. *Animal Behaviour*, **70**, 1429–1437.

Skutch, A.F. (1987) *Helpers at Bird's Nests*. University of Iowa Press, Iowa City.

Steck, N., Wedekind, C. & Milinski, M. (1999) No sibling odor preference in juvenile three-spined sticklebacks. *Behavioral Ecology*, **10**, 493–497.

Svensson, O., Magnhagen, C., Forsgren, E. & Kvarnemo, C. (1998) Parental behaviour in relation to the occurrence of sneaking in the common goby. *Animal Behaviour*, **56**, 175–179.

Swaney, W., Kendal, J., Capon, H., Brown, C. & Laland, K. (2001) Familiarity facilitates learning of foraging behaviour in the guppy. *Animal Behaviour*, **62**, 591–598.

Utne-Palm, A. & Hart, P. (2000) The effects of familiarity on competitive interactions between threespined sticklebacks. *Oikos*, **91**, 225–232.

VanHavre, N. & FitzGerald, G.J. (1987) The adaptive significance of cannabalism in stickle-backs (*Gasterosteidae, Pisces*) *Behavioural Ecology and Sociobiology*, **20**, 125–128.

VanHavre, N. & FitzGerald, G.J. (1988) Shoaling and kin recognition in the threespine stickleback (*Gasterosteus aculeatus* L.). *Biology of Behaviour*, **13**, 190–201.

Waldman, B. (1988) The ecology of kin recognition. *Annual Review of Ecology and Systematics*, **19**, 543–571.

Warburton, K. (2003) Learning of foraging skills by fish. *Fish and Fisheries*, **4**, 203–215.

Warburton, K. & Lees, N. (1996) Species discrimination in guppies: learned responses to visual cues. *Animal Behaviour*, **52**, 371–378.

Ward, A. & Hart, P. (2003) The effects of kin and familiarity on interactions between fish. *Fish and Fisheries*, **4**, 348–358.

Ward, A. & Krause, J. (2001) Body length assortative shoaling in the European minnow, *Phoxinus phoxinus*. *Animal Behaviour*, **62**, 617–621.

Ward, A., Botham, M., Hoare, D., James, R., Broom, M., Godin, J. & Krause, J. (2002) Association patterns and shoal fidelity in the three-spined stickleback. *Proceedings of the Royal Society of London, Series B*, **269**, 2451–2455.

Ward, A., Axford, S. & Krause, J. (2003) Cross-species familiarity in shoaling fishes. *Proceedings of the Royal Society of London, Series B*, **270**, 1157–1161.

Ward, A., Hart, P. & Krause, J. (2004) The effects of habitat- and diet-based cues on association preferences in three-spined sticklebacks. *Behavioral Ecology*, **15**, 925–929.

Ward, A., Holbrook, R., Krause, J. & Hart, P. (2005) Social recognition in sticklebacks: the role of direct experience and habitat cues. *Behavioral Ecology and Sociobiology*, **57**, 575–583.

Warner, R. (1988) Traditionality of mating-site preferences in a coral reef fish. *Nature*, **335**, 719–721.

Watts, D. & Strogatz, S. (1998) Collective dynamics of 'small-world' networks. *Nature*, **393**, 440–442.

Wegner, K.M., Reusch, T.B.H. & Kalbe, M. (2003a) Multiple parasite species are driving major histocompatibility complex polymorphism in the wild. *Journal of Evolutionary Biology*, **16**, 233–241.

Wegner, K.M., Kalbe, M., Kurtz, J., Reusch, T.B.H. & Milinski, M. (2003b) Parasite selection for immunogenetic optimality. *Science*, **301**, 1343.

Winberg, S. & Olsén, K.H. (1992) The influence of rearing conditions on the sibling odour preference of juvenile Arctic charr *Salvelinus alpinus* (L.). *Animal Behaviour*, **44**, 157–164.

Wisenden, B.D. (2000) Olfactory assessment of predation risk in the aquatic environment. *Philosophical Transactions of the Royal Society of London, B*, **355**, 1205–1208.

Wisenden, B. & Smith, R. (1998) A re-evaluation of the effect of shoalmate familiarity on the proliferation of alarm substance cells in ostariophysan fishes. *Journal of Fish Biology*, **53**, 841–846.

Wootton, R.J.A. (1985) *Functional Biology of Sticklebacks*. Croom Helm, London.

Chapter 9

Social Organization and Information Transfer in Schooling Fishes

*Iain D. Couzin, Richard James, David Mawdsley,
Darren P. Croft and Jens Krause*

9.1 Introduction

As fishes move through spaces they sense their environment, process information and frequently react to both internal and external stimuli (Odling-Smee &
Braithwaite 2003; and see Chapter 7). Even blind cave fish (*Anoptichthys jordani*,
Characidae), using their lateral line in the absence of vision, can build up some
contextual representation of the geometrical space in which they live (von Campenhausen *et al.* 1981; Burt de Perera 2004). Thanks to integration of behaviour through
interaction between individuals, in schooling species the way in which spatial representation, and other experiences are acquired depends not only on an individual's
information processing, but also on that of other group members. Even if groups are
not strictly coherent entities relative to the timescale over which certain ecologically
important factors occur (such as movement from one source of food to another)
schooling still biases motion, and thus the potential for information acquisition by
individuals. Thus the social behaviour of fishes has the potential to influence how,
and what, information is acquired by members of a population.

By definition, in mobile animal groups, be they composed of bacteria (Ben-Jacob
et al. 1994), fishes or herding vertebrates (Couzin & Krause 2003), individuals must
(to a greater or lesser degree) bias their own directed motion with that of others.
This social context affects the way we need to consider both how information about
the environment can be acquired, and also how it can be processed both within
individuals and collectively. Social learning research has made significant advances
in our understanding of information processing by individuals (see Chapter 10),
and, importantly, in the case of fishes has revealed the potential for relatively complex cognitive properties, suggesting that the individual-level capacities of these
individuals have previously been underestimated (Bshary *et al.* 2002), a topic that
is addressed comprehensively elsewhere in this volume (see Chapter 12). Here we
propose that the information-processing and problem-solving capacity associated
with integrated behaviour in fish schools has also been largely neglected. Improving
our understanding of these aspects of grouping may be equally important. Although
much has been written on the use of 'inadvertent social information' or 'public
information' (for a recent review, see Danchin *et al.* 2004, and the informative
subsequent discussion of this topic, Bednekoff 2004; Dall 2004; Laland *et al.* 2004;

Lotem & Winkler 2004), less is known about issues relating to the collective capacity for information acquisition, the filtering of environmental and/or sensory noise, information storage or collective decision-making in animal groups. It is particularly unclear what kind of information processing can be achieved by the collective that is not possible, or is extremely difficult, for individuals in isolation. Such collective processing, previously little recognized, could provide additional insights, especially if integrated into current social learning theory.

We shall also consider the potential for information transfer in wild populations of a fish species, the guppy (*Poecilia reticulata*, Poeciliidae), where studies of the networks of social interactions can be informative. The number of possible pairwise interactions becomes enormous in such networks, yet, as this chapter will demonstrate, guppy behaviour results in non-random associations correlated both with individual sex, phenotype and environmental structure that may facilitate effective transfer of pertinent information between individuals.

9.2   Integrated collective motion

Many fish schools perform highly integrated collective motion (Radakov 1973; Partridge 1982; Reynolds 1987; Couzin *et al.* 2002); so much so that they can appear like some animate fluid as they change volume, direction and shape with density waves propagating across the group (Radakov 1973), yet still remaining a cohesive whole. The survival of individuals in such groups is strongly dependent on social interactions (Radakov 1973; Partridge 1982; Parrish 1989; Romey 1995; Krause & Ruxton, 2002) and an intrinsic feature is that important information about the environment (e.g. the location of predators or resources) can be acquired by responding to the positions and orientations of others. Much of the collective behaviour of such groups can only be understood by considering these interactions and it is useful for our purposes here to consider the fundamental social tendencies that individuals exhibit when within a group: first, that individuals avoid collisions with others if they move too close (Ikawa & Okabe 1997; Krause & Ruxton 2002); and second, that if individuals are not performing avoidance behaviour, they will tend to be both attracted to, and to align with, others (Okubo 1980; Partridge 1982; Couzin *et al.* 2002).

Despite the simplicity of this description of behaviour it is impossible for us to comprehend how a group of more than a few individuals behaves when following these general behavioural rules using verbal argument, or through experiments, alone. As Dimitrii Radakov wrote in his groundbreaking 1972 (translated into English 1973) synthesis of fish schooling, 'The use of mathematics to express quantitative patterns manifested in fish schools seems to us not only justified but necessary, in particular because the patterns are often manifested in the actions of a large number of separate units – fishes. In such a situation it is scarcely possible to do without mathematics, when we must not be content with the ecologists' accustomed verbal description of the qualitative expression of a phenomenon but must explain and express precisely its quantitative characteristic' (Radakov 1973; p. 26). The limitation of verbal description is a restrictive, and generic, problem in a wide range of biological disciplines because a fundamental issue is understanding how functional complexity (such as the functioning of a biological tissue) results from the actions and interactions between the individual components (such as the cells forming the tissue). It is the recursive nature of these interactions that makes them particularly hard to investigate without the use of mathematics and computer simulations; the behaviour of an individual influences that of others (for example near neighbours), which, in turn, influence the focal individual, and so on.

Computer models, therefore, have become an integral part of systems biology, being used extensively in applications such as genomics, cell signalling and developmental biology (Kitano 2002). They are also becoming an increasingly powerful tool for the study of collective animal behaviour (Parr 1927; Breder 1954; Camazine *et al.* 2001; Couzin & Krause 2003; Sumpter 2005). We can specify individual rules (such as those expressed above) in mathematical terms, creating 'equations of motion' (Parr 1927; Breder 1954; Okubo 1980; Grünbaum & Okubo 1994) and by iterating these equations in time (usually simultaneously for all individuals in discrete, but sufficiently small, time steps) using computer simulation, collective motion can be both visualized and analysed numerically (and of course compared to the behaviour of the natural systems they represent; Grünbaum *et al.* 2004; Hoare *et al.* 2004; Viscido *et al.* 2004; Hensor *et al.* 2005). Thus we can make, and test, logically consistent predictions about information processing and social learning within animal groups such as fish schools. For a discussion of the principles behind this approach see Camazine *et al.* (2001), Couzin & Krause (2003) and Sumpter (2005).

9.3   Collective motion in the absence of external stimuli

In the absence of external organizing forces, such as an environmental attractor (e.g. a fish aggregating device or food source), simple rules, like those above, can account for the types of collective motion seen in natural fish schools, such as loose, unpolarized shoals if individuals exhibit only attraction and relatively close-range repulsion (Fig. 9.1a; Radakov 1973; Grünbaum *et al.* 2004); highly polarized schools when there is a relatively large range of alignment (Fig. 9.1c; Radakov 1973; Partridge 1982); and milling behaviour (in a doughnut-like shape, or torus), where individuals rotate around an empty core when the alignment range is relatively short (Fig. 9.1b, as exhibited by species such as barracuda, jack and tuna; Parrish & Edelstein-Keshet 1999; Couzin *et al.* 2002). For details of the model used in Figure 9.1 see Couzin *et al.* 2002 (see also Suzuki & Sakai 1973, Okubo 1980 and Shimoyama *et al.* 1996 for similar approaches). The important thing to note for our purposes here is that the patterns can result from individuals following relatively simple generative behavioural rules that do not explicitly describe this pattern nor rely on an external attractor; there is nothing explicit in the rules that generate the torus (Fig. 9.1b) that suggests 'move in a circle' (although, of course, in other cases attractors may facilitate a similar behaviour).

A further property of such groups, where discrete collective behaviours can be seen, is that it may be possible for memory of group shape to be stored in the group's structure, even if individuals themselves do not have any memory of group structure (Couzin *et al.* 2002). This does not contradict the fact that individual fish have memory, but does act to demonstrate the logical plausibility of a type of as yet undiscovered 'collective memory' that may be present in certain fish schools.

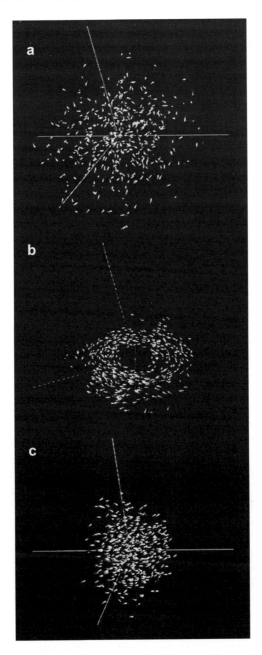

Fig. 9.1   Collective motion in fishes. (a) 'Swarm-like', loose and unpolarized aggregation. (b) 'Torus' formation. (c) 'Dynamic polarized group' where individuals become polarized, but can dynamically change position. For details of the model see Couzin *et al.* 2002.

9.4   Response to internal state and external stimuli: information processing within schools

9.4.1   *Collective response to predators*

Until now we have considered fish behaviour in the absence of specific environmental factors, a useful place to start if considering how collective behaviour results from individual rules. Fishes do, of course, also respond to external stimuli and modify their behaviour both with respect to others and also their environment. The group shape transitions discussed above could be important ways by which they assume an appropriate collective strategy appropriate to their internal state, such as when herring adopt a loose swarm-like formation when feeding, but a highly polarized group if predators are detected (Radakov 1973; and see Fig. 9.1a,c). Similarly, fishes that exhibit the torus formation (Fig. 9.1b) may do so to allow the group to remain stationary (which may be beneficial when resting) yet facilitate individual motion to allow flow of water over the gills, and perhaps energetic savings from every individual being in the slipstream of another. A further possible benefit results from the local alignment of individuals in this formation.

Alignment behaviour of grouping fishes facilitates collective directed motion and also information transfer over distances to a much greater extent than individuals' direct range of perception (Nikol'skii cited in Radakov 1973; Radakov 1973; Couzin & Krause 2003). A particularly clear example of this is seen when large fish schools are attacked by predators and waves of density and turning propagate across the group. Radakov (1973) studied these empirically (see below), observing that they shared some properties with waves in physical media and even cancelled out if they met mid-school. The highly integrated social interactions among fishes in such schools mean that, in some sense, they can be considered a 'social medium', which can rapidly propagate local information (such as the location and direction of a predator attack) simply through the tendency of individuals to group and to align with near neighbours. Escape behaviour can thereby be transmitted through the entire shoal.

In some cases the transmission speed of the information about the predator can be faster than the approach speed of the predator (Radakov 1973; Godin & Morgan 1985), thus providing fishes with an early warning system called the Trafalgar effect (Treherne & Foster 1981). Radakov (1973), for example, found that the speed of propagation of waves of turning reached 11.8–15.1 m s^{-1} in *Atherinomorus* (Atherinomorus stipes, hardhead silverside) (Atherinidae; the maximum burst swimming speed of this species of silverside was found to be around 1 m s^{-1}). Godin *et al.* (1988) determined the relationship between shoal size and predator detection and demonstrated the extent to which individuals benefit from the detection abilities of shoal members.

The dramatic turning associated with predator avoidance could potentially be used by fishes to learn, socially, about dangers not previously encountered (Leshcheva 1970 cited in Radakov 1973; Brown 2003; Brown & Laland 2003; Kelley *et al.* 2003; and see Chapter 10). Radakov (1973), for example, describes Kasimov's study in which juvenile fishes that had been reared on farms were found to make up approximately 90% of the food of predators when they were released, whereas an equal number of young fishes that had developed under natural conditions accounted for only about 10%. Thompson (cited in Radakov 1973), by training young chinook and coho salmon to associate a model trout with an electric shock, demonstrated that untrained fishes were 2.5 times more likely to be eaten by natural predators when released (see also Mirza & Chivers 2000; Brown & Laland 2001).

Gerasimov (cited in Radakov 1973) demonstrated that fishes could rapidly learn to associate a harmless stimulus (a light) with strong avoidance behaviour by reproducing the behaviour of individuals they could see through a glass partition being shocked electrically when the light was turned on. This 'imitative reflex' took very few trials (just one or two associations being required, whereas the unconditioned 'electrodefence reflex' took from four to 20 trials) and was considered 'extremely stable'. The higher the ratio of demonstrator to observer fishes (3:2, 2:1, 5:2) the stronger the influence of the demonstrators, but the response of the demonstrators was influenced by the response of the observers. It would be fascinating to replicate these studies (with a humane alarm stimulus) using one-way glass partitions to remove the recursive nature of the response.

In a related experiment, Chivers & Smith (1994) showed that it was possible to train fathead minnows (*Pimephales promelas*, Cyprinidae) to avoid harmless fishes (such as goldfish) if they had been combined with alarm substance on a number of occasions. However, the experiments showed that socially learnt information regarding real predators was retained for longer.

9.4.2 *Mechanisms and feedbacks in information transfer*

The way in which information is propagated in groups such as fish schools is poorly understood. Current models have investigated a variety of possibilities, mostly focusing on focal individuals biasing their own motion with the average of many, or a subset of, detectable neighbours. Responding to the motion of near neighbours is clearly important to the functionality of schools and the range over which they interact has important effects on collective behaviour. If this range is relatively small, an individual will behave more or less independently of those around it and changes in direction spread poorly or cannot spread (nor do groups remain cohesive). As this range increases, however, individuals respond to a greater number of neighbours, cohesive groups can form, and such groups become capable of transferring directional information, facilitating collective response to external environmental stimuli. If this range becomes too large, however, individual response includes distant individuals whose direction and position may not encode relevant information. Thus the focal individual may not respond in an optimal manner (see also Inada & Kawachi 2002). Consequently, in the case where the range of perception is not restricted by shadowing effects (as it is in some pelagic schools; Partridge 1982), individuals may deliberately restrict the range over which they respond to others (Couzin & Krause 2003; Hoare *et al.* 2004). Even where global information is available, it may benefit individuals to respond only to local information.

The assumptions made about the way individuals react to one another in schools in current models are likely to be generally correct because they explain many features of the internal structure of groups, such as group shape (Couzin *et al.* 2002; Viscido *et al.* 2005) and the dynamics of intergroup interaction (Niwa 1998a and see below). Testing such models, however, is restricted by a dearth of data from real groups. Consequently it is unclear how the adaptive dynamical features of fish schools (amplification of signals versus damping of noise, for example) are encoded in individual rules.

There may be important collective feedback systems whereby individuals change their probability of responding to certain neighbours in a context-dependent fashion. As we consider in more detail below, a few individuals may strongly bias the motion of an entire group. Even in the absence of specified 'leaders', if a certain number of fishes in close proximity turn together (even if this is just by chance, rather than as a response to a signal or cue) this may result in a wave of turning across the whole group (Radakov 1973).

Such 'informational cascades', to use the term from economics (Bikhchandani *et al.* 1992; Giraldeau *et al.* 2002), can result in an autocatalytic response with a defined response threshold. This type of response is common in a wide range of collective systems (Ame *et al.* 2004; Jeanson *et al.* 2005; Sumpter 2005), allowing a rapid adaptive response to changing conditions. In the case where fishes are in a state of heightened response (such as if alarm substance is detected, or following an attack), chance turning events can frequently lead to spontaneous turning by the group, even in the absence of further attacks (Radakov 1973). This heightened response presumably allows the group to be extremely sensitive to weak or ambiguous external stimuli (essentially increasing the 'system gain') but also means it becomes more susceptible to noise (stochastic effects) and false alarms. At the opposite extreme, such as when fishes adopt a poorly polarized group structure (Fig. 9.1a), signals (changes in the direction of a subset of individuals) are more readily damped; the signal-to-noise ratio becomes relatively high, but relatively low gain means the group takes longer to respond, or may even not respond at all, to true signals. The way in which individuals adopt behaviours adaptively to tune the sensitivity of collective response, adjusting gain and thus the signal-to-noise ratio would make a very interesting study and there may well be an important learned component to such responses.

9.4.3   *Information transfer during group foraging and migration*

Alignment and cohesion is of importance not only for evasion of predators, but can also facilitate information transmission in other contexts, such as during foraging or migration. Grünbaum (1997) demonstrated that schooling can facilitate more accurate taxis in noisy environments (such as when responding to thermal or nutritional gradients). Schools act as an integrated array of sensors (Kils 1986), allowing individuals to increase the 'effective' range of their interaction with the environment, and also to average errors in individual motion (error being a fundamental property of sensory systems). Thus schools can dampen small-scale fluctuations to allow perception of gradients not possible for individuals in isolation. For a conceptually similar model for large fish schools see Niwa (1998b). Grouping in this way could allow individuals to find more effectively, and subsequently remember, areas of appropriate habitat, such as food or spawning grounds. This property of schools has also recently inspired new search algorithms in computer science, termed 'particle swarm optimization', allowing for efficient and effective search of complex multi-dimensional function space (Kennedy *et al.* 2001), and further investigations of the collective computational ability of groups in this context may provide new insights into schooling ecology.

Radakov (1973) performed experiments on juvenile pollock (*Pollachius pollachius*, Gadidae) to investigate social transfer of information. A group of 25–35 specimens were kept in an aquarium and he investigated how information about the location of a discrete food source (a patch of gammarids), only visible to a small proportion of the fishes, spread throughout the group. In some trials he used an opaque partition (which extended from the roof of the aquarium to near the base, leaving a 20-cm gap in 100 cm of water), under which fishes could swim to move between sections. By adding food to only one section he could control which fishes could directly see the introduced food. On average, six fishes directly detected food, whereas the rest did not. At 15 s after its introduction, however, on average, 14 fishes had moved into the section with food. Tracking the motion of fishes from film, he could quantify the response of 'observers' to those 'informed individuals' they could see under the partition. Forage-area copying is a common behaviour in many fish species and the head-down posture is frequently used as an indicator that a con-specific has found food (Pitcher & Parrish 1993).

9.5 Informational status, leadership and collective decision-making in fish schools

The environment through which fish schools move is often complex and changing; resources and predators, therefore, are distributed probabilistically in space and time. Because individuals also move between groups (a topic we will return to later), aggregates of social animals often consist of individuals that can differ with respect to their informational status (such as when only some individuals have pertinent information about a currently viable resource), or internal state (such as degree of hunger). In the case of fish schooling, where crowding can severely restrict the range over which individuals can detect one another at any given moment (in pelagic schools individuals tend to be less than one body-length apart, Partridge 1982), or where individuals frequently change groups, it remains questionable whether individuals can actually recognize each other.

Fishes are capable of associating specific locations with rewards (Laland & Williams 1997), and can be trained to move to a known food source at specific times of day (Reebs 2000). By schooling with such 'informed' individuals, those that are naive can more readily find food and learn about their environment (Huse *et al.* 2002). Thus, by learning within a social context, individuals can become informed themselves, subsequently biasing the motion of other individuals that are naive with respect to this information, and allowing knowledge to spread within a population (Radakov 1973; Reader & Laland 2000). Radakov (1973), Reebs (2000) and Swaney *et al.* (2001) all found that relatively few informed individuals are capable of guiding a group of naive individuals, and the latter study also revealed that the degree to which informed individuals were trained influenced the guidance ability. Well trained individuals tend to lose the group, whereas poorly trained individuals remain in the group, but might be less capable of locating the target.

Couzin *et al.* (2005) developed a generic model of this type of guidance in animal groups. Following on from classic (Miller 1944; Brown 1948; Tinbergen 1952) and more recent (McClure *et al.* 2004; Usher & McClelland 2004; Livnat & Pippinger 2005) studies, they proposed that individuals exhibit internal conflict whereby they may have to reconcile different behavioural tendencies, such as schooling, with a desire to move in a specific (or general) direction. Naive individuals, which have no particular directional preference, predominantly school, whereas informed individuals balance schooling with a desire to move in a preferred direction (such as towards the location of a food source, or along a section of a migration route). The degree to which this balance is maintained is controlled through a simple weighting, termed ω. If $\omega = 0$, individuals are completely naive (and therefore have no desire to move in any specific direction). If individuals develop a preference to move in a particular direction (through learning), ω becomes greater than 0 and individuals increasingly bias their tendency to school with others against their competing tendency to direct their motion in a relatively specific desired direction. A single term, therefore, can represent different strengths of directional preference within a group, and thereby represent naive individuals ($\omega = 0$) and those that are informed to various degrees ($\omega > 0$).

Using this framework, Couzin *et al.* (2005) demonstrated that informed individuals can guide those that are naive in an effective manner without requiring either signalling or the ability for individuals to establish who has, and who does not have, information (Thomas D. Seeley of Cornell University has termed this the 'subtle guide hypothesis'). Furthermore, a small (group-size dependent) proportion of informed individuals can guide groups accurately, and as group size increases the required proportion of informed individuals needed to achieve a given accuracy decreases. This could help to explain certain pelagic fish migrations where experienced individuals may influence the motion of those that have not previously performed a migration (Huse *et al.* 2002; Hubbard *et al.* 2004; and see Chapter 14). In the subtle guide model, when only a few individuals within a group were informed, the accuracy of guidance increased with increasing ω, but this came at a cost resulting from a trade-off between increasing accuracy of group motion and the increasing probability that the group splits (informed individuals leaving those that are naive behind). This could be analogous to the fragmentation of experimental groups, where informed individuals were well trained (equivalent to a relatively high value of ω) described in Swaney *et al.* (2001).

Where informed individuals differ in preference, Couzin *et al.* (2005) demonstrated that, even though informed individuals do not know if they are in the majority or minority (they cannot assess the informational status of others), groups can still make consensus decisions, and collectively select the direction associated with the majority. The generic, and simple, nature of this model means that this collective computational ability could be a fundamental property of cohesive fish schools.

Furthermore, groups were also shown to be able to discriminate on the basis of the quality of information (without requiring any modification of the model), selecting collectively the direction associated with least error, even though the individuals themselves were not aware of how their information compared with that of others, or even if there were any other informed individuals. Consequently the model predicts that individuals can respond spontaneously to those with information and that informational status can change freely as individuals acquire information from their environment, either socially or asocially, or as their needs change.

Although this is currently a deliberately simplified version of reality, including a limited number of preferred directions, the framework is highly flexible and can easily be extended to consider more complex cases, or the ability for individuals to modify dynamically both multiple directional preferences and the ability to learn about the environment. In their initial study, Couzin *et al.* (2005) deliberately investigated the case where groups would typically remain cohesive (to simplify the analysis). This is, however, only one area of parameter space in this model. When ω is high, for example, informed individuals are less willing to compromise and will leave the group to move in their desired direction, often taking a subset of naive individuals with them. Under such conditions, groups frequently split if there is informational heterogeneity. Similarly, when the strength of attraction is reduced, groups will readily split and fuse. If combined with memory of new resource locations, this can allow information to spread through populations.

It is the simplicity, and lack of species specificity, of this model that is, in fact, its strength, because it demonstrates the principles underlying a form of collective decision-making that requires only limited cognitive ability. This is not to say that real fishes are this simple, but rather that this kind of collective computation is likely to exist in addition to the complex decision-making capacity of individuals, providing an important mechanism by which grouping individuals can more effectively acquire pertinent information, and consequently learn, about their environment within a social context. Experimental tests of the predictions of this model, currently underway, will also be essential to develop further the theory underlying information transfer in these types of fish schools.

9.6   The structure of fish schools and populations

In addition to informational status, individual differences, such as in speed, interaction range and turning rate, can spontaneously assort individuals both within (Couzin *et al.* 2002) and between groups (Radakov 1973; Couzin & Krause 2003). Therefore phenotypic assortment in populations can be an inevitable mechanism of schooling behaviour with individuals 'self-sorting' into phenotypically similar groups without requiring them to be able to recognize one another or know how their own phenotype compares with others (Couzin & Krause 2003). Fishes may also be capable of active assortment, whereby individuals are capable of assessing others and choosing to bias the strength of their associations (e.g. by modifying their weighting of attraction towards some phenotypes, or actively repelling themselves from others).

Differences among individuals, such as age, nutritional status or sex, may all influence the position adopted by an individual within a group. Krause (1993), for example, demonstrated that starved roach (*Rutilus rutilus*, Cyprinidae) tend to occupy frontal positions where they are presumably more likely to encounter food. Such positions may be relatively dangerous, however, with such individuals more likely to encounter sit-and-wait predators. In large, densely packed pelagic schools, which may consist of tens of thousands, or millions, of individuals, and extend over kilometres, it is likely to be impossible for fishes to know where they are with respect to the group centre. Through simple modifications of local interactions, however, such individuals can change positions with respect to the front, or centre, of the group, thereby modifying their response in a context specific way without necessitating knowledge of their absolute position (Couzin *et al.* 2002; Couzin & Krause 2003).

Habitat preferences may also tend to be associated with phenotypic characteristics, such as body size and/or sex (Croft *et al.* 2003), and these too can produce nonrandom associations in populations. A further property of groups that can be controlled by modification of simple local rules is group size. As shown with comparison to experimental data, individuals can accurately change their probability of being in a group of a certain size at a certain time (Hoare *et al.* 2004). At population-level, individual interactions result in a fission–fusion system (Okubo 1980; Bonabeau & Dagorn 1995; Gueron & Levin 1995; Niwa 1998a; Bonabeau *et al.* 1999), where rates of amalgamation (fusion) and splitting (fission), themselves resulting from individual interactions, define the group-size distribution in the population. Hoare *et al.* (2004), for example, consider how individual killifish (*Fundulus diaphanus*, Cyprinodontidae), by modifying simple local interactions (in this case the range over which they respond to one another), can manipulate their probability of being in groups of a certain size in response to environmental stimuli such as alarm odour and food odour.

The group-size distribution in fish populations tends to be stable (Gueron & Levin 1995; Bonabeau & Dagorn 1995; Niwa 1998a; Bonabeau *et al.* 1999) with the distribution depending on the processes of amalgamation (fusion) and splitting (fission) of groups. Where fusion rates are high relative to fission, the number of groups with relatively few individuals tends to decrease (larger groups can persist for longer) and the group size distribution assumes a long tail. This is characteristic of some pelagic fish schools [skipjack tuna (*Katsuwonus pelamis*, Scombridae), Bayliff 1998; yellowfin tuna (*Thunnus albacares*, Scombridae), Klimley & Holloway 1999], where the half-life of schools can be in the order of weeks. In such populations the timescale for processes such as information-transfer about resources may be completely different to freshwater fishes, where fusion rates tend to be low relative to fission and groups tend to be much more unstable (the group size distribution tends to be more rapidly decreasing). Krause *et al.* (2000), for example, found that in wild populations of killifish the rate of exchange of individuals between schools was high and that schools encountered one another very frequently; on average every 1.1 min. In wild guppy populations, Croft *et al.* (2003) found that encounters occurred approximately every 17 s (see also Seghers 1981 and Helfman 1984 for further evidence of such behaviour). When considering information transfer at population level one has to consider the timescales of biological relevance. For pelagic groups that range widely, food patches may be separated by distances in the order of kilometres. In freshwater fishes the analogous spatial properties may be an order of magnitude less. Consequently one needs to rescale assumptions about information transfer and individual memory accordingly.

It is important to consider the group-size distribution as the result of a dynamic process; it represents a probability distribution of a given individual being in a group of a certain size at a certain time. Many features, such as internal state and sex, can influence how an individual modifies (in an adaptive fashion) its interactions with neighbours to change its probability of being in a group of a given size (such as increasing the probability of being in a large group if a predator is detected). In many cases the distribution of group-size can be recorded in a population (see Couzin & Krause 2003 for a review) but this provides little information about how information transmission could occur within a wild population. There are, however, some long-standing, and more recently developed, methods for assessing the social structure of animal populations that can be used to make predictions for information transmission within and between populations.

It has been well documented that interactions in animals rarely occur randomly (Whitehead & Dufault 1999). As discussed above, particular fish shoals are frequently assorted by a number of phenotypic characteristics including body size, sex, species and parasite load (Radakov 1973; Krause *et al.* 2000). Such non-random interactions may have implications for the transmission of social information, with information being more likely to be transmitted between individuals of a similar phenotype (Hoare & Krause 2003). In order truly to understand the transmission of information in a population we must consider 'who learns from whom'. One starting point is to consider 'who associates with whom', as social learning may be more likely between individuals that have strong social ties. Other factors, however, may also have a role to play, such as dominance (although see Radakov 1973 for a discussion of why dominance is largely absent in many schooling fishes), size, sex, etc. (Laland 2004). Recent progress has been made in investigating the fine-scale structure of animal populations by representing social interaction between individuals as a social network, a method that allows us to study the social structure of a population as a whole.

9.7   Social networks and individual identities

A social network for fishes may be constructed from information about who shoals with whom, with ties in the network representing individuals that have co-occurred in the same shoal. To collect the information required to construct social networks, it is necessary to know the identities of the fishes (in a group or population), particularly if the information transmission, in the context of social learning, requires repeated contacts between the same individuals. The structure of a social network is likely, therefore, to affect the probability and speed at which information will spread through a population.

The past few years have seen significant progress with various marking techniques (e.g. fluorescent elastomer) that allow such individual marking even of relatively small fishes of just 15–20 mm body length (e.g. Croft *et al.* 2004). Such a technique can be used to mark fishes in both the laboratory and field (Croft *et al.* 2004). Once all the individuals in a whole population are marked in this way, we can observe the social contacts (individuals within the same shoal are usually defined as those within four body lengths of one another) between individuals on a nearly continuous basis by using a digital tracking system (Balch *et al.* 2001) or at given time intervals (point-sampling; Martin & Bateson 1993). The latter is logistically simpler and usually the only option when collecting data in the field; for example in an investigation on guppies, Croft *et al.* (2004) marked all adult individuals within a wild population ($n = 199$). The population was then sampled by capturing entire shoals once per day over a period of 15 days, recording the composition of the shoals. Based on this information, a social network of fish association patterns was constructed (Fig. 9.2a). Social networks provide us with information on who had social contact with whom and how often and can form the basis for predictions regarding information flow within a population or even between populations.

A useful first step is to characterize the structure of the social network using simple descriptive statistics such as mean path length (L), mean clustering coefficient (C) and mean degree of connectedness (k) (Albert & Barabási 2002; Newman 2003). L is calculated as the mean number of connections in the shortest path between two individuals in the network (see Newman 2003) and describes a global property of the

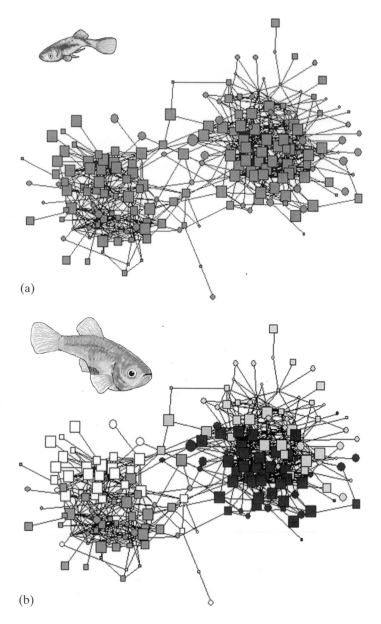

(a)

(b)

Fig. 9.2   (a) A social network of a guppy population in Trinidad. All guppies from two interconnected pools were marked and released. Over the next 2 weeks, approximately 20 shoals were captured daily and fish that belonged to the same shoal were connected in the network. Over time a completely connected network developed that comprised 197 fish. Each circle represents an individual male fish and each square an individual female. The size of the symbol is indicative of the body length of the fish. Individuals interconnected by lines were found at least twice together. (b) Five distinct communities (indicated by different levels of grey) were identified in the guppy network.

network that gives a simple indicator of how quickly social information will spread in an animal population (information can be expected to spread more quickly for lower values of L). C is a measure of the mean cliquishness of the network, calculated for each individual as the fraction of connections that exist between animals in its immediate network neighbourhood (see Newman 2003). C describes an average local property of the network, and is of particular interest because it measures the extent to which two of one's network neighbours are themselves neighbours. In social animals this local structure may result from active associations (e.g. phenotypic assortment or associations between familiar individuals). The degree of connectedness of an individual (from which k is derived) is simply the number of direct social connections (edges) that an individual has (Fig. 9.2). The mean degree for the network (k) can then be calculated as the average of the individual degrees. Together, values of L, C and k can be used to give a broad indication of the structural properties of the network (Watts & Strogatz 1998).

Recent work on both fishes (Croft *et al.* 2004) and dolphins (Lusseau 2003) suggests that wild animal populations are characterized by a non-random social network structure. In particular, networks appear to have relatively short path lengths (small L values), while being highly structured (large C values; Croft *et al.* 2004). Such findings are consistent with the idea of a 'small world' network (Watts & Strogatz 1998), whereby the observed path length (L) is almost as short as that expected in a random network, while the observed cluster coefficient (C) is much larger than expected in a random network. The small world effect is of particular interest because the short path length in a highly structured network may be associated with the rapid transmission of information through a population (Watts & Strogatz 1998).

Standard network theory (as used above for calculating L, C and k) is primarily based on unweighted associations between individuals and thus does not allow the analysis of one of the biological features of most interest, which is the occurrence of stable pairs of fish in wild populations. To measure the persistence of pair-wise associations, Croft *et al.* (2004) used an 'association strength' (AS), which indicates the number of days that a pair of fishes was caught together in a shoal, and compared this value to a randomization test that preserved the shoal size distribution and the number of recaptures for individual fishes. In guppies the social networks were all found to have a non-random structure and exhibited 'social cliquishness' (Croft *et al.* 2005). A number of factors were observed to contribute to this structuring. First, social network structure was influenced by body length and shoaling tendency, with individuals interacting more frequently with conspecifics of similar body length and shoaling tendency. Second, individuals with many social contacts were found to interact with each other more often than with other conspecifics, a phenomenon known as a 'positive degree correlation'. Finally, repeated interactions between pairs of individuals occurred within the networks more often than expected by random interactions. Strong associations between pairs of fish might be a good predictor of information flow between individual fishes within the network and/or cooperative relationships between fishes. A number of studies have looked at individual identities of fishes and their social interactions in the field [yellow perch (*Perca flavescens*, Percidae), Helfman 1984; yellowfin tuna, Klimley & Holloway 1999; three-spined stickleback (*Gasterosteus aculeatus*, Gasterosteidae), Ward *et al.* 2002; guppy, Croft *et al.* 2004). Apart from Helfman's study on the yellow perch, they all found good support for a significant co-occurrence of particular individuals with one another, suggesting active shoalmate choice. However, the only paper that directly tested this idea is Croft *et al.*'s study (2004) on the guppy, providing support for the hypothesis that this co-occurrence of pairs of fishes is indeed based on active social preference rather than similar swimming speeds or microhabitat preferences.

It is currently not known on what such active preferences of individuals for each other are based. Can they really recognize individual fishes (see Chapter 8), or do we observe preferences for familiar individuals (that could be based on microhabitat odours and thus would not require individual recognition; see Ward *et al.* 2004)? It has been suggested that fishes learn preferentially from familiar conspecifics (Brown *et al.* 2003; and see Chapters 8 and 10), which means that strong associations might be indicative of likely pathways of information flow within a population. In a similar way we might expect that other factors that lead to phenotypic assortment might play a role for information transmission within populations, and network analysis can be useful in identifying pathways for information transmission.

9.8   Community structure in social networks

When considering the social structure of group-living animals, one generally thinks of individuals forming groups and various numbers of groups making up a population. Recent developments in network theory, however (see for example, Lusseau & Newman 2004; Newman 2004; Newman & Girvan 2004), have enabled us to explore the possibility that there are meaningful communities, at a level between the group and the population, which may play an important role in determining the social organization, and thus the routes for information transmission, in animal populations. R. James and D. Mawdsley (unpublished observations) investigated the community structure of a population of wild guppies (Fig. 9.2b). There is evidence that the communities are structured by body length differences and, to a lesser extent, by sex. Assortment of fish communities by phenotypic characters (such as body size, colour, pattern, parasite load, etc.) is well established (Mesyatsev cited in Radakov 1973; Radakov 1973; Krause *et al.* 2000; Couzin & Krause 2003). However, so far studies on assortment have principally focused on group composition (although see Radakov 1973 and Couzin & Krause 2003 for exceptions). In contrast, what the community analysis suggests is that individuals can be freely exchanged between groups provided that they belong to the same community. Furthermore, we would predict that information transfer is more rapid within communities than between them. This is a yet untested prediction and should provide an interesting field for future experimental work.

Another important consequence of this type of analysis is that we can identify individuals and their position in the network; for instance, individuals that interconnect communities should play a key role in information transmission in the population as a whole. In the guppy network there is an indication that fishes of intermediate body lengths that cannot be clearly assigned to any particular community take up this position. Whether these individuals are indeed pivotal to information transmission remains to be ascertained.

All of the social network analysis we have presented has assumed that network to be a static entity, derived from the accumulation of data over a period of time. The reality, especially in fluid social systems, is much more dynamic and interesting.

Although the structure of the static network gives an indication of the information-carrying capacity of a population, there are certain to be features that depend strongly on the order in which pairwise connections are made and broken (Moody 2002).

9.9   Conclusions and future directions

Much of the functional complexity of fish schools arises from repeated interactions among individuals. Because they can be readily observed and manipulated, fishes present an excellent opportunity to quantify the behaviour of individuals and to develop new mathematical approaches that link the behaviour of individuals to the higher-order properties at the group and population level that result from these interactions. This approach is still at a relatively early stage of development, but, as we have discussed here, it is likely to provide new insights into how information is acquired, processed and stored. Combining this approach with current ideas in the field of social learning is therefore likely to be an important direction for future research.

Linking experimental studies of fishes to those in natural populations will also be an important goal. Here a key challenge will be to scale from individual motion and interactions to the temporal scale of practical observation in the field (samples may only be taken once per day, yet groups can interact over much shorter timescales: of the order of seconds). One way this may be possible is to create intermediate mesoscale experiments in the laboratory, where automated digital tracking (Balch *et al.* 2001) could allow scaling from subsecond intervals to that of several hours. This could allow detailed studies designed to reveal the important timescales, and temporal ordering, over which social networks form and function.

Scaling will also be a fundamental issue in how we study individual, and collective, decision-making in fish populations. In principle the tightly controlled experiments possible with fishes may provide a way to link the neural processes underlying individual decision-making to how groups resolve situations where informed individuals differ in preference. Future studies of fishes, therefore, may better link mechanistic and evolutionary models of grouping and allow us to make important advances in our understanding of social organization in biology.

9.10   Acknowledgements

We thank Kevin Laland for stimulating discussions. Financial support was provided by the Pew Charitable Trust and NSF to I.D.C., the EPSRC to J.K. and I.D.C., the Leverhulme Trust to J.K. and the FSBI to D.P.C.

9.11   References

Albert, R. & Barabási, A.L. (2002) Statistical mechanics of complex networks. *Reviews of Modern Physics,* **74**, 47–97.

Ame, J.M., Rivault, C. & Deneubourg, J.-L. (2004) Cockroach aggregation based on strain odour recognition. *Animal Behaviour,* **68**, 793–801.

Balch, T., Khan, Z. & Veloso, M. (eds) (2001) Automatically tracking and analyzing the Behavior of Live Insect Colonies. *Proceedings of the Fifth International Conference on Autonomous Agents,* pp. 521–528, Montreal, Quebec, Canada.

Bayliff, W.H. (1998) Integrity of schools of skipjack tuna, *Katsuwonus pelamis*, in the eastern Pacific Ocean, as determined from tagging data. *Fishery Bulletin*, **86**, 631– 643.

Bednekoff, P.A. (2004) Response to Danchin *et al.* (2004). *Science*, **308**, 354.

Ben-Jacob, E., Shochet, O., Tenenbaum, A., Cohen, I., Czirok, A. & Vicsek, T. (1994) Generic modeling of cooperative growth patterns in bacterial colonies. *Nature*, **368**, 46–49.

Bikhchandani, S., Hirshleifer, D. & Welch, I. (1992) A theory of fads, fashions, custom, and cultural changes as informational cascades. *Journal of Political Economics*, **100**, 992–1026.

Bonabeau, E. & Dagorn, L. (1995) Possible universality in the size distribution of fish schools. *Physical Review E*, **51**, 5220–5223.

Bonabeau, E., Dagorn, L. & Freon, P. (1999) Scaling in animal group-size distributions. *Proceedings of the National Academy of Sciences of the USA*, **96**, 4472– 4477.

Breder, C.M. (1954) Equations descriptive of fish schools and other animal aggregations. *Ecology*, **35**, 361–370.

Brown, C. & Laland, K.N. (2001) Social learning and life skills training for hatchery reared fish. *Journal of Fish Biology*, **59**, 471– 493.

Brown, C. & Laland, K.N. (2003) Social learning in fishes: a review. *Fish and Fisheries*, **4**, 280 –288.

Brown, C., Laland, K.N. & Krause, J. (eds) (2003) Learning in fishes: why they are smarter than you think. *Fish and Fisheries*, **4**.

Brown, G.E. (2003) Learning about danger: chemical alarm cues and local risk assessment in prey fishes. *Fish and Fisheries*, **4**, 227–234.

Brown, J.S. (1948) Gradients of approach and avoidance responses and their relation to level of motivation. *Journal of Comparative and Physiological Psychology*, **41**, 450 –465.

Bshary, R., Wickler, W. & Fricke, H. (2002) Fish cognition: a primate's eye view. *Animal Cognition*, **5**, 1–13.

Burt de Perera, T. (2004) Fish can encode order in their spatial map. *Proceedings of the Royal Society of London, Series B*, **271**, 2131–2134.

Camazine, S., Deneubourg, J.-L., Franks, N.R., Sneyd, J., Theraulaz, G. & Bonabeau, E. (2001) *Self-organization in Biological Systems*. Princeton University Press, Princeton, New Jersey.

von Campenhausen, C., Reiss, L. & Weissert, R. (1981) Detection of stationary objects by the blind cave fish, *Anoptichthys jordani* (Characidae). *Journal of Comparative Physiology A*, **143**, 369 –374.

Chivers, D.P. & Smith, R.J. (1994) Fathead minnows, *Pimephales promelas*, acquire predator recognition when alarm substance is associated with the sight of unfamiliar fish. *Animal Behaviour*, **48**, 597–605.

Couzin, I.D. & Krause, J.K. (2003) Self-organization and collective behavior in vertebrates. *Advances in the Study of Behavior*, **32**, 1–75.

Couzin, I.D., Krause, J., James, R., Ruxton, G.D. & Franks, N.R. (2002) Collective memory and spatial sorting in animal groups. *Journal of Theoretical Biology*, **218**, 1–11.

Couzin, I.D., Krause, J., Franks, N.R. & Levin, S.A. (2005) Effective leadership and decision making in animal groups on the move. *Nature*, **433**, 513–516.

Croft, D.P., Arrowsmith, B.J., Bielby, J., Skinner, K., White, E., Couzin, I.D., Magurran, A.E., Ramnarine, I. & Krause, J. (2003) Mechanisms underlying shoal composition in the Trinidadian guppy, *Peocilia reticulata*. *Oikos*, **100**, 429– 438.

Croft, D.P., Krause, J. & James, R. (2004) Social networks in the guppy (*Poecilia reticulata*). *Proceedings of the Royal Society of London, Series B*, **271**, S516–S519.

Croft, D.P., James, R., Ward, A.J.W., Mawdsley, D. & Krause, J. (2005) Assortative interactions and social networks in fish. *Oecologia*, **143**, 211–219.

Dall, S.R.X. (2004) Response to Danchin *et al.* (2004). *Science*, **308**, 353.

Danchin, E., Giraldeau, L.-A., Valone, T.J. & Wagner, R.H. (2004) Public information: from nosy neighbours to cultural evolution. *Science*, **305**, 487– 491.

Gerasimov, V.V. (1964) Specificity to imitation in fish. Ibid. Vol. 2, No. 5(9).

Giraldeau, L.-A., Valone, T.J. & Templeton, J.J. (2002) Potential disadvantages of using socially acquired information. *Philosophical Transactions of the Royal Society of London, Series B*, **357**, 1559–1566.

Godin, J.-G.J. & Morgan, M.J. (1985) Predator avoidance and school size in a cyprinodontid fish, the banded killifish (*Fundulus diaphanus Lesueur*). *Behavioral Ecology and Sociobiology*, **16**, 105–110.

Godin, J.-G.J., Classon, L.J. & Abrahams, M.V. (1988) Group vigilance and shoal size in a small characin fish. *Behaviour*, **104**, 29–40.

Grünbaum, D. (1997) Schooling as a strategy for taxis in a noisy environment. In: J.K. Parrish, & W.M. Hamner (eds), *Animal Groups in Three Dimensions*, pp. 257–281. Cambridge University Press, Cambridge.

Grünbaum, D. & Okubo, A. (1994) Modelling social animal aggregations. In: S.A. Levin (ed.), *Frontiers in Mathematical Biology, Lecture Notes in Biomathematics*, pp. 296–325. Springer-Verlag, Heidelberg.

Grünbaum, D., Viscido, S. & Parrish, J.K. (2004) Extracting interactive control algorithms from group dynamics of schooling fish. In: V. Kumar, N.E. Leonard & A.E. Morse (eds), *Proceedings, Block Island Workshop on Cooperative Control*. Springer-Verlag, Heidelberg.

Gueron, S. & Levin, S.A. (1995) The dynamics of group formation. *Mathematical Biosciences*, **128**, 243–264.

Helfman, G.S. (1984) School fidelity in fishes: the yellow perch pattern. *Animal Behaviour*, **32**, 663–672.

Hensor, E.M.A., Couzin, I.D., James, R. & Krause, J. (2005) Modelling density-dependent fish shoal distributions in the laboratory and field. *Oikos*, **110**, 344–352.

Hoare, D.J. & Krause, J. (2003) Social organization, shoal structure and information transfer. *Fish and Fisheries*, **4**, 269–279.

Hoare, D.J., Couzin, I.D., Godin, J.-G.J. & Krause, J. (2004) Context-dependent group size choice in fish. *Animal Behaviour*, **67**, 155–164.

Hubbard, S., Babak, P., Sigurdsson, S.T. & Magnússon, K.G. (2004) A model of the formation of fish schools and migrations of fish. *Ecological Modelling*, **174**, 359–374.

Huse, G., Railsback, S. & Fernö, A. (2002) Modelling changes in migration pattern of herring: collective behaviour and numerical domination. *Journal of Fish Biology*, **60**, 571–582.

Ikawa, T. & Okabe, H. (1997) Three-dimensional measurements of swarming mosquitoes: a probabilistic model, measuring system, and example results. In: J.K. Parrish, & W.M. Hamner (eds), *Animal Groups in Three Dimensions*. Cambridge University Press, Cambridge.

Inada, Y. & Kawachi, K. (2002) Order and flexibility in the motion of fish schools. *Journal of Theoretical Biology*, **214**, 371–387.

Jeanson, R., Rivault, C., Deneubourg, J.-L., Blanco, S., Jost, C. & Theraulaz, G. (2005) Self-organized aggregation in cockroaches. *Animal Behaviour*, **69**, 167–180.

Kasimov, R. Yu. (1958) Conditioned reflexes in the *Acipenseridae*. *Zoologicheskii Zhurnal*, **37**, No. 9.

Kelley, J.L., Evans, J.P., Ramnarine, I.W. & Magurran, A.E. (2003) Back to school: can antipredator behaviour in guppies be enhanced through social learning? *Animal Behaviour*, **65**, 655–662.

Kennedy, J., Eberhart, R.C. & Shi, Y. (2001) *Swarm Intelligence*. Academic Press, San Diego.

Kils, U. (1986) Verhaltensphysiologische Untersuchungen an pelagischen Schwärmen, Schwarmbildung als Strategie zur Orientierung in Umweltgradienten, Bedeutung der Schwarmbildung in der Aquakultur. *Berichte aus dem Institut für Meereskunde an der Christian-Albrechts-Universitaet Kiel*, **163**, 1–168.

Kitano, H. (2002) Computational systems biology. *Nature*, **420**, 206–210.

Klimley, A.P. & Holloway, C.F. (1999) School fidelity and homing synchronicity in yellowfin tuna, *Thunnus albacares*. *Marine Biology*, **133**, 307–317.

Krause, J. (1993) The relationship between foraging and shoal position in a mixed shoal of roach (*Rutilus rutilus*) and chub (*Leuciscus leuciscus*): a field study. *Oecologia*, **93**, 356–359.

Krause, J. & Ruxton, G.D. (2002) *Living in Groups*. Oxford University Press, Oxford.

Krause, J., Butlin, R.K., Peuhkuri, N. & Pritchard, V.L. (2000) The social organization of fish shoals: a test of the predictive power of laboratory experiments for the field. *Biological Reviews*, **75**, 477–501.

Laland, K.N. (2004) Social learning strategies. *Learning & Behavior*, **32**, 4–14.

Laland, K.N. & Williams, K. (1997) Shoaling generates social learning of foraging information in guppies. *Animal Behaviour*, **53**, 1161–1169.

Laland, K.N., Coolen, I. & Kendal, R. (2004) Response to Danchin *et al.* (2004). *Science*, **308**, 354.

Leshcheva, T.S. (1970) Formation of a defense reflex in young fish. – Tezisy dokladov molodezhnoi nauchoi konferentsii, posvyashchennoi 100 – letiyu so dnya rozhdeniya V.I. Lenina. Institut Evolyutsionnoi Morfologii 1 Ekologii Zhivotnykh im. A.N. Severtsova. Moscow.

Livnat, A. & Pippenger, N. (2006) An optimal brain can be composed of conflicting agents. Proceedings of the National Academy of Sciences USA, **103 (9)**, 3198–3202.

Lotem, A. & Winkler, D.W. (2004) Response to Danchin *et al.* (2004). *Science*, **308**, 354.

Lusseau, D. (2003) The emergent properties of a dolphin social network. *Proceedings of the Royal Society of London, Series B*, **270 (2)**, S186–S188.

Lusseau, D. & Newman, M.E.J. (2004) Identifying the role that animals play in their social networks. *Proceedings of the Royal Society of London, Series B*, **271**, S477–S481.

Martin, P. & Bateson, P. (1993) *Measuring Behaviour: An Introductory Guide*. Cambridge University Press, Cambridge.

McClure, S.M., Laibson, D.I., Loewenstein, G. & Cohen, J.D. (2004) Separate neural systems value immediate and delayed monetary reward. *Science*, **306**, 503–507.

Metyatsev, I.I. (1937) Structure of shoals of schooling fish. Investiya AN SSSR, Seriya Biologicheskaya, No. 3.

Miller, N.E. (1944) Experimental studies of conflict. In: J. McV. Hunt (ed.), *Personality and the Behavior Disorders*, volume 1, pp. 431–465. Ronald Press Co., New York, USA.

Mirza, R.S. & Chivers, D.P. (2000) Predator-recognition training enhances survival of brook trout: evidence from laboratory and field-enclosure studies. *Canadian Journal of Zoology*, **78**, 2198–2208.

Moody, J. (2002) The importance of relationship timing for diffusion. *Social Forces*, **81**, 25–56.

Newman, M.E.J. (2003) The structure and function of complex networks. *SIAM Review*, **45**, 167–256.

Newman, M.E.J. (2004) Detecting community structure in networks. *European Physical Journal B*, **38**, 321–330.

Newman, M.E.J. & Girvan, M. (2004) Finding and evaluating community structure in networks. *Physical Review E*, **69**, 026113.

Nikol'skii, G.V. (1963) Fish ecology "Vysshaya Shkola" Publishing House, Moscow.

Niwa, H.-S. (1998a) School size statistics of fish. *Journal of Theoretical Biology*, **195**, 351–361.

Niwa, H.-S. (1998b) Migration dynamics of fish schools in heterothermal environments. *Journal of Theoretical Biology*, **193**, 215–231.

Odling-Smee, L. & Braithwaite, V.A. (2003) The role of learning in fish orientation. *Fish and Fisheries*, **4**, 235–246.

Okubo, A. (1980) *Diffusion and Ecological Problems: Mathematical Models, Lecture Notes in Biomathematics*, Vol. 10. Springer-Verlag, New York.

Parr, A.E. (1927) A contribution to the theoretical analysis of the schooling behaviour of fishes. *Occasional Papers of Bingham Oceanographic College*, **1**, 1–32.

Parrish, J.K. (1989) Re-examining the selfish herd: are central fish safer? *Animal Behaviour*, **38**, 1048–1053.

Parrish, J.K. & Edelstein-Keshet, L. (1999) Complexity, pattern and evolutionary trade-offs in animal aggregation. *Science*, **284**, 99–101.

Partridge, B.L. (1982) The structure and function of fish schools. *Scientific American*, **245**, 90–99.

Pitcher, T.J. & Parrish, J.K. (1993) Functions of shoaling behavior in teleosts. In: T.J. Pitcher (ed.), pp. 363–439. Chapman and Hall, London.

Radakov, D.V. (1973) *Schooling in the Ecology of Fish*. Translated from the Russian by H. Mills. A Halsted Press Book, John Wiley & Sons Inc., New York.

Reader, S.M. & Laland, K.N. (2000) Diffusion of foraging innovations in the guppy. *Animal Behaviour*, **60**, 175–180.

Reebs, S.G. (2000) Can a minority of informed leaders determine the foraging movements of a fish shoal? *Animal Behaviour*, **59**, 403–409.

Reynolds, C.W. (1987) Flocks, herds, and schools: a distributed behavioural model. *Computer Graphics*, **21**, 25–33.

Romey, W.L. (1995) Position preferences within groups: do whirligigs select positions which balance feeding opportunities with predator avoidance. *Behavioural Ecology and Socio-biology*, **37**, 195–200.

Seghers, B.H. (1981) Facultative schooling behaviour in the spottail shiner (*Notropis hudsonius*): possible costs and benefits. *Environmental Biology of Fishes*, **6**, 21–24.

Shimoyama, N., Sugawara, K., Mitzuguchi, T., Hayakawa, Y. & Sano, M. (1996) Collective motion in a system of motile elements. *Physical Review Letters*, **76**, 3870–3873.

Sumpter, D.J.T. (2006) The principals of collective animal behaviour. *Philosophical Transactions of the Royal Society B: Biological Science*, **361** (1465), 5–22.

Suzuki, R. & Sakai, S. (1973) Movement of a group of animals. *Biophysics*, **13**, 281–282.

Swaney, W., Kendall, J., Capon, H., Brown, C. & Laland, K.N. (2001) Familiarity facilitates social learning of foraging behaviour in the guppy. *Animal Behaviour*, **62**, 591–598.

Thompson, R.B. (1966) Effect of predator avoidance conditioning on the past-related survival rate of artificially propagated salmon in fisheries. *College Fish. Res. Inst. Univ. Washington*, Contrib., No. 240.

Tinbergen, N. (1952) 'Derived' activities; their causation, biological significance, origin and emancipation during evolution. *The Quarterly Review of Biology*, **27**, 1–32.

Treherne, J.F. & Foster, W.A. (1981) Group transmission of predator avoidance in a marine insect: the Trafalgar effect. *Animal Behaviour*, **29**, 911–917.

Usher, M. & McClelland, J.L. (2004) Loss aversion and inhibition in dynamical models of multialternative choice (proofs version). *Psychological Review*, **111**, 757–769.

Viscido, S.V., Parrish, J.K. & Grünbaum, D. (2004) Individual behavior and emergent properties of fish schools: a comparison between observation and theory. *Marine Ecology Progress Series*, **273**, 239–249.

Viscido, S.V., Parrish, J.K. & Grünbaum, D. (2005) The effect of population size and number of influential neighbors on the emergent properties of fish schools. *Ecological Modelling*, **183**, 347–363.

Ward, A.J.W., James, R., Broom, M., Godin, J.-G.J., Botham, M.S. & Krause, J. (2002) Association patterns and shoal fidelity in the three-spined stickleback. *Proceedings of the Royal Society London, Series B*, **269**, 2451–2455.

Ward, A.J.W., Thomas, P., Hart, P.J.B. & Krause, J. (2004) Correlates of boldness in three-spined sticklebacks (*Gasterosteus aculeatus*). *Behavioural Ecology and Sociobiology*, **55**, 561–568.

Watts, D.J. & Strogatz, S.H. (1998) Collective dynamics of 'small-world' networks. *Nature*, **393**, 440–442.

Whitehead, H. & Dufault, S. (1999) Techniques for analyzing vertebrate social structure using identified individuals: review and recommendations. *Advances in the Study of Behavior*, **28**, 33–74.

Chapter 10
Social Learning in Fishes

Culum Brown and Kevin Laland

10.1   Introduction

In making decisions, such as how to find food and mates or how to avoid predators, many animals utilize information that is produced by others. Such individuals are referred to as 'observers' in the social learning literature (Heyes & Galef 1996) or 'eavesdroppers' in the signal-receiver literature (McGregor 1993). Socially transmitted information may simply be an inadvertent by-product of the 'demonstrating' individual's behaviour, or a signal targeted towards a particular individual or audience. Any learning that involves the use of socially provided information is termed 'social learning'. Social learning refers to any incidence in which individuals acquire new behaviour or information about their environment via observation of, or interaction with, other animals or their products.

It is generally assumed that social learning is beneficial because naive individuals can acquire locally adaptive behaviour quickly and efficiently from more knowledgeable individuals (Boyd & Richerson 1985), for instance, without having to incur the costs of exploration or the risks of learning about predators. Social learning is sometimes assumed to be more common in, of a more sophisticated form in, or even restricted to, 'intelligent' or 'large-brained' taxa. However, research over the past 50 years has demonstrated that social learning is common among fishes, birds and mammals, and should now be regarded as a regular feature of vertebrate life (Lefebvre & Palameta 1988; Heyes & Galef 1996). Moreover, recent studies have revealed that individuals, including fishes, will not always utilize socially available information, and will switch between reliance on social and asocial sources of information according to the costs and benefits (Kendal *et al.* 2005).

Many of the processes that underpin social learning lend themselves particularly well to shoals of fishes, although they are by no means restricted to shoaling species; for example, social learning through 'local enhancement' or 'stimulus enhancement' occurs when the behaviour (or simply the presence) of one individual attracts the attention of another individual to a particular location or stimulus about which the naive individual subsequently learns something; for instance, observers may learn that food is found at that location. 'Social facilitation' occurs when the behaviour of one individual induces an identical behaviour in another individual and the latter then learns something via the expression of this behaviour; for instance, it learns about the consequences of producing the behaviour in that context. 'Guided learning' or 'exposure' refer to instances in which, by following or being with a knowledgeable animal, an individual is exposed to similar features of the environment and comes to learn the same behaviour. 'Observational conditioning' occurs when the response of a demonstrator to a stimulus elicits a matching response on the part of an observer, who simultaneously perceives the original stimulus, and effectively learns that the response is an appropriate response to it; for instance, rhesus monkeys (*Macaca mulatta*) can acquire a fear of snakes when they witness experienced monkeys behaving fearfully in the presence of one (Mineka & Cook 1988). A variety of other processes, some assuming more sophisticated psychological underpinnings, can also facilitate social learning in animals, described variously as imitation, goal emulation, copying, etc. (see Whiten & Ham 1992 and Heyes 1994 for reviews). We know of no laboratory evidence that fishes are capable of imitation, that is, learning to produce particular bodily movements through observation of others. However, suggestive circumstantial evidence of imitation is provided by Mazeroll & Montgomery's report (1995) that in the local migrations of Brown surgeonfish (*Acanthurus nigrofuscus*, Acanthuridae), followers not only take the route of leaders but reproduce their postural changes (e.g. dips and rolls). However, it is a feature of animal social learning that simple processes are sufficient both to allow individuals to acquire adaptive information, and to mediate behavioural traditions in populations.

Over the past 20 years the study of social learning has expanded greatly and is now a major topic of research in ethology, behavioural ecology and comparative psychology (Galef & Giraldeau 2001; Shettleworth 2001). Documented cases of social learning in fishes are now commonplace and our growing understanding of the underlying mechanisms is enabling further identification and clarification of such phenomena. This chapter reviews evidence that social learning plays a role in fish in terms of anti-predator behaviour; migration and orientation; foraging; mate choice; and aggression, all of which may be facilitated by eavesdropping.

10.2   Anti-predator behaviour

It is well-known that the anti-predator behaviour of many fishes has a learned component, and countless species of fish improve their anti-predator response with experience (Kieffer & Colgan 1992; and see Chapter 3). However, learning about predators is a risky business, and there is little room for mistakes or extensive training (Kelley & Magurran 2003). Given the large risks and limited opportunities for asocial learning, it is easy to see the potential benefits of learning about predators from others. Not surprisingly, social learning of predator presence or predator identity appears widespread in fishes (Suboski & Templeton 1989; Brown & Laland 2001) as well as in other animals (reviewed by Griffin 2004).

The use of socially transmitted information enables individuals to respond to threats without having to verify the presence of danger independently. Fishes are particularly well equipped for rapid information transfer via sight and the lateral-line system. Such information may pass through a shoal much more quickly than the speed of an approaching predator or the diffusion of an alarm substance (Webb 1980; Potts 1984), resulting in what appears to be a synchronous response of shoal members. While such responses are not in themselves social learning (being merely manifestations of communication between shoal members), in the process young and inexperienced individuals frequently learn to identify predators, acquire appropriate anti-predator responses, or refine these responses; for example, Kelley *et al.* (2003) found that wild-caught guppies (*Poecilia reticulata*, Poeciliidae) captured in low predation areas altered their response to predators when paired with conspecifics from high, but not from low, predation locations in the presence of a predator model. These fishes from areas of low predation, when paired with their more experienced conspecifics, increased their shoaling cohesion and inspected the predator model from a greater distance.

Members of fish shoals are able to make decisions about predators by observing changes in the behaviour of shoalmates. When fishes return from a predator inspection visit it appears that the behaviour of the rest of the shoal changes in response to the level of threat perceived by the inspectors, as manifest in their behaviour (Pitcher *et al.* 1986). When fishes are attacked or startled by the presence of a predator they frequently display a fright response. It has long been known that the fright responses of many fishes can be precipitated by a chemical alarm substance emitted by other fishes (von Frisch, cited in Pfeiffer 1974; Brown 2003; and see Chapter 4). However, von Frisch discovered that anosmic fishes joined the fright reactions of conspecifics and subsequent research has established that such fright responses can also be visually transmitted (Verheijen 1956). The fright response provides an obvious visual cue that a predator is in the vicinity; for example, European minnows (*Phoxinus phoxinus*, Cyprinidae) increased the frequency of flight responses after observing the flight response of conspecifics in a neighbouring tank that had been threatened by a predator (Magurran & Higham 1988). Not only is the speed of communication considerably increased over the diffusion of a chemical, but predator presence can be communicated to and by fishes that have no direct experience with the predator. Social learning of fright responses has also been reported in mixed-species shoals, where the fright response of one species elicits a similar response in a second (Krause 1993; Mathis *et al.* 1996).

Conceivably the elevated anti-predator responses of fishes that have recently witnessed conspecifics and heterospecifics behaving fearfully in the presence of a predator could be the result of a number of processes. Central to the story would seem to be the formation of an association between the fright reaction of conspecifics and the predator. Observers may learn the identity of a predator by avoiding anything that elicits fright responses in conspecifics. Evidence for this comes from Suboski *et al.* (1990), who demonstrated that minnows could learn a fright response to olfactory cues from a novel predator if these cues were presented at the same time as either the visual fright reactions or the alarm substance of conspecifics. A similar mechanism can help fish to learn about risky habitats (Chivers & Smith 1995). The above examples demonstrate observational conditioning. It would seem that many fishes exhibit an unlearned fright response to the alarm substance (von Frisch, cited in Pfeiffer 1974), which would be sufficient for it to act as an unconditioned stimulus and would, through Pavlovian conditioning, explain the conditioned response to the predator when the alarm substance and predator are paired. It is less clear how the fright reactions of conspecifics should play this role. Perhaps they too elicit an unlearned fear response, but more plausibly they could have acquired higher-order conditioning properties through prior association with the alarm substance, or with other predators. In addition, observers may learn to flee, seek refuge, freeze or shoal tightly when others do so, and that these behaviour patterns are appropriate responses to the predator. Furthermore, observers may have their motivational or physiological state changed as a result of perception of fearful conspecifics, in a manner that leaves them more likely to learn about all aspects of their environment. Clearly a complete understanding of this apparently simple example of social learning awaits further analysis.

With the exception of the Kelley *et al.* (2003) study described above, there is, as yet, little direct evidence that fishes learn anti-predator responses to natural predators through social learning (i.e. that they learn to respond differently to different threats, such as shoaling versus hiding). Nevertheless, investigations into the avoidance responses of guppies to artificial predators (Sugita 1980; Brown & Laland 2002a) suggest that this is possible. Sugita (1980) found that guppies learn to avoid an electric shock by following demonstrator fishes into one of two safe compartments in a shuttle box. Brown & Laland (2002a) exposed naive fishes to a model trawl apparatus, which simulates a predator. Half of the naive fishes were placed with demonstrators trained to take one escape route and the others with demonstrators trained to take an alternative route. Both groups remained faithful to the demonstrated route and escaped more frequently than control groups while the demonstrators were present. However, once the demonstrators were removed, although the experimental groups were still far more efficient at escaping than control groups that had not been with demonstrators, their fidelity to the demonstrated route degraded. These results suggest that the fishes learnt an appropriate escape response by following the example set by demonstrators. While, in the absence of clear demonstration, functional aspects of traditional behaviour were maintained (i.e. to escape by swimming through a hole in the trawl), more arbitrary components were lost (i.e. utilization of the previously demonstrated route when equally viable alternatives are available). Similar experiments carried out on natural populations of Trinidadian guppies *in situ* found that even in the absence of demonstrators the route preference was maintained (Reader *et al.* 2003; Fig. 10.1), suggesting that social learning about anti-predator responses is heightened in wild populations compared to pet-shop strains. Population variation in social learning performance has yet to

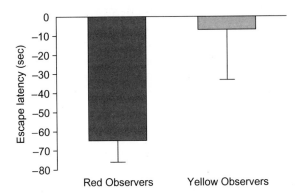

Fig. 10.1   The mean (SE) difference in escape latency (red route escape latency – yellow escape route latency) in wild guppies placed with demonstrators trained to escape via the red and yellow escape routes. Negative scores indicate a preference to escape via the red route. (Redrawn based on Reader *et al.* 2003.)

receive much attention from researchers, but may shed further light on the evolution of social learning by revealing the ecological contexts in which it is favoured (see below for further discussion).

Given the mounting evidence that social learning about predators appears to occur readily in fish populations, there is increased interest in the possibility that social learning may be used as a tool to train naive, hatchery-reared fishes to recognize predators (Brown & Laland 2001; Brown & Day 2002). Vilhunen and colleagues have examined social learning of anti-predator responses in endangered, land-locked salmonids. These populations are fully maintained by reintroductions from hatchery stocks after hydroelectric schemes destroyed their primary spawning grounds. Initial studies revealed that socially acquired avoidance of predator odours is a particularly effective means of training naive fishes to recognize predators (Vilhunen *et al.* 2005), and these early exposures to predators do lead to improved survival upon subsequent contact with live predators (Vilhunen 2006).

10.3   Migration and orientation

A number of studies have explored how, through social learning, fishes can learn to orientate around their environment and, in the process, learn the route to food sites, resting sites, schooling sites or mating sites.

One of the most elegant demonstrations of social learning in a natural population of fishes was carried out by Helfman & Shultz (1984). Helfman *et al.* (1982) had discovered that specific resting sites in coral reefs contained groups of French grunts (*Haemulon flavolineatum*, Haemulidae) that made daily migrations to feeding grounds. These groups appeared to be joined occasionally by newly recruited juveniles that had followed older individuals and seemingly had subsequently learned the migration path. Helfman & Shultz (1984) tested this by transplanting individuals between resting locations and then recorded their path towards foraging grounds. In the experimental condition the transplanted juveniles were allowed to follow the residents for 2 days before the residents were removed. In a control condition the resident population was removed prior to transplanting. While the experimental fishes learnt the same migration path as the resident adults, the control fishes continued to use paths appropriate to their original resting site.

Bluehead wrasse (*Thalassoma bifasciatum*, Labridae) show similar migratory traditions. These fishes have mating-site locations that remain constant over many generations. Warner (1988, 1990) removed entire populations and replaced them with transplanted populations. Not only did the new fishes establish entirely new mating grounds, but these new locations remained constant over subsequent generations. In the 12 years of studying 22 patches on the reef, not once was a new mating site established or lost, despite fluctuations in the wrasse population size. Combined with the observation that reef populations are not subject to significant genetic differentiation (Warner 1988), this finding provides strong evidence of cultural variation.

In laboratory experiments, Laland & Williams (1997, 1998) found that naive 'observer' guppies could learn a route to a foraging patch by following more knowledgeable 'demonstrators'. Observers were placed in the presence of demonstrators that had previously been trained to take one of two routes to feed. Observers typically shoaled with the demonstrators and fed at the feeding site. When the demonstrators were removed, the observers continued to utilize the same route to feed, despite the availability of an alternative route. Traditions were established in small populations in which experienced fishes were repeatedly replaced with naive conspecifics, and yet the route preferences remained. Laland & Williams (1997) found that as the number of demonstrators increased, the more likely it was that observers would remain faithful to the route. This is consistent with the hypothesis that animal social learning is commonly 'conformist', with the rate of information transmission increasing with the number of individuals displaying the behaviour. The strength of social learning in Laland & Williams's experiment (1998) was such that guppies would even maintain traditions for maladaptive[1] circuitous routes for brief periods. This conformity hypothesis is particularly important in group-living species that are under heavy selective pressure from predators to look and behave similarly to other group members (Brown & Laland 2002a).

What are the mechanisms underlying such natural and laboratory-based migratory traditions? Some recent experimental and theoretical studies shed light on this issue. Reebs (2000, 2001) has carried out experimental studies of leading and following in fishes, which demonstrate how small numbers of individuals can direct the movements of entire groups. Among shoals of 12 golden shiners (*Notemigonus crysoleucas*, Cyprinidae), some individuals were trained to swim to a specific location in a large tank where they were fed at the same time each day. The tank was divided into two areas, one of which was shaded and preferred by the fishes, the other was brightly lit but was also where food was delivered. In experiments with shoals composed of a mix of trained and naive fishes, in which the trained fishes were in the minority (down to a single trained individual in the shoal), Reebs demonstrated that even a few trained fishes were capable of leading the entire shoal to the feeding area at the time that food was delivered.

Reebs's empirical findings receive theoretical support from an agent-based model developed by Couzin *et al.* (2005). These researchers programmed virtual individuals with an aggregation rule, which endowed them with a tendency to be close to and aligned with neighbouring individuals, and provided a subset of individuals with information about a preferred direction, representing, for example, the direction to a known resource or segment of a migration route. Couzin *et al.* showed that this information could be transferred within the group without signalling and when group members do not know which individuals have information. Moreover, the larger the group, the smaller the proportion of informed individuals needed to guide the group (see also Chapter 9). The model suggests that, merely by shoaling, a small group of demonstrator fishes that are highly motivated to swim to a particular location can lead a much larger group of naive individuals to that location, even if the naive individuals do not pay particular attention to the demonstrators or even know who they are. Previous hypotheses put forward to account for learned migratory traditions in animals often assumed that young or naive individuals paid particular attention to, or learned to follow, older or more experienced conspecifics. The Couzin *et al.*

---

[1] Note this experiment provides evidence for the social transmission of maladaptive *information* (i.e. take the long route), and for a suboptimal behavioural *tradition* (for taking the long route), but neither the *behaviour* of the fish (where it pays to shoal for protection from predators) nor the general *capacity* for social learning (which is typically advantageous) should be described as maladaptive.

(2005) explanation is attractive because it relies on what would appear to be a much simpler mechanism than such accounts, and helps to explain why blocking does not prevent naive individuals from learning an association between spatial cues and reinforcement.

Further insight comes from a state-dependent model of foraging by Rands *et al.* (2003, 2004) which found that, assuming some advantage to foraging together, the individual with lower reserves determines when a pair of individuals should forage. They suggest that group coordination may be resolved through the spontaneous emergence of temporary 'leaders' and 'followers', owing to the build up of differences in energetic state. In support of this, studies of foraging in groups of fishes have shown that leadership decisions may often be made by individuals with lower reserves (Krause *et al.* 1998). The analysis, once again, implies that simple rules of thumb, which require no detailed knowledge of the state of other individuals, can lead to coordinated behaviour on the part of a group.

There is also laboratory experimental evidence that the natural shoaling tendency exhibited by many fishes can generate a simple form of conformity, which enhances the stability of group behaviour. This is illustrated by an experiment by Day *et al.* (2001) investigating how shoal size affects foraging efficiency in fishes. It is generally accepted that fishes in large shoals find food sooner than those in small shoals, largely as a consequence of having more socially transmitted information available to them. However, Day *et al.* (2001) discovered that in cases where the food patch is visually isolated from the shoal, smaller groups of fishes discover it more quickly than larger groups (Fig. 10.2). Day *et al.* interpret this apparent contradiction as stemming from the fact that fishes are generally reluctant to leave the safety of the shoal and forage on their own at a patch that is hidden. The larger the shoal, the stronger the apparent compulsion to remain with it. In small shoals, the protection conferred by the shoal is low and individuals are more likely to break away and search for food. This reason-

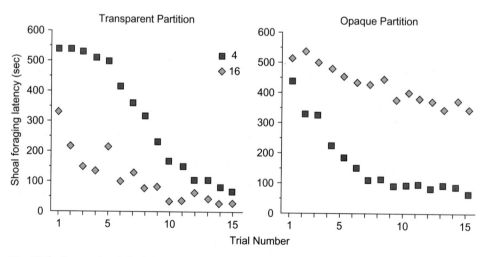

Fig. 10.2   Large shoals find the location of a foraging patch faster when the patch is located behind a transparent partition. However, the reverse is true when the patch is located behind an opaque partition. (Redrawn after Day *et al.* 2001.)

ing is the logical inverse of the observation that fishes will be more likely to join a large than a small shoal (Lachlan *et al.* 1998). Supporting this interpretation is Day *et al.*'s finding that individuals in large groups located food behind an otherwise identical transparent barrier faster than individuals in small groups (Day *et al.* 2001). We anticipate that the conformity illustrated above might operate to maintain the stability of migratory traditions in fishes, because each individual would be penalized by increased vulnerability to predation if it were to take an alternative route to the rest of the shoal, even if it were, in some respects, more efficient.

The simple shoaling and learning processes suggested by the above findings probably underlie the traditions of coral reef fishes, and perhaps the migrations of walleye (*Stizostedion vitreum*, Percidae; Olson *et al.* 1978) and surgeonfish (Mazeroll & Montgomery 1995). It is still largely unknown if fishes use social learning to aid in navigation during large-scale migrations, such as those seen in salmonids and eels (Odling-Smee & Braithwaite 2003; Tsukamoto *et al.* 2003; and see Chapter 7).

10.4 Foraging

Given the interest among behavioural ecologists in prey choice (Stephens & Krebs 1986; Warburton 2003; and see Chapter 2), there has been surprisingly little attention given to how social learning might facilitate the acquisition of dietary preferences. However, social learning is implicated in patch choice, where the use of socially transmitted information is well established in fishes (Pitcher & House 1987; Ryer & Olla 1991). 'Forage area copying' (Barnard & Sibly 1981) appears to operate through a local enhancement process. Once discovered, the foraging behaviour of the finder of a patch acts as a cue to the rest of the shoal, which quickly joins it. Groups of fishes find food faster than individuals because the probability of detecting a patch varies directly with group size, while the time spent on vigilance is inversely related to group size (Morgan & Colgan 1987). In open water, goldfish (*Carassius auratus auratus*, Cyprinidae), bluntnose minnows (*Pimephales notatus*, Cyprinidae), Alaskan pollock (*Theragra chalcogramma*, Gadidae), three-spined sticklebacks (*Gasterosteus aculeatus*, Gasterosteidae) and guppies all forage more efficiently in social groups than alone (Pitcher *et al.* 1982; Pitcher & House 1987; Morgan 1988; Ryer & Olla 1992; Peuhkuri *et al.* 1995; Day *et al.* 2001).

Increased foraging efficiency in a social context is not only restricted to shoaling species. Juvenile Atlantic salmon (*Salmo salar*, Salmonidae) take up benthic foraging stations from whence they dart to the surface to intercept prey items before returning to the riverbed. Brown & Laland (2002b) set out to determine whether the darting motion sends a message to other fishes that food is available, and whether this cue could be utilized by naive fishes to learn to forage on novel prey items. They found that, over the course of the experiment, 100% of individuals paired with previously trained demonstrators learned to accept the novel prey. Naive fishes paired with equally naive individuals actually perform worse (50% of individuals learned to feed on novel prey items) than individuals learning in isolation (73%). The presence of an inactive conspecific provided negative feedback to both individuals and thus both were reluctant to begin feeding: a finding that was labelled 'social inhibition'.

Social enhancement of foraging has also been reported in hatchery-reared juvenile chum salmon (*Oncorhynchus keta*, Salmonidae), Alaskan pollock and rock bass (*Ambloplites rupestris*, Centrarchidae) (Templeton 1987; Ryer & Olla 1991, 1992). Sundstrom & Johnsson (2001) found an increase in foraging performance when in visual contact with another feeding conspecific in wild but not hatchery-reared brown trout (*Salmo trutta*, Salmonidae), suggesting that the conditions under which hatchery fishes are raised may diminish their ability to exploit social cues.

Finally, there is experimental evidence that fishes can learn novel foraging behaviour through observation of conspecifics. Anthouard (1987) reports that juvenile European sea bass (*Dicentrarchus labrax*, Moronidae) learned to press a lever to receive a food reward through observation of proficient, trained demonstrators. These self-feeders are now commonly employed in aquaculture facilities (Shima *et al.* 2003).

10.5   Mate choice

The role of learning in mate choice decisions is discussed in detail elsewhere (see Chapter 5). This chapter provides only an overview of the evidence of social learning in mate choice. There is increasing evidence that social factors play a role in mate choice in many fishes (Westneat *et al.* 2000), including mollies (*Poecilia latipinna*, Poeciliidae; Witte & Ryan 2002); guppies (Dugatkin 1992); gobies (*Pomatoschistus microps*, Gobiidae; Reynolds & Jones 1999); and the Japanese rice fish (*Oryzias latipes*, Adrianichthyidae; Grant & Green 1996). 'Mate-choice copying' is said to have occurred when the probability of an individual selecting another as a sexual partner increases because other individuals (of the same sex) have selected the same partner (Gibson & Hoglund 1992). In the paradigm experiment (e.g. Dugatkin 1992), two males are secured at the ends of an aquarium, one with a demonstrator female nearby. The observer, another female, placed centrally, watches the other female interact with one of the males. When, after the demonstrator has been removed, the observer is allowed to choose between the two males, she consistently chooses the male that had the female nearby. In mollies, similar observations consistent with mate-choice copying have been reported for males (Schlupp & Ryan 1997). There is also evidence suggesting that mate-choice copying occurs in the wild (Witte & Ryan 2002).

One interpretation of these finding is that the observing female utilizes the presence of the female near a male as an indication of his quality, and biases her choice of male accordingly (Dugatkin & Godin 1992). It is assumed that, like other forms of social learning, mate-choice copying reduces the time individuals spend sampling potential mates before coming to a decision on which mate is of the highest quality. However, there are a number of other possible interpretations of the data (e.g. shoalmate choice), and researchers have struggled to replicate some prominent findings (Brooks 1996, 1999; Lafleur *et al.* 1997). If fishes choose mates on the basis of major histocompatability complex compatibility (Milinski 2003), choosing mates based on the choice of others seems to make little sense, because optimal matches of major histocompatability complex genotypes between partners will vary from individual to individual (Milinski *et al.* 2005).

Mate-choice copying is by no means the only way that socially influenced mate choice may occur; for example, in a number of species, females prefer males with a greater number of eggs present in their nest: stickleback (Goldsmidt *et al.* 1993); bullhead (*Cottus gobio*, Cottidae; Bisazza & Marconato 1988); fantail darter (*Etheostoma flabellare*, Percidae; Knapp & Sergeant 1989); fathead minnow (*Pimephales promelas*, Cyprinidae; Unger & Sergent 1988); blenny (*Aidablennius sphinx*, Blenniidae; Kraak & Weissing 1996); sand goby (*Pomatoschistus minutus*, Gobiidae; Forsgren *et al.* 1996); and others. Although the observing female may not have witnessed the laying of these eggs, their presence may act as a cue suggestive of mating success, or prior female choice. Once again, there are other explanations. In the case of the sand goby, females seem to use egg number as an estimate of a male's ability to defend their nest rather than prior female choice (Lindstrom & Kangas 1996). In sticklebacks, males steal eggs from neighbouring nests in an attempt to bolster their attractiveness (Largiader *et al.* 2001). However, females might select nests with large numbers of eggs already present for reasons that have nothing to do with mate choice *per se*; for instance, predator risk dilution (Jamieson 1995). While these other processes can generate mating patterns similar to mate-choice copying, they do not necessarily constitute social learning. Isolating the exact mechanism responsible for such behavioural patterns remains a challenge for future investigation. For further discussion see Lafleur *et al.* (1997), Westneat *et al.* (2000), and also Chapter 5.

10.6 Aggression

Male Siamese fighting fish (*Betta splendens*, Osphronemidae), monitor aggressive interactions between neighbouring conspecifics and use the information on relative fighting ability in subsequent aggressive interactions with the males they have observed (Oliveira *et al.* 1998). Similar observations have been made in rainbow trout (*Oncorhynchus mykiss*, Salmonidae; Johnsson & Akerman 1999). This exploitation of communicated signals in a network has become known as 'eavesdropping' (McGregor 1993). Essentially, information from a transmitter that is directed at a particular conspecific, the receiver, can be 'overheard' by peripheral individuals or 'bystanders'. The bystander then uses that information in later interactions with either the signaller or receiver. Oliveira *et al.*'s findings suggest that the level of aggression that eavesdroppers observe in interactions between a pair of demonstrators strongly affects their subsequent agonistic interactions. Similarly, green swordtails (*Xiphophorus helleri*, Poeciliidae) were less likely to initiate fights, escalate fights or win against the winners of the contests they had observed (Earley & Dugatkin 2002). Eavesdropping provides a method whereby individuals can gain information about the social status of others without having to expend energy or risk injury in social contests. Theoretical modelling suggests that the advantages of such a system are most apparent when the potential costs of combat (death or severe injury) are high (Johnstone 2001). For further discussion of the role of prior experience in agonistic interactions see Chapter 6.

Early studies assumed that the signaller and receiver were unaware of the eavesdropper (hence the terminology), indeed most investigations in the signal–receiver literature tend to focus on the dyad alone, ignoring the social context. Further investigation in more realistic social surroundings, however, revealed that the signal can be deliberately altered to take the presence of bystanders into account. Male Siamese fighting fishes alter their threat displays depending on the audience, because females also eavesdrop on male–male displays. In the presence of female observers, males reduce the number of aggressive components (e.g. bites) in their display and tailor it more towards a sexual display (Doutrelant *et al.* 2001). Thus the focus of research has swiftly switched from dyads to the examination of communication networks.

Much of the work on eavesdropping has tended to concentrate on agnostic or competitive displays; however, Johnstone & Bshary (2004) highlight the fact that the theory is equally applicable to a wide range of social contexts where individuals are likely to keep track of the 'social image' of conspecifics or heterospecifics. It could, for instance, be equally applicable to altruistic or cooperative encounters like those seen in cleaner–client relationships (Bshary 2002). Despite the obvious correspondence in subject matter, unfortunately eavesdropping has been comparatively neglected within the social learning literature.

10.7   Trade-offs in reliance on social and asocial sources of information among fishes

The belief that social learning is restricted to, or of particularly importance to, large-brained species of vertebrates is widespread, but is inconsistent with the data presented here. Many researchers have suggested that social learning abilities may be more strongly associated with ecology than taxonomy (Klopfer 1959; Lefebvre & Palameta 1988), and we endorse this argument. Another common assumption is that social learning will be particularly prevalent in highly social animals. This assumption would anticipate that social learning is likely to be restricted to shoaling fishes. However, this again is not the case; for instance, Atlantic salmon parr are known for their highly aggressive nature, yet are capable of social learning (Brown & Laland 2002b).

We suggest that consideration of the costs and benefits associated with reliance on social and asocial sources of information is more likely to explain variation in reliance on social learning within and between species than crude inferences based on assumptions of cleverness or sociality. If individuals use social information when personal information is costly, unreliable or likely to be out of date, then there may be differing propensities for social learning in populations for which survival demands vary along these dimensions. Populations at greater risk of predation when they collect personal information will be more likely to use social information than others less at risk.

A good example is provided by Coolen *et al.*'s study (2003) of public information use in sticklebacks. Both three-spined sticklebacks and nine-spined sticklebacks (*Pungitius pungitius*, Gasterosteidae) can use cues provided by others (public information) to locate foraging patches, but only nine-spined sticklebacks could assess patch quality (Coolen *et al.* 2003; Fig. 10.3). This difference in the information these two closely related species obtained by observing conspecifics may depend on the costs of relying on asocial learning to obtain information about the environment.

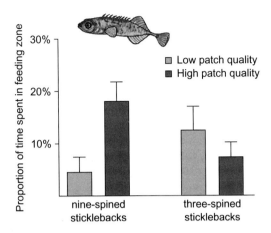

Fig. 10.3 Nine-spined sticklebacks were able to assess the quality of a foraging patch by observing conspecifics, whereas three-spined sticklebacks could not. (Redrawn after Coolen *et al.* 2003.)

Nine-spined sticklebacks lack body armour and have far smaller spines than does the three-spined species and they are therefore more vulnerable to predation while sampling food patches. It is likely that, because of their increased vulnerability, nine-spined sticklebacks rely more heavily on observing the behaviour of others from the safety of cover before deciding in which patch to forage. While this assumption is yet to be empirically tested, it does suggest that the propensity to rely on public information varies considerably between species, and perhaps even within species, depending on the relative costs and benefits associated with an individual's motivational state.

Van Bergen *et al.* (2004) revealed that nine-spined sticklebacks switch between reliance on public and private information (information gained by personally sampling the environment) depending on its costs. When private information was reliable and recently acquired, sticklebacks ignored public information and based their foraging decision on information gained from their own personal experiences. However, when private information was less reliable or outdated, sticklebacks tended to switch to social learning. Seemingly these fishes prefer to base their decisions on information recently gained from personal experience, but reliance on this information decays over time as it becomes outdated. Sticklebacks switched to reliance on social learning if their own information was greater than 7 days old. These studies are consistent with the hypothesis that the relative costs and benefits of reliance on social and asocial sources of information will explain a substantial proportion of the variance in social learning across fishes.

10.8 Conclusions

In summary, there is now unequivocal evidence, from laboratory and field studies, that a variety of different species of fishes are capable of social learning, including learning how to find food and which foods to eat; recognizing predators; and assessing mate and rival quality. When viewed in the context of the burgeoning literature on fish cognition (Bshary *et al.* 2002; and see Chapter 12), it is quite apparent that the abilities and complexity of social behaviour of this group have previously been seriously underestimated.

We end by noting that the widespread use of social learning by fishes may have important implications for conservation and fisheries reintroductions (Suboski & Templeton 1989; Brown & Laland 2001; Brown & Day 2002). Typically, over 95% of all fishes released from hatcheries die from predation or starvation in the first few weeks following release (Brown & Laland 2001): an enormous waste of resources. It is conceivable that hatchery-reared fishes could be trained *en masse* to recognize predators and prey using social learning protocols. The evidence presented above suggests that it may be possible to cut post-release mortality figures dramatically by allowing hatchery fishes to learn from more experienced or wild conspecifics. The manners in which the cognitive abilities of fishes can be exploited are discussed in greater detail in Chapter 14).

10.9   Acknowledgements

We are grateful to the Biotechnology and Biological Sciences Research Council and Royal Society for financial support and Stephen Reebs for his helpful comments on the manuscript.

10.10   References

Anthouard, M. (1987) A study of social transmission in juvenile *Dicentrarchus labrax* (Pisces, Serranidae), in an operant-conditioning situation. *Behaviour*, **103**, 266–275.

Barnard, C. & Sibly, R. (1981) Producers and scroungers: a general model and its application to feeding flocks of house sparrows. *Animal Behaviour*, **29**, 543–550.

van Bergen, Y., Coolen, I. & Laland, K.N. (2004) Nine-spined sticklebacks exploit the most reliable source when public and private information conflict. *Proceedings of the Royal Society of London, Series B – Biological Sciences*, **271**, 957–962.

Bisazza, A. & Marconato, A. (1988) Female mate choice, male–male competition and parental care in the river bullhead, *Cotus gobio*. *Animal Behaviour*, **36**, 1352–1360.

Boyd, R. & Richerson, P.J. (1985) *Culture and the Evolutionary Process*. Chicago University Press, Chicago.

Brooks, R. (1996) Copying and the repeatability of mate choice. *Behavioural Ecology and Sociobiology*, **39**, 323–329.

Brooks, R. (1999) Mate choice copying in guppies: females avoid the place where they saw courtship. *Behaviour*, **136**, 411–421.

Brown, C. & Day, R. (2002) The future of stock enhancements: bridging the gap between hatchery practice and conservation biology. *Fish and Fisheries*, **3**, 79–94.

Brown, C. & Laland, K.N. (2001) Social learning and life skills training for hatchery reared fish. *Journal of Fish Biology*, **59**, 471–493.

Brown, C. & Laland, K.N. (2002a) Social learning of a novel avoidance task in the guppy, *P. reticulata*: conformity and social release. *Animal Behaviour*, **64**, 41–47.

Brown, C. & Laland, K.N. (2002b) Social enhancement and social inhibition of foraging behaviour in hatchery-reared Atlantic salmon. *Journal of Fish Biology*, **61**, 987–998.

Brown, G.E. (2003) Learning about danger: chemical alarm cues and local risk assessment in prey fishes. *Fish and Fisheries*, **4**, 227–234.

Bshary, R. (2002) Biting cleaner fish use altruism to deceive image-scoring client reef fish. *Proceedings of the Royal Society of London, Series B – Biological Sciences*, **269**, 2087–2093.

Bshary, R., Wickler, W. & Fricke, H. (2002) Fish cognition: a primate's eye view. *Animal Cognition*, **5**, 1–13.

Chivers, D.P. & Smith, J.F. (1995) Chemical recognition of risky habitats is culturally transmitted among fathead minnows, *Pimephales promelas* (Osteichthyes, Cyprinidae). *Ethology*, **99**, 286–296.

Coolen, I., van Bergen, Y., Day, R.L. & Laland, K.N. (2003) Species difference in adaptive use of public information in sticklebacks. *Proceedings of the Royal Society of London, Series B – Biological Sciences*, **270**, 2413–2419.

Couzin, I.D., Krause, J., Franks, N.R. & Levin, S.A. (2005) Effective leadership and decision making in animal groups on the move. *Nature*, **433**, 513–516.

Day, R., MacDonald, T., Brown, C., Laland, K. & Reader, S.M. (2001) Interactions between shoal size and conformity in guppy social foraging. *Animal Behaviour*, **62**, 917–925.

Doutrelant, C., McGregor, P.K. & Oliveira, R.F. (2001) The effect of an audience on intrasexual communication in male Siamese fighting fish, *Betta splendens*. *Behavioral Ecology*, **12**, 283–286.

Dugatkin, L.A. (1992) Sexual selection and imitation: females copy the mate choice of others. *American Naturalist*, **139**, 1384–1489.

Dugatkin, L.A. & Godin, J.-G.J. (1992) Reversal of mate choice by copying in the guppy (*Poecilia reticulata*). *Proceedings of the Royal Society of London, Series B – Biological Sciences*, **249**, 179–184.

Earley, R.L. & Dugatkin, L.A. (2002) Eavesdropping on visual cues in green swordtail (*Xiphophorus helleri*) fights: a case for networking. *Proceedings of the Royal Society of London, Series B – Biological Sciences*, **269**, 943–952.

Forsgren, E., Karlsson, A. & Kvarnemo, C. (1996) Female sand gobies gain direct benefits by choosing males with eggs in their nests. *Behavioral Ecology and Sociobiology*, **39**, 91–96.

Galef, B.G. Jr & Giraldeau, L.-A. (2001) Social influences on foraging in vertebrates: causal mechanisms and adaptive functions. *Animal Behaviour*, **61**, 3–15.

Gibson, R.M. & Hoglund, J. (1992) Copying and sexual selection. *Trends in Ecology and Evolution*, **7**, 229–232.

Goldsmidt, T., Bakker, T.C.M. & Feuth-de Bruijn, E. (1993) Selective choice in copying of female sticklebacks. *Animal Behaviour*, **45**, 541–547.

Grant, J.W. & Green, L.D. (1996) Mate copying versus preferences for actively courting males by female Japanese medaka (*Oryzias latipes*). *Behavioural Ecology*, **7**, 165–167.

Griffin, A.S. (2004) Social learning about predators: a review and prospectus. *Learning & Behavior*, **32**, 131–140.

Helfman, G.S. & Schultz, E.T. (1984) Social tradition of behavioural traditions in a coral reef fish. *Animal Behaviour*, **32**, 379–384.

Helfman, G.S., Meyer, J. L. & McFarland (1982) The ontogeny of twilight migration patterns in grunts (Pisces: Haemulidae). *Animal Behaviour*, **30**, 379–384.

Heyes, C.M. (1994) Social learning in animals: categories and mechanisms. *Biological Reviews*, **69**, 207–231.

Heyes, C.M. & Galef, B.G. (1996) *Social Learning in Animals: the Roots of Culture*. Academic Press, London.

Jamieson, I. (1995) Do female fish prefer to spawn in nests with eggs for reasons of mate choice copying or egg survival? *American Naturalist*, **145**, 824–832.

Johnsson, J.I. & Akerman, A. (1999) Watch and learn: preview of the fighting ability of opponents alters contest behaviour in rainbow trout. *Animal Behaviour*, **56**, 771–776.

Johnstone, R.A. (2001) Eavesdropping and animal conflict. *Proceedings of the National Academy of Science of the USA*, **98**, 9177–9180.

Johnstone, R.A. & Bshary, R. (2004) Evolution of spite through indirect reciprocity. *Proceedings of the Royal Society of London, Series B – Biological Sciences*, **271**, 1917–1922.

Kelley, J.L. & Magurran, A.E. (2003) Learning of predator recognition and anti-predator responses in fishes. *Fish and Fisheries*, **4**, 216–226.

Kelley, J.L., Evans, J.P., Ramnarine, I.W. & Magurran, A.E. (2003) Back to school: can antipredator behaviour in guppies be enhanced through social learning? *Animal Behaviour*, **65**, 655–662.

Kendal, R.L., Coolen, I., van Bergen, Y. & Laland, K.N. (2005) Tradeoffs in the adaptive use of social and asocial learning. *Advances in the Study of Behaviour*, **35**, 333–380.

Kieffer, J.D. & Colgan, P.W. (1992) The role of learning in fish behaviour. *Reviews in Fish Biology and Fisheries*, **2**, 125–143.

Klopfer, P.H. (1959) Social interactions in discrimination learning with special reference to feeding behaviour in birds. *Behaviour*, **14**, 282–299.

Knapp, R.A. & Sergeant, R.C. (1989) Egg mimicry as a mating strategy in the fantail darter, *Etheostoma flabellare*: females prefer males with eggs. *Behavioural Ecology and Socio-biology*, **25**, 321–326.

Kraak, S.B.M. & Weissing, F.J. (1996) Female preference for nests with many eggs: a cost–benefit analysis of female choice in fish with paternal care. *Behavioral Ecology*, **7**, 353–361.

Krause, J. (1993) Transmission of fright reaction between different species of fish. *Behaviour*, **127**, 37–48.

Krause, J., Reeves, P. & Hoare, D. (1998) Positioning behaviour in roach shoals: the role of body length and nutritional state. *Behaviour*, **135**, 1031–1039.

Lachlan, R.F., Crooks, L. & Laland, K.N. (1998) Who follows whom? Shoaling preferences and social learning of foraging information in guppies. *Animal Behaviour*, **56**, 181–190.

Lafleur, D.L., Lozano, G.A. & Sclafani, M. (1997) Female mate-choice copying in guppies, *Poecilia reticulata*: a re-evaluation. *Animal Behaviour*, **54**, 579–586.

Laland, K.N. & Williams, K. (1997) Shoaling generates social learning of foraging information in Guppies. *Animal Behaviour*, **53**, 1161–1169.

Laland, K.N. & Williams, K. (1998) Social transmission of maladaptive information in the guppy. *Behavioural Ecology*, **9**, 493–499.

Largiader, C.R., Fries, V. & Bakker, T.C.M. (2001) Genetic analysis of sneaking and egg-thievery in a natural population of the three-spined stickleback (*Gasterosteus aculeatus* L.). *Heredity*, **48**, 459–468.

Lefebvre, L. & Palameta, B. (1988) Mechanisms, ecology and population diffusion of socially-learned, food-finding behavior in feral pigeons. In: T.R. Zentall & B.G. Galef (eds), *Social Learning: Psychological and Biological Perspectives*, pp. 141–164. Lawrence Erlbaum Associates, New Jersey.

Lindstrom, K. & Kangas, N. (1996) Egg presence, egg loss, and female mate preferences in the sand goby (*Pomatoschistus minutus*). *Behavioral Ecology*, **7**, 213–217.

Magurran, A.E., & Higham, A. (1988) Information transfer across fish shoals under predator threat. *Ethology*, **78**, 153–158.

Mathis, A., Chivers, D.P. & Smith, R.J.F. (1996) Cultural transmission of predator recognition in fishes, intraspecific and interspecific learning. *Animal Behaviour*, **51**, 185–201.

Mazeroll, A.I. & Montgomery, W.L. (1995) Structure and organization of local migrations in brown surgeonfish (*Acanthurus nigrofuscus*). *Ethology*, **99**, 89–106.

McGregor, P.K. (1993) Signaling in territorial systems – a context for individual identification, ranging and eavesdropping. *Philosophical Transactions of the Royal Society of London, Series B – Biological Sciences*, **340**, 237–244.

Milinski, M. (2003) The function of mate choice in sticklebacks: optimizing MHC genetics. *Journal of Fish Biology*, **63**, 1–16.

Milinski, M., Griffiths, S., Wegner, K.M., Reusch, T.B.H., Haas-Assenbaum, A. & Boehm, T. (2005) Mate choice decisions of stickleback females predictably modified by MHC peptide ligands. *Proceedings of the National Academy of Sciences of the USA*, **102**, 4414–4418.

Mineka, S. & Cook, M. (1988) Social learning and the acquisition of snake fear in monkeys. In: T.R. Zentall & B.G. Galef (eds), *Social Learning: Psychological and Biological Perspectives*, pp. 51–74. Lawrence Erlbaum Associates, New Jersey.

Morgan, M.J. (1988) The influence of hunger, shoal size and predator presence on foraging in bluntnose minnows. *Animal Behaviour*, **36**, 1317–1322.

Morgan, M.J. & Colgan, P.W. (1987) The effects of predator presence and shoal size on bluntnose minnows, *Pimephales notatus*. *Environmental Biology of Fishes*, **20**, 105–111.

Odling-Smee, L. & Braithwaite, V.A. (2003) The role of learning in fish orientation. *Fish and Fisheries*, **4**, 235–246.

Oliveira, R.F., McGregor, P.K. & Latruffe, C. (1998) Know thine enemy: fighting fish gather information from observing conspecific interactions. *Proceedings of the Royal Society of London, Series B – Biological Sciences*, **265**, 1045–1049.

Olson, D.E., Schupp, D.H. & Macins, V. (1978) A hypothesis of homing behaviour of walleyes as related to observed patterns of passive and active movement. *American Fisheries Society Special Publication*, **11**, 52–57.

Peuhkuri, N., Ranta, E., Juvonen, S.K. & Lindstrom, K. (1995) Schooling affects growth in the 3-spined stickleback, *Gasterosteus aculeatus*. *Journal of Fish Biology*, **46**, 221–226.

Pfeiffer, W. (1974) Pheromones in fish and amphibia. In: M.C. Birch (ed.), *Pheromones*, pp. 269–296. North-Holland, Amsterdam.

Pitcher, T.J. & House, A. (1987) Foraging rules for group feeders: area copying depends upon density in shoaling goldfish. *Ethology*, **76**, 161–167.

Pitcher, T.J., Magurran, A.E. & Winfield, I.J. (1982) Fish in larger shoals find food faster. *Behavioral Ecology and Sociobiology*, **10**, 149–151.

Pitcher, T.J., Green, D.A. & Magurran, A.E. (1986) Dicing with death: predator inspection behaviour in minnow shoals. *Journal of Fish Biology*, **28**, 439–448.

Potts, W.K. (1984) The chorus-line hypothesis of manoeuvre coordination in avian flocks. *Nature*, **309**, 344–345.

Rands, S.A., Cowlishaw, G., Pettifor, R.A., Rowcliffe, J.M. & Johnstone, R.A. (2003) Spontaneous emergence of leaders and followers in foraging pairs. *Nature*, **423**, 432–434.

Rands, S.A., Pettifor, R.A., Rowcliffe, J.M. & Cowlishaw, G. (2004) State-dependent foraging rules for social animals in selfish herds. *Proceedings of the Royal Society of London, Series B – Biological Sciences*, **271**, 2613–2620.

Reader, S.M., Kendal, J.R. & Laland, K.N. (2003) Social learning of foraging sites and escape routes in wild Trinidadian guppies. *Animal Behaviour*, **66**, 729–739.

Reebs, S.G. (2000) Can a minority of informed leaders determine the foraging movements of a fish shoal? *Animal Behaviour*, **59**, 403–409.

Reebs, S.G. (2001) Influence of body size on leadership in shoals of golden shiners, *Notemigonus crysoleucas*. *Behaviour*, **138**, 797–809.

Reynolds, J.D. & Jones, J.C. (1999) Female preference for preferred males is reversed under low oxygen conditions in the common goby (*Pomatoschistus microps*). *Behavioural Ecology*, **10**, 149–154.

Ryer, C.H. & Olla, B.L. (1991) Information transfer and the facilitation and inhibition of feeding in a schooling fish. *Environmental Biology of Fishes*, **30**, 317–323.

Ryer, C.H. & Olla, B.L. (1992) Social mechanisms facilitating exploitation of spatially variable ephemeral food patches in a pelagic marine fish. *Animal Behaviour*, **44**, 69–74.

Schlupp, I. & Ryan, M.J. (1997) Male sailfin mollies (*Poecilia latipinna*) copy the mate choice of other males. *Behavioural Ecology*, **8**, 104–107.

Shettleworth, S.J. (2001) Animal cognition and animal behaviour. *Animal Behaviour*, **61**, 277–286.

Shima, T., Yamamoto, T., Furuita, H. & Sumiki, N. (2003) Effect of the response interval of self-feeders on the self-regulation of feed demand by rainbow trout (*Oncorhynchus mykiss*) fry. *Aquaculture*, **224**, 181–191.

Stephens, D.W. & Krebs, J.R. (1986) *Foraging Theory*. Princeton University Press. Princeton.

Suboski, M.D. & Templeton, J.J. (1989) Life skills training for hatchery fish: Social learning and survival. *Fisheries Research*, **7**, 343–352.

Suboski, M.D., Bain, S., Carty, A.E., McQuoid, L.M., Seelen, M.I. & Seifert, M. (1990) Alarm reaction in acquisition and social transmission of simulated predator recognition by zebra danio fish (*Brachydanio rerio*). *Journal of Comparative Psychology*, **104**, 101–112.

Sugita, Y. (1980) Imitative choice behaviour in guppies. *Japanese Psychological Research*, **22**, 7–12.

Sundstrom, L.F. & Johnsson, J.I. (2001) Experience and social environment influence the ability of young brown trout to forage on live novel prey. *Animal Behaviour*, **61**, 249–255.

Templeton, J. (1987) Individual differences in the behaviour of juvenile rock bass (*Ambloplites rupestris*): causes and consequences. Masters Thesis, Queens University.

Tsukamoto, K., Aoyama, J. & Miller, M.J. (2003) Migration, speciation, and the evolution of diadromy in anguillid eels. *Canadian Journal of Fisheries and Aquatic Science*, **59**, 1989–1998.

Unger, L.M. & Sergent, R.C. (1988) Alloparental care in the fathead minnow, *Pimephales promelas*: females prefer males with eggs. *Behavioural Ecology and Sociobiology*, **23**, 27–32.

Verheijen, F.J. (1956) Transmission of a flight reaction among a school of fish and the underlying sensory mechanisms. *Experientia*, **12**, 202–204.

Vilhunen, S., Hirvonen, H. & Laakkonen, M.V.M. (2005) Less is more: social learning of predator recognition requires a low demonstrator to observer ratio in Arctic charr (*Salvelinus alpinus*). *Behavioral Ecology and Sociobiology*, **57**, 275–282.

Vilhunen, S. (2006) Repeated anti-predator conditioning: a pathway to habituation or to better avoidance? *Journal of Fish Biology*, **68**, 25–43.

Von Fisch, K. (1941) U" ber einen Shreckstoff der Fischhaut und seine biologishe Bedeutung. *Z. vergl. Physiol.*, **29**, 46–145.

Warburton, K. (2003) Learning of foraging skills by fish. *Fish and Fisheries*, **4**, 203–215.

Warner, R.R. (1988) Traditionality of mating-site preferences in a coral reef fish. *Nature*, **335**, 719–721.

Warner, R.R. (1990) Male versus female influences on mating-site determination in a coral-reef fish. *Animal Behaviour*, **39**, 540–548.

Webb, P.W. (1980) Does schooling reduce fast-start response latencies in teleosts? *Comparative Biochemistry and Physiology A: Comparative Physiology*, **65**, 231–234.

Westneat, D.F., Walters, A., McCarthy, T.M., Hatch, M.I. & Hein, W.K. (2000) Alternative mechanisms of nonindependent mate choice. *Animal Behaviour*, **59**, 467–476.

Whiten, A. & Ham, R. (1992) On the nature and evolution of imitation in the animal kingdom: reappraisal of a century of research. *Advances in the Study of Behaviour*, **21**, 239–283.

Witte, D.J. & Ryan, M.J. (2002) Mate-choice copying in the sailfin molly, *Poecilia latipinna*, in the wild. *Animal Behaviour*, **63**, 943–949.

Chapter 11
Cooperation and Cognition in Fishes

Michael S. Alfieri and Lee Alan Dugatkin

11.1   Introduction

'The theory of evolution is based on the struggle for life and the survival of the fittest. Yet cooperation is common between members of the same species and even between members of different species' (Axelrod & Hamilton 1981, p. 1390). In this simple, but powerful quote, Robert Axelrod (a professor of political science) and William Hamilton (a professor of evolutionary biology) illustrate a principal difficulty in understanding cooperation in light of evolutionary theory. Why should any organism help another at risk to itself if there is no apparent benefit in doing so? This question has perplexed evolutionary biologists since the inception of the field. Indeed, Charles Darwin initially struggled to explain how sterile insects, individuals that sacrifice reproduction to contribute to the production of the hive or colony without gaining obvious benefits from their cooperative behaviour, could fit into his theory of natural selection (Darwin 1859). A key question regarding cooperative behaviour and natural selection faced Darwin: namely, how could natural selection favour cooperation if the cooperation does not increase the fitness of those that express this trait? Darwin posited one possible solution when he noted, 'this difficulty, though appearing insuperable, is lessened, or, as I believe, disappears, when it is remembered that selection may be applied to the family, as well as to the individual, and may thus gain the desired end' (Darwin 1859, p. 237). Here Darwin described one of the main paths by which cooperation can arise and be maintained in a population, later described and formalized as kin selection or inclusive fitness by W.D. Hamilton (1964a,b). This chapter will briefly describe kin selection, plus three other categories of cooperation, and will suggest the necessary cognitive prerequisites for cooperation to occur, and provide empirical examples that illustrate each category of cooperation in fishes.

There are four commonly recognized categories invoked to explain the origin and maintenance of cooperation (Table 11.1): namely, kin selection (Hamilton 1964a,b); reciprocity (Trivers 1971); byproduct mutualism (West Eberhard 1975; Brown 1983); and trait-group selection (Wilson 1980) (categories reviewed in Dugatkin *et al.* 1992; Dugatkin 1997). These categories are not mutually exclusive and more than one category may accurately describe cooperative behaviour within or among species. This is evident by our example of predator inspection that appears twice, in two different categories of cooperation in this chapter. Because cooperation is often envisioned as a cognitively complex trait, the cognitive abilities that are needed for cooperation to persist in populations have recently been examined (Dugatkin 1997; Gadagkar 1997; Bshary 2002a; Dugatkin & Alfieri 2002; Sachs *et al.* 2004), and at least

Table 11.1 Categories and cognition prerequisites of selected cooperative behaviour in fishes.

Category of cooperation	Definition of cooperative category	Possible cognitive prerequisites	Representative examples and species	References of examples
Kin selection Hamilton (1964a,b)	An act of cooperation is directed towards kin with costs to the cooperator	Categorical recognition of kin	Cooperative breeding in *Lamprologus brichardi*	Taborsky (1984, 1985)
			Territorial defence in Atlantic salmon and rainbow trout	Brown & Brown (1993)
Reciprocity Trivers (1971)	A costly act of cooperation is repaid to the cooperator by the recipient at a future time	Memory of past events and individual recognition	Egg trading in black hamlet fish	Fischer (1980, 1988)
			Predator inspection in three-spined sticklebacks and guppies	Milinski (1987) and Dugatkin (1988)
			Cleaning between cleaner wrasse and their client groupers	Trivers (1971), Tebbich *et al.* (2002)
By-product mutualism West Eberhard (1975) Brown (1983)	Individuals act together to achieve a beneficial outcome that could not have been achieved as efficiently by any single individual	Categorical recognition of environment (e.g. harsh vs mild)	Cooperative foraging in blue tang surgeonfish and wrasse	Foster (1985, 1987)
Trait-group selection D.S. Wilson (1975, 1980)	Within-group costs to cooperator are less than between-group benefits	Categorical recognition of others (e.g. cooperator vs non-cooperator)	Predator inspection in three-spined sticklebacks and guppies	Dugatkin & Mesterton-Gibbons (1996)

some of the controversy surrounding the study of cooperative behaviour has focused on these abilities (Dugatkin 1997; Clutton-Brock 2002; Stevens & Hauser 2004). Disagreement about the cognitive abilities of fishes in particular may have begun in some of the earlier works on animal social behaviour. In his pioneering book, *Sociobiology: The New Synthesis*, E.O. Wilson addresses the evolutionary mechanisms behind several social behaviours in the animal kingdom, including cooperation (E.O. Wilson 1975). Wilson suggests that the 'lack of intelligence' (among other possibilities) in lower vertebrates, including fishes, is a reason why he did not see specific cooperative behaviour in this group. In the 30 years since the publication of this book, the evidence is now quite clear that fishes possess impressive cognitive abilities worthy of study in the field of cooperative behaviour (Dugatkin 1997; Bshary 2002a; Dugatkin & Alfieri 2002; Brown & Laland 2003; Griffiths 2003). However, as is evident from the ongoing debates (Riolo *et al.* 2001; Hammerstein 2002; Stevens & Hauser 2004; Trivers 2004; Pfeiffer *et al.* 2005), more directed empirical studies focusing on the specific cognitive prerequisites necessary for cooperation are needed.

11.2   Why study cooperation in fishes?

The study of cooperative behaviour can benefit greatly from research focusing on fishes. Fishes display enormous diversity in morphology, behaviour, physiology, and life histories, and can be found in almost all aquatic environments (Godin 1997). Among vertebrates, teleost fishes are the most abundant and diverse group, representing over 50% of all known vertebrate species (Diana 2004). Half of all fish species spend at least part of their lives in groups (Shaw 1978), allowing for at least the potential for cooperative interactions to evolve under many diverse conditions (although social groups are not a necessary prerequisite for cooperative behaviour, as seen in the example between cleaner and client fish below). Examples of cooperation in the animal kingdom are widespread and many of these examples include fishes (Dugatkin 1997). Recently, fishes have been examined as a model system for the study of cognition (Bshary 2002a; Laland & Hoppitt 2003), and, additionally, as models for the study of cooperation and the role of cognition in cooperative behaviour (Dugatkin & Mesterton-Gibbons 1996; Dugatkin 1997; Tebbich *et al.* 2002). However, even with our ever improving understanding of fish systems, the need for rigorous, directed, empirical studies examining the cognitive abilities necessary for cooperation in fishes is still great.

This chapter presents selected examples of cooperation from the fish literature that highlight the categories of cooperation and the potential cognitive abilities needed for each (Table 11.1). Each example may be the result of a single or multiple categories of cooperation; however, this framework is used to illustrate the diverse cognitive abilities and varied methods by which cooperation can be found in fishes.

11.3   Cooperation and its categories

The phrase 'to cooperate' has many meanings. It may, for example, imply 'to achieve cooperation', a result that is realized at the group level. Alternatively, to cooperate may refer to the actions of the individual. That is, to behave in a manner that makes cooperation possible, although cooperation may not be realized unless other members of a group also act cooperatively; for example, imagine two individuals trapped in a cave where the exit is blocked by a large boulder. We may speak of cooperation as the outcome of the two individuals working together to move the boulder, or may define cooperation by the act of helping to move the boulder (something an individual does). In our review of cooperation, we refer to the latter, the behavioural act that allows for cooperation. That is, cooperation results when two or more individuals behave in a coordinated manner and the outcome is that participants gain some type of benefit (Dugatkin 1997). We define 'benefit' as a positive contribution towards one's fitness (see Wilson & Dugatkin 1992 for a discussion of relevant definitions). The benefit may be direct (for example, gaining a meal, a valuable piece of information, or protection from predators), or indirect (for example, improving the survival of kin or increasing the possibility of future benefits). Recently, authors have defined cooperation in a similar manner (Sachs *et al.* 2004; Stevens & Hauser 2004).

11.3.1 *Category 1 – Kin selection*

Theoretically, kin selection is perhaps the most intuitive category of cooperation. To see why, we need to take a 'gene's eye' perspective on cooperation. By definition, blood kin share many genes that are identical by descent; that is, they are derived from some common ancestor. A gene that codes for helping one's blood kin is increasing the probability that copies of itself that reside in blood kin are transmitted to the next generation (Hamilton 1964a,b). 'Hamilton's rule' posits that cooperation will be selected for in a population when $rb - c > 0$, where b is the benefits to the recipient of a cooperator's action, r is the coefficient of relatedness (i.e. the proportion of genes shared between two individuals from a common ancestor), and c is the cost to the cooperator of cooperating. Phrased in the cold language of natural selection, relatives are worth helping in direct proportion to their genetic (blood) relatedness.

11.3.1.1 *Cognition and kin selection*

When an individual's blood kin are scattered throughout a given environment, kin recognition allows the benefits of cooperation to be differentially allocated to such blood kin. Such recognition may be based on some behavioural attribute or on so-called 'recognition genes', which may allow identification of kin based on odour, morphological marks, etc. A large body of research has demonstrated that fish are able to recognize others (including kin) as well as alter behaviour based on this recognition (reviewed by Krause *et al.* 2000; Griffiths 2003; Ward & Hart 2003). In their recent approach to studying cooperation, Sachs *et al.* (2004) refer to kin-selected cooperation based on recognition of kin, learned or heritable, as 'kin choice'. Individuals need not, however, be able to recognize others as kin for cooperation via kin selection to occur. When an individual is always (or almost always) surrounded by close relatives, then strategies to treat any encounter as an encounter with kin may evolve. These cases of kin-selected cooperation have been referred to as 'kin fidelity' (Sachs *et al.* 2004). In such cases, cooperation via kin selection is possible without the cognitive abilities to recognize an individual as kin. This 'rule of thumb' (i.e. treat others as kin) is, however, vulnerable to invasion, as non-kin may receive the cooperative benefits reserved for kin.

11.3.1.2   Example of kin-selected cooperation: cooperative breeding

In cooperative breeding societies, groups are composed of breeding individuals and helpers that either delay their own breeding or completely forego reproduction to assist breeding pairs (Brown 1987). Although the literature on cooperative breeding is dominated by examples involving birds, insects and mammals, a few species of fish have been studied in light of cooperative breeding. Michael Taborsky's research (Taborsky 1984, 1985) on the cooperative breeding cichlid (*Lamprologus brichardi*, Cichlidae) provides a case example of kin-selected cooperation in fishes. In *L. brichardi*, sexually mature offspring stay at the nest and help maintain and defend eggs despite the fact that they pay a cost of staying and helping. This cost may be realized through reduced growth rates or lost breeding opportunities. After several other possible hypotheses were tested as to why helpers stay and help at the nest (including the potential benefits of gaining experiences for future success in raising their own young, safety in the natal territory, possibility of taking over the parental territory, improved diet as a result of cannibalism), kin-selected benefits were determined to be the best explanation for this behaviour in *L. brichardi* (reviewed in Dugatkin 1997). Recently, Griffin & West (2003) used meta-analysis to determine the relative importance of kin selection among 18 cooperatively breeding vertebrates (birds and mammals). They found a significant pattern of kin recognition and preferential treatment of closely related kin among helpers. Although their study did not include fish, it highlights the importance of kin recognition and kin selection in other cooperatively breeding vertebrates (for alternative explanations see Clutton-Brock 2002).

Recent research on cooperative breeding in fishes has focused on the cichlid (*Neolamprologus pulcher*, Cichlidae; Werner *et al.* 2003; Bergmuller & Taborsky 2005; Bergmuller *et al.* 2005; Brouwer *et al.* 2005). In this species smaller (younger) helper fish are more closely related to both the breeding pair and their brood than are larger (older) helpers (Stiver *et al.* 2004; Bergmuller & Taborsky 2005; Bergmuller *et al.* 2005; Brouwer *et al.* 2005). As such, kin selection predicts that cooperation should be more common in smaller helpers than in larger helpers. Brouwer *et al.* (2005) present compelling experimental evidence supporting kin-selected cooperation in smaller fishes (see Bergmuller & Taborsky 2005; Brouwer *et al.* 2005 for explanations of cooperation among larger fish).

11.3.1.3   Example of kin-selected cooperation: conditional territory defence

Another cooperative interaction in fishes that is probably based on kin selection is conditional territory defence behaviour. A territory should be defended if the benefits of owning an area (e.g. food, shelter) are greater than the costs of defending it (e.g. energy expenditure, potential injury; Brown 1964; and see Grant 1997 for costs and benefits of territoriality in fishes). Brown & Brown (1993) found that both Atlantic salmon (*Salmo salar*, Salmonidae), and rainbow trout (*Oncorhynchus mykiss*, Salmonidae), two species in which kin recognition has been documented (Brown & Brown 1992), displayed reduced aggressive behaviour, increased tolerance, and reduction in territory size when the holder of the neighbouring territory was kin compared to non-kin. In this example of kin-selected cooperation, both indirect fitness benefits (the improved survival of a close relative) and direct fitness benefits (reduction in energy and potential harm occurred in aggressive defence) are achieved. Similar kin-selection benefits, including territory sharing and latency to acquire food patches, were found in Atlantic salmon in a study by Griffiths & Armstrong (2002).

11.3.2   *Category 2 – Reciprocity*

A second category of cooperation is reciprocity, also called direct reciprocity, directed reciprocation or reciprocal altruism (Dugatkin 1997; Sachs *et al.* 2004). In his 1971 work, Robert Trivers describes a model by which cooperative behaviour can persist in a population both in the absence of, or in conjunction with, kin selection. During reciprocity, an act of cooperation is repaid to the cooperator by the recipient at a future time. Cooperation via reciprocity is vulnerable to 'cheating'; that is, after receiving the beneficial act from a cooperator, a recipient receives a higher payoff from simply not returning the favour. Trivers addresses the 'cheater problem' by discussing cooperation in light of a game theory model, the Prisoner's Dilemma (Fig. 11.1; Luce & Raiffa 1957; Rapoport & Chammah 1965, cited in Trivers 1971). During the game, two individuals are faced with the choice to either cooperate or defect (i.e. not cooperate). During a single encounter, each player receives a greater payoff if they defect. To see why, consider the payoff matrix (Fig. 11.1). If player 2 cooperates, player 1 receives the greatest payoff if it defects ('T' the temptation to cheat) than if it also cooperates ('R' the reward for mutual cooperation). By defecting, player 1 receives the benefit of a cooperative act towards itself but does not pay any of the costs associated with cooperating. Alternatively, if player 2 defects and

| | Player 2 | |
	Cooperate	Defect
Cooperate	R = 3	S = 0
Defect	T = 5	P = 1

(Player 1 labels the rows)

Fig. 11.1   The Prisoner's Dilemma game. Each cell represents the payoff to player 1 given its interaction with player 2. The game is constructed so that the Temptation to cheat (T) > Reward for mutual cooperation (R) > the Punishment for mutual defection (P) > Sucker's payoff for cooperating when your opponent defects (S).

player 1 cooperates, player 1 pays all the costs of cooperation and receives none of the benefits. In this scenario player 1 receives the lowest possible payoff ('S' the sucker's payoff). Thus, if player 2 defects player 1 should also defect and both players receive ('P') the punishment for mutual defection. While playing the Prisoner's Dilemma game, then, it appears that the strategy to defect is the best strategy for each player – so where is the 'dilemma'? The dilemma exists in the fact that if both players cooperate they each receive a reward for mutual cooperation ('R') that is greater than the punishment for mutual defection ('P'). To achieve cooperation, Trivers (1971) suggests that the game is not played only once but iterated games must be considered between the same two individuals so that each has an opportunity to respond to the others to 'mimic real life'.

In 1981, Axelrod and Hamilton used the iterated Prisoner's Dilemma (iPD) to examine numerous strategies in a computer tournament in which they invited experts to submit a set of behavioural rules that would interact with other sets of rules to determine how cooperation can arise and be maintained in a population. In two separate tournaments, the strategy that outcompeted all others and allowed for cooperation to thrive in a population (if the probability of meeting a given partner was above a critical threshold) was Tit-for-Tat, submitted by Anatol Rapoport. Tit-for-Tat (TFT) is a simple set of rules that directs a player to cooperate on one's first move, and subsequently copy an opponent's last move (Axelrod & Hamilton 1981; Axelrod 1984). The iPD game continues to be used as a focal model to study reciprocity (Axelrod 1984; Dugatkin 1997; Dugatkin & Reeve 1998).

11.3.2.1   Cognition and reciprocity

In the previous section we described reciprocity and how the Prisoner's Dilemma has been used to examine the conflict between cooperation and the temptation to defect (i.e. not cooperate). To play TFT (which includes copying ones partner's last move) an individual must be able to recognize who they are paired with, as well as how that partner acted during their last encounter. As such, the cognitive requirements associated with playing TFT are the ability to recognize partners and remember the outcome of previous encounters. These, however, are not strict pre-requisites as individual recognition would not be required when individuals interact with only one partner for a longer time (Axelrod & Hamilton 1981). The consistency between partners may result from spatial constraints during interactions (e.g. individuals are not very mobile), or a lack of alternative partners in a population. Recent authors have further stressed the importance of memory and learning when individuals are playing the iPD game; for example, Milinski & Wedekind (1998) showed that in humans constraints on memory can affect strategies used when playing the iPD game. Additionally, Gutnisky & Zanutto (2004) present a model highlighting scenarios during which operant learning can benefit individuals while playing the iPD game.

The study of cooperation via reciprocity has recently been explored in the wild. These works suggest that in some natural populations of freshwater fish species, the conditions necessary for cooperation via reciprocity are found (Ward *et al.* 2002; Croft *et al.* 2004; Croft *et al.* 2005; Croft *et al.* 2006); for example, Ward *et al.* (2002) and Croft *et al.* (2004, 2005, 2006) have demonstrated that sticklebacks

(*Gasterosteus aculeatus*) and guppies (*Poecilia reticulata*, Poeciliidae), respectively, form stable social affiliations, an important condition for cooperation via reciprocity (Dugatkin 1997).

11.3.2.2   Example of reciprocity: egg trading

Teleost fishes represent the only known vertebrates that are capable of simult-aneous hermaphroditism; that is, possession by a single individual of both eggs and sperm at the same time. While this is most prevalent in deep-sea fishes (Smith 1975), it also occurs in the Serranidae, a shallow-water family of fishes that includes the subfamily Serraninae, the sea basses (Fischer & Petersen 1987; Fischer 1988). Reciprocity via egg-trading has been well studied in black hamlet fish (*Hypoplectrus nigricans*, Serranidae; Fischer 1980, 1981, 1987), zebra goby (*Lythrypnus zebra*, Gobiidae; St. Mary 1996), belted sandfish (*Serranus subligarius*, Serranidae; Oliver 1997; Cheek 1998), tobacco fish, *Serranus tabacarius*, Serranidae; Petersen 1995), and chalk bass (*Serranus tortugarum*, Serranidae; Fischer 1984; Petersen & Fischer 1996).

Fischer (1980) described simultaneous hermaphroditism in the black hamlet fish. During the last 2 hours of the day before sunset, fish come together in pairs at the reef edge or slope to spawn, usually away from their foraging territories. After several courtship displays, one fish initiates the spawn by releasing eggs that are externally fertilized by its partner. Eggs are much more expensive to produce than sperm, and so 'cheaters' could benefit by limiting their role in spawning to that of sperm donor. All available eggs are not released, but instead are parcelled out over four to five releases on average during a spawning period. The parcelling of eggs allows fish to alternate their role as either male or female (i.e. egg trading), with one individual providing eggs to be fertilized in exchange for eggs from a partner to fertilize. Fischer (1988) refers to this exchange as 'delayed reciprocity'. But why should not a partner defect (not cooperate) – that is, not switch roles – in this system? Kin selection is not a likely explanation, as these fishes are obligate outbreeders and eggs are planktonic (Fischer 1988). In the black hamlet, cooperation appears to be maintained because these fish are playing the iPD game, in which a cooperative act is parcelling eggs, defection is not providing eggs to a partner, and iterations are the exchange of parcels of eggs during the spawning period. Fischer (1988) argues that the costs and benefits associated with cooperation and defection during egg trading are consistent with the payoff matrix in the iPD game. Additionally, he presents evid-ence that there is a temptation to cheat and this temptation is met with retaliation. Fischer (1980) found that fish wait significantly longer to provide eggs to a partner that has failed to reciprocate, compared to a partner that provided eggs on the pre-vious move. This evidence suggests that the black hamlet are using a strategy similar to TFT in the iPD game; a strategy that is more forgiving than TFT called 'Generous Tit-for-Tat' (Nowak & Sigmund 1992).

Although this evidence is compelling, alternative explanations may exist. One alternative to the TFT strategy has been suggested by Connor (1992), namely pseudoreciprocity (Connor 1986). In pseudoreciprocity (also called byproduct mutualism), unlike in reciprocity, there is no incentive to cheat as the benefits of mutual cooperation are greater than the temptation to cheat and the cognitive requirements of pseudoreciprocity do not include individual recognition or memory.

Connor (1992) argues that fish engaging in egg trading are benefiting themselves and only as an incidental outcome also benefiting others. This remains to be tested.

11.3.2.3   Example of reciprocity: predator inspection

Small fish have been observed swimming away from the relative safety of their shoal and moving towards a potentially dangerous predator. This behaviour has been called predator inspection (Pitcher *et al.* 1986) and approaching (Dugatkin & Godin 1992a) and has been observed in many fish species, including guppies, sticklebacks, European minnows (*Phoxinus phoxinus*, Cyprinidae), paradise fish (*Macropodus opercularis*, Anabantidae), damselfish (*Stegastes planifrons*, Pomacentridae), blue-gill sunfish (*Lepomis macrochirus*, Centrarchidae) and mosquitofish (*Gambusia affinis*, Poeciliidae) (see Dugatkin & Godin 1992a; Pitcher 1992; Smith 1997 for reviews on predator inspection in fishes). The phenomenon of prey inspecting a potential predator is not limited to fish and has been observed in other taxa, including mammals (Walther 1969, Cheney & Seyfarth 1990; Fitzgibbon 1994) and birds (Altmann 1956; Curio 1978; Olendorf *et al.* 2004). The advantages of predator inspection include signalling to the predator that it has been seen and that an attack would be unsuccessful; gathering information about the potential threat that may be transmitted back to the shoal; and advertising individual quality to conspecifics (see Dugatkin & Godin 1992b for a review of costs and benefits of predator inspection). If multiple individuals inspect together, it can be considered a form of cooperation. In fact, predator inspection in fishes has become one of the most popular experimental systems to study reciprocity (see Dugatkin 1997 and Stevens & Hauser 2004 for controversies surrounding predator inspection and cooperation).

Debates not withstanding, it appears that predator inspection behaviour fits the assumed payoffs from the Prisoner's Dilemma, namely, $T > R > P > S$. Specifically, a fish benefits the most if it allows its partner to move closer to a potential predator to either signal to the predator it has lost the element of surprise and/or to gain information about the threat. As such, the temptation to defect (T), that is staying out of harm's way while another individual inspects, is greater than the payoff for both fishes if they inspect together and gain the reward for mutual cooperation (R). However, the fishes receive a greater reward if they both inspect than if they both remain in the shoal and do not approach the predator, thereby suffering the punishment for mutual defection (P). Finally, it would be most dangerous (i.e. least reward) for a fish to approach a predator alone, receiving the sucker's payoff (S). Early works by Milinski (1987) and Dugatkin (1988) on sticklebacks and guppies, respectively, have provided a strong foundation to suggest that these fish are not only cooperating via reciprocity, but are also probably using the TFT strategy, or a similar strategy, during predator inspections. Milinski (1987) and Dugatkin (1988) used a series of experiments during which cooperation or defection was simulated by placing a mirror parallel to (cooperation) or at an angle away from (defection) an individual inspecting a predator. In the case of simulated cooperation, the mirror image of the inspecting fish stayed next to the subject during its movement toward the predator. However, the mirror placed at an angle made the image appear to swim away from the subject as it moved toward the predator, so it mimicked an act of defection. Both Milinski (1987) and Dugatkin (1988) found that in the simulated cooperation trials, fish approached a predator more closely when an inspector perceived its inspecting partner (its mirror image) cooperating as defecting from the inspection. This suggests that fish are copying a partner's last move as predicted by the TFT strategy. In addition to this, Milinski (1987) and Dugatkin (1988) tested specific predictions of the TFT strategy; for example, they found that fish retaliated against defectors (fish moved away from the predator when their partner appeared to defect), and were forgiving (continued to inspect once their partner reappeared close to the subject) (see Dugatkin 1997 for a review).

As stated earlier, the cognitive abilities required for cooperation via reciprocity and the iPD include recognition of partners and remembering their previous move. These prerequisites were tested in both sticklebacks (Milinski *et al.* 1990a,b) and guppies (Dugatkin & Alfieri 1991a,b). These works present evidence that sticklebacks and guppies both show preferences for partners that were more likely to cooperate with them during previous inspection bouts, suggesting that individuals are able to discriminate between past partners as well as remember their previous encounters with them. Recently, the work of Dugatkin & Alfieri (1991a) has been repeated but with a current focus on the preference of an observer guppy that was restricted to only watching, but not interact with potential cooperators and defectors (Brosnan *et al.* 2003). This work did not find any significant preferences between observer guppies and either perceived cooperators or defectors and suggests that the repeated act of inspection between co-inspectors, as opposed to merely watching a conspecific inspect, is necessary for cooperation to grow between inspectors.

There has been much debate in the literature over the role of reciprocity during predation inspection in fishes. Although the scope of this discussion is too extensive to address fully in this chapter, aspects of the debate are evident from the exchange between Connor (1996), who argues that predator inspection behaviour can be explained via byproduct mutualism, and Dugatkin (1996) and Milinski (1996), who responded to this claim (for more on this topic see Dugatkin 1997).

11.3.2.4   *Example of reciprocity: interspecific cleaning behaviour*

Trivers (1971) provided an early example of cooperation via reciprocity by describing the interactions between a cleaner fish, such as the cleaner wrasse (*Labroides dimidiatus*, Labridae) and their clients, such as the grouper (*Epinephelus striatus*, Serranidae). During this cooperative interaction, 'cleaner' fish repeatedly interact with 'client' fish, during which time cleaners feed on the parasites and unhealthy tissue of clients. Clients often swim to a specific location, a cleaning station, where a specific cleaner fish will probably swim into the gill chambers and mouth of a client or host fish to feed on ectoparasites (Trivers 1971). Cleaning stations are often found in the same place, with the same cleaners and specific clients repeatedly returning to these locations, so that there are repeated interactions between specific clients and cleaners. There is also a substantial cost (e.g. lost time, increased predation) to both cleaners and clients for repeatedly establishing new pairs. For this to be an example of cooperative behaviour via reciprocity, cleaner–client pairs must repeatedly interact and therefore must either have some form of individual recognition or have only very limited opportunity to interact with different partners at a cleaning stations (Axelrod & Hamilton 1981). Trivers presents compelling evidence that these assumptions are met in at least some cleaner–client systems and that this is a primary example of reciprocal altruism (Trivers 1971), although his example has been met with much debate; for example, Gorlick *et al.* (1978) directly question Trivers' example. They describe behaviour of several *Labroides* spp. that do not meet the assumptions set forth by Trivers, including evidence that cleaners will feed on the healthy tissue, mucus, and fins of clients in addition to ectoparasites; clients may not feed on cleaners not because of a cooperative interaction but rather because cleaners often avoid the mouth of some piscivorous clients as well as avoid clients in search of food; and some cleaners are distasteful to clients and are avoided as a food item (Gorlick *et al.* 1978 and references within). Further examination of some of these issues including biting by and diet preferences of cleaner fish have been examined in studies of cooperative behaviour (Bshary 2002b; Bshary & Grutter 2002; Grutter & Bshary 2004).

Recently, the study of cleaning fishes has focused on specific aspects of cooperative behaviour; for example, Tebbich *et al.* (2002) studied individual recognition and preference for familiar versus unfamiliar fishes in the cleaner wrasse and its client, the striated surgeonfish (*Ctenochaetus striatus*, Acanthuridae). They provide evidence that the cleaner fish spend significantly more time with familiar versus unfamiliar clients, suggesting an important role of individual recognition in cleaning behaviour. Tebbich *et al.* (2002) did not, however, find a preference for either familiar or unfamiliar cleaners in the client fish. They suggest that this may be a result of experimental artefacts including low statistical power and/or low motivation on the part of the client resulting from insufficient time to allow client and cleaners to establish a 'significant relationship'. Additionally, they suggest that there is not a need for recognition of individuals in this system but rather a need for recognition of a site where clients return and thereby facilitate repeated interactions with the same cleaners (Tebbich *et al.* 2002). The roles that the development of cleaner–client relationships (Bshary 2002c) and the nature of tactile stimulations (Bshary & Wurth 2001) play on cooperative interactions in this system continue to be studied.

11.3.3   *Category 3 – Byproduct mutualism*

Byproduct mutualism is a form of cooperation during which two or more individuals act together to achieve an outcome that could not have been achieved as efficiently (or at all) by any single individual (West Eberhard 1975; Brown 1983; Connor 1995) and there is no temptation for either individual not to cooperate. Using the definition we have described here, byproduct mutualism has also been called no-cost cooperation (Dugatkin 1997), pseudoreciprocity (Connor 1986), selfish cooperation (Stevens & Hauser 2004), and two-way byproducts (Sachs *et al.* 2004). In this category of cooperation, cheaters (i.e. non-cooperators) receive a lower pay-off than cooperators. Although this makes byproduct mutualism conceptually very easy to understand, it has been argued that it should not be considered cooperation at all because the temptation to cheat and the cost to act cooperatively are absent. However, given coordinated actions are needed between individuals to achieve an outcome of a greater reward than any one individual could obtain, we feel byproduct mutualism does indeed fulfil our definition of cooperation. Cooperation via byproduct mutualism occurs when environmental situations dictate that acting together yields greater rewards than acting alone. In this model, the environment is categorized as either 'harsh', in which case the best strategy is to cooperate, or 'mild' in which case the best strategy is not to cooperate (Mesterton-Gibbons & Dugatkin 1992).

11.3.3.1   Cognition and byproduct mutualism

Cognitive requirements are often less demanding in byproduct mutualism compared to other categories of cooperation because memory and recognition of individuals are not needed. The cognitive requirement of byproduct mutualism is a categorical recognition of one's environment; if you are in a 'harsh' environment, then co-operate, if in a 'mild' environment, do not cooperate (Dugatkin 1997; Dugatkin & Alfieri 2002). However, in situations where populations have been subjected to a particular type of environmental condition (harsh or mild) for long periods of time, the ability to recognize the environmental type may not be necessary. As a result of the lesser cognitive requirements associated with byproduct mutualism (as com-pared to reciprocity), it has been suggested that byproduct mutualism is a more common form of cooperation in the animal kingdom. Stevens & Hauser (2004) state that 'selfish cooperation' (i.e. byproduct mutualism) is in fact common in animal societies mostly because the cognitive requirements are severe enough to act as a barrier to the evolution of most types of reciprocity.

11.3.3.2   Example of by-product mutualism: cooperative foraging

The feeding habits of adult blue tang surgeonfish (*Acanthurus coeruleus*, Acanthuridae) present an example of cooperation via byproduct mutualism. Blue tang surgeonfish feed on algae and form feeding schools (Foster 1985). Often, highly desirable algal mats are defended by territorial dusky damselfish (*Stegastes dorso-punicans*, Pomacentridae). A solitary blue tang cannot overcome the defence of a damselfish; however, feeding schools can overcome a territorial damselfish and feed on the algal resource (Foster 1985). In this example, the 'harsh' environment that stimulates cooperation via byproduct mutualism in the surgeonfish is the territorial defence of the damselfish.

Dugatkin (1997) and Dugatkin & Mesterton-Gibbons (1996) describe similar examples of cooperative foraging through byproduct mutualism, including work by Foster (1987) on the wrasse (*Thalasomma lucasanum*, Labridae). In this example, the sergeant major damselfish (*Abudefduf troschelli*, Pomacentridae) defends its embryos successfully from small groups of wrasses (fewer than 30 individuals); how-ever, it cannot prevent the predation of the embryos by larger groups (hundreds of individuals). Here again, the harsh environment is defined by territorial defence and only through cooperative actions can a reward not obtainable by one (or a few) be obtained by the group. Interestingly, large groups of wrasses were only seen when damselfish were nesting with embryos, suggesting that group size formation in the wrasse is a direct response to achieving a cooperative reward. Similar examples of byproduct mutualism cooperation resulting from the harsh environment of a territ-ory holder are cited in Dugatkin (1997).

Bshary (2002a) presents interesting anecdotal evidence of complex cooperative hunting behaviour between giant moray eels (*Gymnothorax javanicus*, Muraenidae) and two different groupers, the red sea coral groupers (*Plectropomus pessuliferus*, Serranidae) and lunartail groupers (*Variola louti*, Serranidae). Groupers have been observed following eels and octopuses while they hunt in an attempt to capture the prey that the eels and octopuses flush out of hiding places in corals (Diamant & Shpigel 1985). Bshary (2002a) describes his observations of the two grouper species mentioned above approaching a resting moray eel and shaking their bodies within close proximity of the eel. In half of these close encounters, the eel and grouper would then swim off together in close proximity to hunt for prey. The eel would swim into a coral while the grouper waited above the coral. In one instance, Bshary (2002a) witnessed a grouper wait outside a coral head for an escaped prey item to emerge, swim away from the coral, then return with a grey moray eel (*Siderea grisea*, Muraenidae). Although he did not witness capture of prey by any eels or groupers, he suggests that groupers are recruiting eels to hunt cooperatively and that groupers and eels assume different roles during a hunt (Bshary 2002a). Interactions between eels and groupers are interesting examples of complex, byproduct mutualism cooperative foraging behaviour, assuming the harsh environment is the difficulty in capturing a prey item.

11.3.4 *Category 4 – Trait group selection*

Until the 1960s, the term 'group selection' was associated with Wynne-Edwards' 'for the good of the species' argument (Wynne-Edwards 1962). Wynne-Edwards argued that only groups with cooperators that benefited their species (i.e. self-sacrificing individuals that would control population size to safeguard against overexploitation) would be selected, as the benefits to the group outweighed the costs to cooperators. In general, this view of selection has not been supported by theoretical or empirical works, and has never gained favour within the scientific community, as evident from criticisms by Williams (1966). Williams argued that individuals that possessed the trait for self-sacrifice (e.g. restricting reproduction for the good of the group) would be selected against, as a result of natural selection favouring individuals that were not sacrificing producing relatively more offspring than sacrificing individuals, and passing on this trait to a larger percentage of subsequent generations. Over evolutionary time, self-sacrificing individuals would be outcompeted within populations (Alcock 2001). However, more complex models of group selection have been proposed. Modern or 'trait group selection' (D.S. Wilson 1975) models describe cooperative behaviour in populations by examining fitness based on the productivity of local groups or 'trait groups'. Here, the effects of cooperative acts are examined, both at the level of the individuals within the trait group (where a cooperator pays a cost that non-cooperators do not), and at the level of the trait groups within the global population or deme (Dugatkin & Mesterton-Gibbons 1996). Cooperation is possible (even if there is a cost to the cooperator) if the within-group costs are less than between-group benefits, so that groups with cooperators are more productive than groups without cooperators (Sober & Wilson 1998). (For more on this see Wilson 1990; Mesterton-Gibbons & Dugatkin 1992; Wilson & Sober 1994; Dugatkin 1997; Sober & Wilson 1998.)

11.3.4.1   Cognition and trait-group selection

Although it is possible to have cooperation via trait-group selection without individual recognition or memory of events (Wilson 1980), trait group cooperation would be favoured if individuals recognized and associated with other cooperators (Peck 1993; Wilson & Dugatkin 1997; Roberts & Sherratt 1998). Such recognition would allow for the formation of trait groups with many cooperators, and such groups should have greater productivity than groups with proportionally more cheaters. This would require individuals to recognize others as belonging to a general category of cooperators or cheaters but could also include individual recognition (Dugatkin & Alfieri 2002).

11.3.4.2   Example of trait-group selected cooperation:
predator inspection

Predator inspection (see description above) involves the movement of a few individuals (inspectors) away from a larger group towards a potential predator. For trait group cooperation to be the mechanism of cooperation during predator inspection one needs to document costs to individuals who cooperate when they leave the group to inspect, document benefits accrued by groups from having inspectors, and show that groups with inspectors should have an advantage over groups without inspectors. Dugatkin & Mesterton-Gibbons (1996) review several empirical studies in the guppy that provide evidence for each of these requirements. First, there are costs associated with inspection. Inspectors have been shown to be at greater risk of predation (Dugatkin 1992) and inspectors obtain less food than non-inspectors (Dugatkin & Godin 1992a). Second, groups benefit from the action of inspectors. The information obtained by inspectors is transmitted back to the group and the entire group benefits from this information (Magurran & Higgam 1988). Finally, groups with inspectors may have some fitness advantage over groups without inspectors. Although they did not test this directly, the results from Dugatkin & Godin (1992a) suggest that it is indeed the case that groups with inspectors are attacked less often by predators than groups without inspectors. While more work needs to be undertaken in this system, the within-group costs of inspection may be less than the between-group benefits, suggesting predator inspection as an example of trait-group selection in fishes. At equilibrium, selection would balance the within-group cost to cheating (i.e. cheaters get information about predators, but pay no cost) against the between-group benefit of having many inspectors (e.g. groups are very vigilant against predators).

   A higher proportion of inspectors to cheaters in a group may result in greater productivity, that is, reduced attacks, for the group (Dugatkin & Godin 1992a). Two fish species that have been extensively studied with regard to predator inspection are the guppy, and the stickleback. Empirical studies on both species have shown that guppies and sticklebacks are able to recognize familiar individuals in general (Magurran *et al.* 1994; Barber & Ruxton 2000, respectively), as well as in the specific context of preference for a familiar cooperative conspecific (Dugatkin & Alfieri 1991a,b in guppies and Milinski *et al.* 1990a,b in sticklebacks). In guppies, however, preferential assortment seems to be limited to individuals in small groups (Dugatkin

& Alfieri 1991a), and is absent when group size becomes too large (Dugatkin & Wilson, 2000). Additionally, several studies of guppies and sticklebacks have shown that kin selection is probably not a selective force on cooperative behaviour in these species because of low relatedness in sampled populations (Griffiths & Magurran 1999; Russell *et al.* 2004 in guppies and FitzGerald & Morisette 1992 in sticklebacks; see also Ward & Hart 2003, and Chapter 8 for a general review of familiarity and kin recognition).

The work on predation inspection demonstrates nicely that different categories of cooperation are not mutually exclusive. As we have now seen, cooperation during predator inspection may include elements of both reciprocity and group-selected behaviour.

11.4   Conclusions

We have selected specific examples of empirical studies in fishes that have addressed each of the four categories of cooperation and their cognitive requirements (Table 11.1). Through further investigation, each example may help refine models of cooperation or perhaps inspire the creation of new models. We feel that this is a productive way best to continue our understanding of the relationships between cooperation, cognition and fish behaviour. We hope that this work will stimulate research that directly addresses the cognitive requirements in all forms of cooperative behaviour.

11.5   Acknowledgements

We would like to thank Jennifer Sadowski, Glena Temple and Darren Croft for helpful and thoughtful comments on this chapter.

11.6   References

Alcock, J. (2001) *Animal Behavior an Evolutionary Approach*, 7th edn. Sinauer Associates Inc., Massachusetts.

Altmann, S.A. (1956) Avian mobbing behavior and predator recognition. *Condor*, **58**, 241–253.

Axelrod, R. (1984) *The Evolution of Cooperation*. Basic Books Inc., New York.

Axelrod, R. & Hamilton, W.D. (1981) The evolution of cooperation. *Science*, **211**, 1390–1396.

Barber, I. & Ruxton, G.D. (2000) The importance of stable schooling: do familiar sticklebacks stick together? *Proceedings of the Royal Society of London, Series B*, **267**, 151–155.

Bergmuller, R. & Taborsky, M. (2005) Experimental manipulation of helping in a cooperative breeder: helpers 'pay to stay' by pre-emptive appeasement. *Animal Behaviour*, **69**, 19–28.

Bergmuller, R., Heg, D. & Taborsky, M. (2005) Helpers in a cooperatively breeding cichlid stay and pay or disperse and breed, depending on ecological constraints. *Proceedings of the Royal Society of London, Series B*, **272**, 325–331.

Brosnan, S.F., Earley, R.L. & Dugatkin, L.A. (2003) Observational learning and predator inspection in guppies (*Poecilia reticulata*). *Ethology*, **109**, 823–833.

Brouwer, L., Heg, D. & Taborsky, M. (2005) Experimental evidence for helper effects in a cooperatively breeding cichlid. *Behavioral Ecology*, **16**, 667–673.

Brown, C. & Laland, K.N. (2003) Social learning in fishes: a review. *Fish and Fisheries*, **4**, 280–288.

Brown, G.E. & Brown, J.A. (1992) Do rainbow trout and Atlantic salmon discriminate kin? *Canadian Journal of Zoology*, **70**, 1636–1640.

Brown, G.E. & Brown, J.A. (1993) Social dynamics in salmonid fishes: do kin make better neighbours? *Animal Behaviour*, **45**, 863–871.

Brown, J.L. (1964) The evolution of diversity in avian territorial systems. *Wilson Bulletin*, **76**, 160–169.

Brown, J.L. (1983) Cooperation – a biologist's dilemma. In: J.S. Rosenblatt (ed.), *Advances in the Study of Behaviour*, pp. 1–37. Academic Press, New York.

Brown, J.L. (1987) *Helping and Communal Breeding in Birds*. Princeton University Press, Princeton, New Jersey.

Bshary, R. (2002a) Fish cognition: a primate's eye view. *Animal Cognition*, **5**, 1–13.

Bshary, R. (2002b) Biting cleaner fish use altruism to deceive image-scoring client reef fish. *Proceedings of the Royal Society of London, Series B*, **269**, 2087–2093.

Bshary, R. (2002c) Building up relationships in asymmetric co-operation games between the cleaner wrasse *Labroides dimidiatus* and client reef fish. *Behavioral Ecology and Sociobiology*, **52**, 365–371.

Bshary, R. & Grutter, A.S. (2002) Asymmetric cheating opportunities and partner control in a cleaner fish mutualism. *Animal Behaviour*, **63**, 547–555.

Bshary, R. & Wurth, M. (2001) Cleaner fish *Labroides dimidiatus* manipulate client reef fish by providing tactile stimulation. *Proceedings of the Royal Society of London, Series B*, **268**, 1495–1501.

Cheek, A.O. (1998) Ovulation does not constrain egg parcel size in the simultaneous hermaphrodite *Serranus subligarius*. *Environmental Biology of Fishes*, **52**, 435–442.

Cheney, D.L. & Seyfarth, R.M. (1990) *How Monkeys See the World*. University of Chicago Press, Chicago.

Clutton-Brock, T. (2002) Breeding together: kin selection and mutualism in cooperative vertebrates. *Science*, **296**, 69–72.

Connor, R.C. (1986) Pseudo-reciprocity: investing in mutualism. *Animal Behaviour*, **34**, 1562–1584.

Connor, R.C. (1992) Egg-trading in simultaneous hermaphrodites: an alternative to Tit-for-Tat. *Journal of Evolutionary Biology*, **5**, 523–528.

Connor, R.C. (1995) The benefits of mutualism: a conceptual framework. *Biological Reviews*, **70**, 247–257.

Connor, R.C. (1996) Partner preferences in by-product mutualisms and the case for predator inspection in fish. *Animal Behaviour*, **51**, 451–454.

Croft, D.P., Krause, J. & James, R. (2004) Social networks in the guppy (*Poecilia reticulata*). *Proceedings of the Royal Society of London Letters*, **271**, 516–519.

Croft, D.P., James, R., Ward, A.J.W., Botham, M.S., Mawdsley, D. & Krause, J. (2005) Assortative interactions and social networks in fish. *Oecologia*, **143**, 211–219.

Croft, D.P., James, R., Thomas, P.O.R., Hathaway, C., Mawdsley, D., Laland, K.N. & Krause, J. (2006) Social structure and co-operative interactions in a wild population of guppies (*Poecilia reticulata*). *Behavioural Ecology and Sociobiology*, **59**, 644–650.

Curio, E. (1978) The adaptive significance of avian mobbing. I. Teleonomic hypotheses and predictions. *Zeitschrift für Tierpsychologie*, **48**, 175–183.

Darwin, C. (1859) *On the Origin of Species*. J. Murray, London.

Diamant, A. & Shpigel, M. (1985) Interspecific feeding associations of groupers (Teleostei: Serranidae) with octopuses and moral eels in the gulf of Eilat (Aqaba). *Environmental Biology of Fishes*, **13**, 153–159.

Diana, J.S. (2004) *Biology and Ecology of Fishes*, 2nd edn. Cooper Publishing Group, LLC, USA.

Dugatkin, L.A. (1988) Do guppies play Tit for Tat during predator inspection visits? *Behavioral Ecology and Sociobiology*, **25**, 395–399.

Dugatkin, L.A. (1992) Tendency to inspect predators predicts mortality risk in the guppy, *Poecilia reticulata*. *Behavioral Ecology*, **3**, 124–128.

Dugatkin, L.A. (1996) Tit for Tat, by-product mutualism and predator inspection: a reply to Connor. *Animal Behaviour*, **51**, 455–457.

Dugatkin, L.A. (1997) *Cooperation Among Animals: An Evolutionary Perspective*. Oxford University Press, New York.

Dugatkin, L.A. & Alfieri, M. (1991a) Guppies and the Tit for Tat strategy: preference based on past interaction. *Behavioral Ecology and Sociobiology*, **28**, 243–246.

Dugatkin, L.A. & Alfieri, M. (1991b) Tit for Tat in guppies: the relative nature of cooperation and defection during predator inspection. *Evolutionary Ecology*, **5**, 300–309.

Dugatkin, L.A. & Alfieri, M.S. (2002) A cognitive approach to the study of animal cooperation. In: M. Bekoff & C. Allen (eds), *The Cognitive Animal*, pp. 413–419. M.I.T. Press, Cambridge, Massachusetts.

Dugatkin, L.A. & Godin, J.-G.J. (1992a) Predator inspection, shoaling and foraging under predation hazard in the Trinidadian guppy, *Poecilia reticulata*. *Environmental Biology of Fishes*, **34**, 265–276.

Dugatkin, L.A. & Godin, J.-G.J. (1992b) Prey approaching predators: a cost-benefit perspective. *Annales Zoologici Fennici*, **29**, 233–252.

Dugatkin, L.A. & Mesterton-Gibbons, M. (1996) Cooperation among unrelated individuals: reciprocal altruism, byproduct mutualism, and group selection in fishes. *Biosystems*, **37**, 19–30.

Dugatkin, L.A. & Reeve, H.K. (eds) (1998) *Game Theory and Animal Behavior*. Oxford University Press, Oxford.

Dugatkin, L.A. & Wilson D.S. (2000) Assortative interactions and the evolution of cooperation during predator inspection in guppies (*Poecilia reticulata*). *Evolutionary Ecology Research*, **2**, 761–767.

Dugatkin, L.A., Mesterton-Gibbons, M. & Houston, A.I. (1992) Beyond the prisoner's dilemma: towards models to discriminate among mechanisms of cooperation in nature. *Trends in Ecology and Evolution*, **7**, 202–205.

Fischer, E.A. (1980) The relationship between mating system and simultaneous hermaphroditism in the coral reef fish, *Hypoplectrus nigricans* (Serranidae). *Animal Behaviour*, **28**, 620–633.

Fischer, E.A. (1981) Sexual allocation in a simultaneously hermaphroditic coral reef fish. *American Naturalist*, **117**, 64–82.

Fischer, E.A. (1984) Egg trading in the chalk bass, *Serranus tortugarum*, a simultaneous hermaphrodite. *Zeitschrift fuer Tierpsychologie*, **66**, 143–151.

Fischer, E.A. (1987) Mating behavior in the black hamlet-gamete trading or egg trading? *Environmental Biology of Fishes*, **18**, 143–148.

Fischer, E.A. (1988) Simultaneous hermaphroditism, tit-for-tat and the evolutionary stability of social systems. *Ethology and Sociobiology*, **9**, 119–136.

Fischer, E.A. & Petersen, C.W. (1987) The evolution of sexual patterns in the seabasses. *BioScience*, **37**, 482–489.

FitzGerald, G.J. & Morisette, J. (1992) Kin recognition and choice of shoal mates by threespine sticklebacks. *Ethology Ecology and Evolution*, **4**, 273–283.

Fitzgibbon, C.D. (1994) The costs and benefits of predator inspection behavior in Thomson gazelles. *Behavioral Ecology and Sociobiology*, **34**, 139–148.

Foster, S.A. (1985) Group foraging by a coral reef fish: mechanism for gaining access to defended resources. *Animal Behaviour*, **33**, 782–792.

Foster, S.A. (1987) Acquisition of a defended resource: a benefit of group foraging for the neotropical wrasse, *Thalassoma lucasanum*. *Environmental Biology of Fishes*, **19**, 215–222.

Gadagkar, R. (1997) *Survival Strategies: Cooperation and Conflict in Animal Societies*. Harvard University Press, Cambridge, Massachusetts.

Godin, J.-G.J. (1997) Behavioural ecology of fishes: adaptations for survival and reproduction. In: J.-G.J. Godin (ed.), *Behavioural Ecology of Teleost Fishes*, pp. 1–9. Oxford University Press, New York.

Gorlick, D.L., Atkins, P.D. & Losey, G.S. Jr. (1978) Cleaning stations as water holes, garbage dumps, and sites for the evolution of reciprocal altruism? *The American Naturalist*, **112**, 341–353.

Grant, J.W.A. (1997) Territoriality. In: J.-G.J. Godin (ed.), *Behavioural Ecology of Teleost Fishes*, pp. 81–103. Oxford University Press, New York.

Griffin, A.S. & West, S.A. (2003) Kin discrimination and the benefit of helping in co-operatively breeding vertebrates. *Science*, **302**, 634–636.

Griffiths, S.W. (2003) Learned recognition of conspecifics by fishes. *Fish and Fisheries*, **4**, 256–268.

Griffiths, S.W. & Armstrong, J.D. (2002) Kin-biased territory overlap and food sharing among Atlantic salmon juveniles. *Journal of Animal Ecology*, **71**, 480–486.

Griffiths, S.W. & Magurran, A.E. (1999) Schooling decisions in guppies (*Poecilia reticulata*) are based on familiarity rather than kin recognition by phenotype matching. *Behavioral Ecology and Sociobiology*, **45**, 437–443.

Grutter, A.S. & Bshary, R. (2004) Cleaner fish, *Labroides dimidiatus*, diet preference for different types of mucus and parasitic gnathiid isopods. *Animal Behaviour*, **68**, 583–588.

Gutnisky, D.A. & Zanutto, B.S. (2004) Cooperation in the iterated Prisoner's Dilemma is learned by operant conditioning mechanisms. *Artificial Life*, **10**, 433–461.

Hamilton, W.D. (1964a) The genetical evolution of social behaviour I. *Journal of Theoretical Biology*, **7**, 1–16.

Hamilton, W.D. (1964b) The genetical evolution of social behaviour II. *Journal of Theoretical Biology*, **7**, 17–52.

Hammerstein, P. (2002) Why is reciprocity so rare in social animals? A Protestant appeal. In: P. Hammerstein (ed.), *Genetic and Cultural Evolution of Cooperation*, pp. 83–93. Dahlem University Press, Berlin.

Krause, J., Butlin, R., Peuhkuri, N. & Pritchard, V. (2000) The social organization of fish shoals: a test of the predictive power of laboratory experiments for the field. *Biological Reviews*, **75**, 477–501.

Laland, K.N. & Hoppitt, W. (2003) Do animal's have culture? *Evolutionary Anthropology*, **3**, 150–159.

Luce, R.D. & Raiffa, H. (1957) *Games and Decisions*. Wiley, New York.

Magurran, A.E. & Higgam (1988) Information transfer across fish shoals under predator threat. *Ethology*, **78**, 153–158.

Magurran, A.E., Seghers, B., Shaw, P. & Carvalho, G. (1994) Schooling preferences for familiar fish in the guppy, *Poecilia reticulata*. *Journal of Fish Biology*, **45**, 401–406.

Mesterton-Gibbons, M. & Dugatkin, L.A. (1992) Cooperation among unrelated individuals: evolutionary factors. *Quarterly Review of Biology*, **67**, 267–281.

Milinski, M. (1987) Tit for Tat and the evolution of cooperation in sticklebacks. *Nature*, **325**, 433–435.

Milinski, M. (1996) By-product mutualism, Tit for Tat reciprocity and cooperative predator inspection: a reply to Connor. *Animal Behaviour*, **51**, 458–461.

Milinski, M. & Wedekind, C. (1998) Working memory constrains human cooperation in the Prisoner's Dilemma. *Proceedings of the National Academy of Science of the USA*, **95**, 13755–13758.

Milinski, M., Kulling, D. & Kettler, R. (1990a) Tit for Tat: sticklebacks 'trusting' a cooperating partner. *Behavioral Ecology*, **1**, 7–12.

Milinski, M., Pfugler, D., Kulling, D. & Kettler, R. (1990b) Do sticklebacks cooperate repeatedly in reciprocal pairs? *Behavioral Ecology and Sociobiology*, **27**, 17–23.

Nowak, M.A. & Sigmund, K. (1992) Tit for tat in heterogeneous populations. *Nature*, **355**, 250–252.

Olendorf, R., Getty, T. & Scribner, K. (2004) Cooperative nest defense in red-winged blackbirds: reciprocal altruism, kinship or by-product mutualism? *Proceedings of the Royal Society of London, Series B*, **271**, 177–182.

Oliver, A.S. (1997) Size and density dependent mating strategies in the simultaneously hermaphroditic seabass *Serranus subligarius* (Cope, 1870). *Behaviour*, **134**, 563–594.

Peck, J.R. (1993) Friendship and the evolution of cooperation. *Journal of Theoretical Biology*, **162**, 195–228.

Petersen, C.W. (1995) Reproductive behavior, egg trading and correlates of male mating success in the simultaneous hermaphrodite *Serranus tabacarius*. *Environmental Biology of Fishes*, **43**, 351–361.

Petersen, C.W. & Fischer E.A. (1996) Intraspecific variation in sex allocation in a simultaneous hermaphrodite: the effect of individual size. *Evolution*, **50**, 636–645.

Pfeiffer, T., Rutte, C., Killingback, T., Taborsky, M. & Bonhoeffer, S. (2005) Evolution of cooperation by generalized reciprocity. *Proceedings of the Royal Society of London, Series B*, **272**, 1115–1120.

Pitcher, T.J. (1992) Who dares wins: the function and evolution of predator inspection behaviour in shoaling fish. *Netherlands Journal of Zoology*, **42**, 371–391.

Pitcher, T.J., Green, D.A. & Magurran, A.E. (1986) Dicing with death: predator inspection behaviour in minnow shoals. *Journal of Fish Biology*, **28**, 438–448.

Rapoport, A. & Chammah, A. (1965) *Prisoner's Dilemma*. University of Michigan Press, Ann Arbor, Michigan.

Riolo, R.L., Cohen, M.D. & Axelrod, R. (2001) Evolution of cooperation without reciprocity. *Nature*, **414**, 441–443.

Roberts, G. & Sherratt, T. (1998) Development of cooperative relationships through increasing investment. *Nature*, **394**, 175–179.

Russell, S.T., Kelley, J.L., Graves, A. & Magurran, A.E. (2004) Kin selection and shoal composition dynamics in the guppy, *Poecilia reticulata*. *Oikos*, **106**, 520–526.

Sachs, J.L., Mueller, U.G., Wilcox, T.P. & Bull, J.J. (2004) The evolution of cooperation. *Quarterly Review of Biology*, **79**, 135–160.

Shaw, E. (1978) Schooling fishes. *American Scientist*, **66**, 166–175.

Smith, C.L. (1975) The evolution of hermaphroditism in fishes. In: R. Reinboth (ed.), *Intersexuality in the Animal Kingdom*, pp. 295–310. Springer-Verlag, New York.

Smith, R.J.F. (1997) Avoiding and deterring predators. In: J.-G.J. Godin (ed.), *Behavioural Ecology of Teleost Fishes*, pp. 163–190. Oxford University Press, Oxford.

Sober, E. & Wilson, D.S. (1998) *Unto others: The Evolution and Psychology of Unselfish Behavior*. Harvard University Press, Cambridge, Massachusetts.

St. Mary, C.M. (1996) Sex allocation in a simultaneous hermaphrodite, the zebra goby *Lythrypnus zebra*: insights gained through a comparison with its sympatric congener, *Lythrypnus dalli*. *Environmental Biology of Fishes*, **45**, 177–190.

Stevens, J.R. & Hauser, M.D. (2004) Why be nice? Psychological constraints on the evolution of cooperation. *Trends in Cognitive Sciences*, **8**, 60–65.

Stiver, K.A., Dierkes, P., Taborsky, M. & Balshine, S. (2004) Dispersal patterns and status change in a co-operatively breeding fish *Neolamprologus pulcher*: evidence from microsatellite analyses and behavioural observations. *Journal of Fish Biology*, **65**, 91–105.

Taborsky, M. (1984) Broodcare helpers in the cichlid fish *Lamprologus brichardi*: their costs and benefits. *Animal Behaviour*, **32**, 1236–1252.

Taborsky, M. (1985) Breeder-helper conflict in a child fish with broodcare helpers: an experimental analysis. *Behaviour*, **95**, 45–75.

Tebbich, S., Bshary, R. & Grutter, A.S. (2002) Cleaner fish *Labroides dimidiatus* recognize familiar clients. *Animal Cognition*, **5**, 139–145.

Trivers, R. (1971) The evolution of reciprocal altruism. *Quarterly Review of Biology*, **46**, 35–57.

Trivers, R. (2004) Genetic and cultural evolution of cooperation. *Science*, **304**, 964–965.

Walther, F.R. (1969) Flight behaviour and avoidance of predators in Thomson's gazelles (*Gazella thomsoni* Guenther 1884) *Behaviour*, **34**, 184–221.

Ward, A.J.W. & Hart, P.J.B. (2003) The effects of kin and familiarity on interactions between fish. *Fish and Fisheries*, **4**, 348–358.

Ward, A.J.W., Botham, M.S., Hoare, D.J., James, R., Broom, M., Godin, J.-G.J. & Krause, J. (2002) Association patterns and shoal fidelity in the three-spined stickleback. *Proceedings of the Royal Society of London, Series B*, **269**, 2451–2455.

Werner, N.Y., Balshine, S., Leach, B. & Lotem, A. (2003) Helping opportunities and space segregation in cooperatively breeding cichlids. *Behavioral Ecology*, **14**, 749–756.

West Eberhard, M.J. (1975) The evolution of social behavior by kin selection. *Quarterly Review of Biology*, **50**, 1–33.

Williams, G.C. (1966) *Adaptation and Natural Selection*. Princeton University Press, Princeton, New Jersey.

Wilson, D.S. (1975) A theory of group selection. *Proceedings of the National Academy of Sciences of the USA*, **72**, 143–146.

Wilson, D.S. (1980) *The Natural Selection of Populations and Communities*. Benjamin Cummings, Menlo Park.

Wilson, D.S. (1990) Weak altruism, strong group selection. *Oikos*, **59**, 135–140.

Wilson, D.S. & Dugatkin, L.A. (1992) Altruism. In: E. Fox Keller & E.A. Lloyd (eds), *Keywords in Evolutionary Biology*, pp. 28–33. Harvard University Press, Cambridge, Massachusetts.

Wilson, D.S. & Dugatkin, L.A. (1997) Group selection and assortative interactions. *The American Naturalist*, **149**, 336–351.

Wilson, D.S. & Sober, E. (1994) Re-introducing group selection to the human behavioral sciences. *Behavioral and Brain Science*, **17**, 585–654.

Wilson, E.O. (1975) *Sociobiology: The New Synthesis*. Harvard University Press, Cambridge, Massachusetts.

Wynne-Edwards, V.C. (1962) *Animal Dispersion in Relation to Social Behavior*. Oliver and Boyd, Edinburgh.

Chapter 12
Machiavellian Intelligence in Fishes

Redouan Bshary

12.1 Introduction

The aim of this paper is to present an overview of the social strategic behaviour of fishes. The content was inspired by the 'Machiavellian intelligence hypothesis' (Byrne & Whiten 1988), which proposes that the main cognitive challenge for individual primates is to cope with and exploit the complexity of the social environment in a manner that enhances their fitness. More specifically, an individual has to know all group members and their (genetical and social) relationships in order to find the right coalition partners and to prevent opponents from building successful coalitions. Key cognitive abilities of individuals are thus the ability to understand and remember complex relationships, to cooperate, and skills in manipulation and deception of group members. Other social cognitive abilities like social learning and the formation of traditions are less prominent in the Machiavellian intelligence hypothesis, but play a role in the closely related social brain hypothesis (Dunbar 1992; Barton & Dunbar 1997), which stresses a link between social complexity and neocortex size evolution in mammals. There are positive correlations between group size (a correlate of social complexity) and neocortex ratio (neocortex size regressed against the size of the rest of the brain) in primates, carnivores and bats (Barton & Dunbar 1997). Thus, both the Machiavellian intelligence hypothesis and the social brain hypothesis seem to be applicable to a variety of taxa. This chapter will try to apply this reasoning to social strategies of fishes.

To be able to present the evidence for Machiavellian intelligence in fishes, an important issue has to be clarified. Many primatologists interpret the social behaviour of primates as the result of complex cognitive mechanisms (Byrne & Whiten 1988). The hypothesis that Machiavellian intelligence is linked to neocortex size (or the size of functionally homologous structures) is based on the assumption that social animals not only successfully solve the complexity of their social environment but that they actually evolved some understanding about why a certain behaviour is successful or not. This has spurred a major debate and much research effort into trying to find out whether primates are able, for example, to understand what other individuals want, feel, believe; that is, whether they have a theory of mind (Premack & Woodruff 1978). Despite this effort, the issue has remained quite elusive (Heyes 1998). Certainly, no appropriate tests have ever been conducted on fishes. Therefore, it is important to make a clear distinction between two components that are, unfortunately, mixed up in most definitions of 'cunning' social behaviour, namely the phenomenon and the underlying mechanism. This issue can be illustrated by a definition of tactical deception. On the phenomenological level, tactical deception is defined as the production of a signal out of its normal context, causing context-specific behaviour of the signal's recipient to its own disadvantage and to the signaller's advantage (Hauser 1998). A good example is a false alarm call that yields access to food monopolized by others (Munn 1986). However, everyday language and social scientists would use the term 'tactical deception' only if the actor knows why the signal works and that it has detrimental effects for the recipient. In other words, the signaller has to be conscious about his action and conscious about what the recipient perceives, so the signaller must have a theory of mind. However, theory of mind is not the only mechanism that may produce this behavioural pattern. Animals may learn to produce signals out of context in a much simpler way, via operant conditioning (Thorndike 1911). An initial error (an alarm call that was immediately perceived to be unjustified because, for example, a log was misidentified as a crouching leopard) is positively reinforced (others flee, yielding access to food), which increases the probability that a false alarm call is produced intentionally in the future.

In the absence of knowledge about the underlying mechanism, I propose that it is most fruitful to restrict the definition of behavioural sequences to the phenomeno-logical level. On this level, many different species from different taxa can be compared. As knowledge of underlying mechanisms increases, researchers can exploit this to distinguish between cases: tactical deception based on theory of mind; tactical decep-tion based on operant conditioning; or tactical deception based on a single error. An alternative way of defining animal behaviour in terms of cognition has been promoted by Clayton *et al.* (2003), who faced the problem that although animals may remember the 'what', 'when' and 'where' of particular events, current theories of human epi-sodic memory refer, in addition, to the conscious experience of self that accompanies episodic recall, a state that has no obvious manifestation in non-linguistic behaviour (Tulving 1983; Tulving & Markowitsch 1998). As a solution, Clayton *et al.* (2003) pro-mote the term 'episodic-like memory'. In analogy, one could use 'tactical deception-like' behaviour for cases where the underlying mechanism is not theory of mind. My personal feeling is that adding 'like' may solve some definition problems (in par-ticular those where the definition is linked to language) but not all. Most problematic is the use of 'like' definitions when a phenomenon is described in the absence of definite knowledge about the underlying mechanism. Should we classify the acquisi-tion of tool use in wild chimpanzees as 'social learning-like' as long as we do not have experimental evidence for a social learning mechanism? This chapter will stick to the distinction between phenomenon and underlying mechanism, and will look at evidence for individual recognition, living in individualized groups, cooperation, manipulation, reconciliation, and deception in fishes. These issues will be addressed mainly from a functional perspective, with only partial speculation about underlying mechanisms. Investigating the mechanisms is a key requirement for future studies.

12.2   Cognitive abilities of fishes that form the basis for Machiavellian intelligence

For individuals to be able to show cunning social behaviour, it is necessary that they recognize each other on an individual basis. A further ability would be to know the relationships between other individuals and use this information for making decisions. Such cognitive abilities are only necessary if (a) individuals interact repeatedly with each other, and (b) conflicts of interest occur. These requirements are usually fulfilled if individuals live in stable groups. However, repeated interactions and conflicts may also occur in shoals, between territory neighbours and in interspecific mutualisms like cleaning interactions. Thus, living in stable groups is not a requirement for the evolution of complex strategic social behaviour.

12.2.1 *Individual recognition*

Individual recognition has been shown for a variety of fish species (see Chapter 8 for a full review). Fishes often live in diverse stable groups of varying sizes and sex composition and defend their territories and/or their eggs and larvae. Damselfish (genera *Dascyllus* and *Amphiprion*) of the Red Sea live in stable social assemblages as pairs, harems, or solitary neighbouring individuals, where unknown individuals are treated differently to established neighbours (Fricke 1975). Cichlids in particular are well known for their uniparental or biparental brood care (reviewed by Keenleyside 1991). Individual recognition based primarily on optical cues probably exists in all these stable groups; it has been demonstrated experimentally in a variety of species (Noble & Curtis 1939; Fricke 1973a, 1974; Hert 1985; Balshine-Earn & Lotem 1998). While most studies deal with optical recognition, smell linked to the major histocompatibility complex (Olsen *et al.* 1998; Milinski *et al.* 2005) and sound (Myrberg & Riggio 1985) may also be important. I am not aware of any study where researchers failed to show individual recognition in fishes. Therefore, it seems reasonable to argue that individual recognition will have evolved whenever it was useful, and that behavioural actions of fishes are unlikely to be constrained by an absence of individual recognition. This statement includes fishes that live in aggregations, as indicated by data on partner choice in predator inspection behaviour (Milinski *et al.* 1990a) and in foraging situations (review by Dugatkin 1997). Interspecific individual recognition can also be found in fishes. It has been demonstrated experimentally that the cleaner wrasse (*Labroides dimidiatus*, Labridae) can recognize individual clients (Tebbich *et al.* 2002). Interspecific individual recognition might also occur in interspecific shoals for which interspecific social learning has already been demonstrated (Krause 1993).

12.2.2 *Information gathering about relationships between other group members*

Fishes are known to 'eavesdrop', that is, to use information from observations of interactions between conspecifics (McGregor 1993; and see Chapters 5 and 10). Dugatkin & Godin (1992a) provided experimental evidence for female guppies changing their preferences between two males if they observed another female being courted by the less preferred male. These results have been replicated with sailfin mollies (*Poecilia latipinna*, Poeciliidae), under field conditions (Witte & Noltemeier 2002; Witte & Ryan 2002). Earlier, Schlupp *et al.* (1994) had found that male sailfin mollies may improve their reproductive success by mating with females of the parthenogenetic amazon molly (*Poecilia formosa*, Poeciliidae), because it increases the probability that females of their own species select them as partners. Thus, these females extract information from observed interactions between males and other females. Oliveira *et al.* (1998) showed experimentally that given the opportunity to observe fights between conspecifics, Siamese fighting fish (*Betta splendens*, Osphronemidae) attack 'losers' in a previous fight more vigorously than 'winners'. In this experiment, both observed fishes were actually winners in fights with two other conspecifics that were hidden from the observer's perspective; thus, to the observer it looked like the two winners were interacting, and the one that stopped threat behaviour first was the 'loser'. Similar results were obtained by Earley & Dugatkin on fighting assessment in swordtails (*Xiphophorus helleri*, Poeciliidae) (Earley & Dugatkin 2002; and see Earley & Dugatkin 2005 on other poeciliid fishes). Eaves-dropping should generate selection for behavioural changes in the individuals that are being observed (so-called 'audience effects'), as the outcome of the current inter-action will affect future outcomes. Audience effects could be achieved genetically through a general increase/decrease in the frequency of behaviour like aggression (Johnstone 2001). Siamese fighting fish, however, solve the problem in a smart way, that is, they increase aggression only if observers are present but not if they are absent (Dutreland *et al.* 2001). Observations of interactions between third parties and audi-ence effects also play a major role in cleaning mutualism (Bshary 2002a, and see below). As in the case of individual recognition, it is important to note that there is, as yet, no negative evidence, that is, no species of fish failed to obtain information from interactions between others that would have been relevant to them at a later stage.

12.2.3   *Cooperation and cheating*

Intraspecific cooperation in fishes is discussed in detail in Chapter 11. I will therefore just summarize briefly some key findings that are important for the Machiavellian intelligence hypothesis. The most famous cooperative behaviour described in fishes involves individuals, pairs or several individuals leaving the safety of a shoal to inspect a nearby predator (Pitcher *et al.* 1986). During inspection, pairs of three-spined sticklebacks (*Gasterosteus aculeatus*, Gasterosteidae), and guppies (*Poecilia reticulata*, Poeciliidae), among others, approach the predator in alternating moves (Milinski 1987; Dugatkin 1988). Although the exact game structure is still debated, inspection seems to fit a Prisoner's Dilemma game (Luce & Raiffa 1957), in the sense that cheating a partner by lagging behind seems to be a profitable option and joint inspection therefore an altruistic form of cooperation (reviewed by Dugatkin 1997). Croft *et al.* (2005) found that pairs of guppies that frequently engaged in predator inspection did so in a more cooperative way, exchanging lead position more often than other pairs. Given the increased risk of predation for the lead fish when inspect-ing in a pair, changes in the lead position strongly suggest an act of cooperation based on reciprocity, in a manner predicted by the 'Tit-for-Tat' strategy. Two results yield important insights about the cognitive abilities underlying predator inspection in fishes. First, Milinski *et al.* (1990a) showed that individual sticklebacks prefer specific partners to others. This result implies that (a) school members recognize each other, and (b) there are better (more cooperative) partners than others. Second, partners build up 'trust' in each other if they have cooperated repeatedly (Milinski *et al.* 1990b). The two fishes were actually in different aquaria, and an opaque partition would make the partner disappear all of a sudden (= cheating), while the removal of the partition would make the partner cooperative from the focal individual's point of view. In the decisive experiment, all partners seemed to cheat because the partition was always present. Nevertheless, fishes approached a predator more closely when accompanied by a partner that had 'cooperated' in the past, than when accompanied by a partner that had 'cheated' in the past. Thus, these fishes do not adjust their behaviour simply to the behaviour of co-inspecting individuals, which implies that they are capable of book-keeping (remembering their partners' behaviour during previous interactions), with several partners simultaneously.

Cooperative hunting is another case of cooperation that merits attention and future research with an emphasis on cognition. Cooperative hunting, in the sense that several predators hunt the same prey simultaneously, is widespread in fishes, especially in mackerels (Atlantic makerel *Scomber scombrus*, Scombridae and black skipjack, *Euthynuus yaito*), which have been described herding their prey (Hiatt & Brock 1948; Sette 1950). Schmitt & Strand (1982) even argued that in yellowtails (*Seriola lalandei*, Carangidae), individuals play different roles during such hunts (splitting the school of prey, herding the prey), and refrain from single hunting attempts until the prey is in a favourable position. In addition, Schmitt & Strand (1982) mention that the hunting strategies are variable and that they depend on the prey species. It is of interest that in cooperative hunting each individual has to monitor the movements of partners relative to the prey to bring itself into the best position for an attack. If individuals assume different roles during hunts, it would be helpful to know whether they specialize in different roles and whether there is some reciprocity between group members if different roles yield different success rates (Gazda *et al.* 2005). Interspecific cooperative hunting between Red Sea groupers (*Plectropomus pessuliferus*, Serranidae), and giant moray eels (*Gymnothorax javanicus*, Muraenidae), (Bshary *et al.* 2002) is interesting because of the signalling that is involved. The groupers swim towards the morays (which usually rest in crevices during daytime) and shake their heads repeatedly (Bshary *et al.*, unpublished information). In response, the morays might swim off with the grouper. Morays hunt in the crevices that are inaccessible to the groupers, while the groupers hunt in the water above the reef. Coordinating the hunt thus yields a double predation effect. The key issue is that groupers elicit joint hunting to individuals of a different species without having previously spotted potential prey. Only once the partners swim off together may a fish that does not behave 'properly' be attacked, while the other great numbers of fishes in the reef are ignored. The signalling towards the moray is thus intentional, in the sense that the decision to hunt is made before a prey has been singled out (definition by Boesch & Boesch 1989). The ability to make decisions in the absence of appropriate external stimuli is certainly of advantage when it comes to problems of Machiavellian intelligence.

To summarize so far, fishes have several cognitive abilities that are necessary for cunning Machiavellian-intelligence-like behaviour: they can recognize each other on an individual basis; they monitor relationships between third parties; they cooperate when cheating would be an alternative option; and they may be able to plan behaviour. The following section, will concentrate on two systems that have yielded results that may justify the application of the Machiavellian intelligence hypothesis to fishes. The first example concerns the group-living cichlids in Lake Tanganyika; the second one deals with marine cleaning mutualism.

12.2.4   *Group-living cichlids*

Stable social groups are best known for cichlids of the great African lakes (Keenleyside 1991). Cooperative breeding, that is, the presence of individuals that help the breeding pair raise its offspring, has been described in eight species of fishes up to now (Taborsky 1994), of which six species are Lamprologine cichlids and endemic to Lake Tanganyika. In the best-studied species, the princess of Burundi (*Neolamprologus pulcher/brichardi*, Cichlidae), there are, on average, five helpers of both sexes and of various sizes (Taborsky & Limberger 1981; Balshine *et al.* 2001). Helpers may be related or unrelated to the breeding pair, as breeding individuals are often replaced from outside (Taborsky & Limberger 1981) and because helpers may switch between groups (Stiver *et al.* 2004; Bergmüller *et al.* 2005a; Dierkes *et al.* 2005). Another cichlid (*Neolamprologus multifasciatus*, Cichlidae) endemic to Lake Tanganyika lives in extended family groups (*sensu* Emlen 1997); that is, in stable groups with two or more sexually active members of both sexes (Kohler 1997). In these social species, individual recognition of group members is very likely, and experimentally shown for *N. brichardi* (Hert 1985; Balshine-Earn & Lotem 1998).

From a Machiavellian intelligence perspective, these systems are very interesting because there are several important conflicts between individual group members, which should promote social intelligence. Conflicts occur over:

1   various group tasks such as sand digging, nest and egg/larvae maintenance, and defence against competitors or predators;
2   snail shells or crevices that are used for shelter;
3   most importantly, over reproduction.

In cooperatively breeding species, helpers have some reproductive success at the expense of dominant fish reproductive success (Dierkes *et al.* 1999), and in *Neolamprologus multifasciatus*, males and females may disagree over the optimal group composition. This is because males could benefit from the presence of additional females under certain circumstances, because it would increase their reproductive success (Kohler 1997). Moreover, the costs of helping are measurable. Sexually mature helpers face various costs when delaying dispersal: reduced growth rates (Taborsky 1984; Heg *et al.* 2004); delayed reproduction; and increased energy expenditure as a result of helping and from costly social interactions (Grantner & Taborsky 1998; Taborsky & Grantner 1998). In conclusion, any (genetic or cognitive) factor that helps to reduce these costs or increases reproductive success might be expected to be favoured by natural selection.

A variety of interesting behaviours and strategies have been found in the *Neolamprologus* cichlids. Schradin & Lamprecht (2000) showed experimentally that *Neolamprologus multifasciatus* males actively intervene in female–female aggression in favour of the unfamiliar female, and that this intervention increased the probability that the new females would settle in the group. Males of the dwarf cichlid (*Apistogramma trifasciatum*, Cichlidae) seem to behave similarly (Burchard 1965). A general feature of *Neolamprologus* species is that subordinates frequently show submissive behaviours towards high-ranking group members (Kohler 1997, Bergmüller *et al.* 2005b). This submissive behaviour apparently functions as pre-emptive conflict avoidance (Bergmüller & Taborsky 2005). In *N. pulcher/ brichardi*, strategic options are highly finely tuned and variable.

1 Dominants are more likely to tolerate large helpers if competition for space with other species is very high (Taborsky 1985).
2 Temporarily removed helpers assisted more in territory maintenance and defence and visited the brood chamber more often after they were returned (Balshine-Earn *et al.* 1998) and helpers that could potentially breed independently reduced helping and submissive behaviour in the home territory (Bergmüller *et al.* 2005b).
3 In the field, Balshine-Earn *et al.* (1998) found that residents attacked temporarily removed helpers when the latter were returned to their groups' territories. Thus, there is some social pressure on individuals to contribute to group tasks.
4 Helpers often visit neighbouring territories and may either switch or reduce working load at home if conditions outside the home territory are favourable (Bergmüller *et al.* 2005a).
5 There is a positive correlation between the amount of submissive behaviour performed by helpers and the amount of aggression they receive from dominants. In an experiment, helpers showed less submissive behaviour towards breeders per received aggression if they had defended the group than if they were prevented from doing so (Bergmüller & Taborsky 2005). The authors concluded that helping and submissive behaviour are at least partly interchangeable options for the same goal: appeasement to avoid being punished or evicted from the group.
6 Small helpers showed less submissive behaviour towards large helpers when they helped more, indicating that there are also conflicts between the helpers over individual contributions to group tasks (Bergmüller & Taborsky 2005). Further conflicts between helpers are a result of apparent competition for space (Balshine-Earn *et al.* 1998; Werner *et al.* 2004).

In summary, individual fishes in cooperatively breeding species seem to be capable of monitoring the behaviour of several group members as well as the social dynamics of neighbouring groups. They use this information to fine tune their behaviour. Appeasement behaviour is an interesting component of interactions, as the behaviour itself provides no direct benefits to the recipient, so subordinates seem to use it to manipulate the decisions of dominant group members. But then why do dominants accept the appeasement behaviour of the subordinates? It would be interesting to know whether there are any physiological constraints on submissive behaviour; for example, if high testosterone levels in males block submissive behaviour, subordinate males have to keep testosterone levels low, which might lower sperm production. Thus, a future study should test whether submissive behaviour might reveal some honest information about the helper being in a non-reproductive state. The fact that helpers prevented from helping subsequently increased their helping efforts in the absence of aggression by dominant group members (Bergmüller & Taborsky 2005), suggests that individuals can counter the consequences of idle behaviour before they have to face them. Perhaps most interestingly, it is important to find out how variable workload and individual contributions to the common good are negotiated depending on environmental conditions.

12.2.5   *Machiavellian intelligence in cleaning mutualisms*

In a cleaning mutualism, so-called client fish trade the removal of parasites and dead or infected tissue for an easy meal for so-called 'cleaner fish' (Losey *et al.* 1999; Côté 2000; Bshary & Noë 2003). The best-studied cleaner fish species, *Labroides* sp. and *Elacatinus* sp., have small territories ('cleaning stations') that clients actively visit. Over the last few years, plenty of evidence has accumulated suggesting that these interactions are indeed mutualistic (Grutter 1999; Cheney & Côté 2001; Bshary 2003; Grutter *et al.* 2003). Nevertheless, research on the cleaner wrasse revealed that there are a variety of potential conflicts between cleaners and clients, namely the risk for cleaners of being eaten by predatory clients (Trivers 1971), timing and durations of interactions (Bshary 2001), partner choice (Bshary & Grutter 2002a), and the cleaners' preference for at least some clients' mucus over ectoparasites (Grutter & Bshary 2003, 2004). The latter point is crucial because it shows that interactions are not a simple byproduct mutualism (Brown 1983) but clients need to control the behaviour of cleaner fish in order to prevent being cheated ('cleaner eats mucus') and to make cleaners cooperate ('feeding on ectoparasites'). The control mechanism depends critically on the strategic options of the clients: predators may retaliate by eating a cheating cleaner, while non predatory clients lack this option, and visiting clients with access to several cleaning stations may switch partner if cheated, while resident clients with access to the local cleaner only lack that option. As predicted by a generalized and phenomenon-based Machiavellian intelligence hypothesis, the social conflicts between cleaners and clients and the variety of strategic options have promoted an array of interesting behaviour that raises questions about the underlying cognitive abilities.

12.2.5.1   *Categorization and individual recognition of clients*

Observations on interactions suggest that cleaner wrasse (*Labroides dimidiatus*, Labridae) discriminate between three client categories (Bshary 2001). First, they discriminate between predatory and non-predatory client species, the former being virtually never cheated, in contrast to the latter. Among the non-predatory clients, cleaners further distinguish between resident clients and visiting clients, giving the latter priority of service. This is advantageous because visiting clients would not wait for inspection and would visit a different cleaning station instead, while residents simply have to wait if they want to be serviced at all (Bshary & Schäffer 2002). An important difference between visitors and residents is their response to cleaner fish cheating: visitors simply swim off, while residents chase the cleaners around (Bshary & Grutter 2002a). The aggression functions as punishment (*sensu* Clutton-Brock & Parker 1995). Aggressive chasing bears immediate energetic costs to both client and cleaner but it makes cleaners more cooperative in future interactions. According to theory, punishment can function only if there is individual recognition (Ostrom 1990). Choice experiments with one-way mirrors confirmed that cleaners can recognize clients individually; in the absence of environmental clues, they spent more time near a familiar client than near an unfamiliar client of the same species and of similar size (Tebbich *et al.* 2002). Individual cleaners may interact with more than 100 individual resident clients belonging to various species (R. Bshary, unpublished. observations). Thus, they may well be capable of recognizing more than 100 clients on an individual basis and to remember their last interaction with each of them (see below), although this clearly has to be tested in detail in a future study.

12.2.5.2   Building up relationships between cleaners and resident clients

If cleaner wrasse are experimentally translocated to a new site, resident clients first chase them around (Bshary 2002b). The cleaners spend most of the first day exhibiting very peculiar behaviour, which has been observed so far only among cleaners of the genus *Labroides*: they ride on the clients' backs, snouts pointing into the blue, while applying tactile stimulation to the clients' backs with the pelvic and pectoral fins (Potts 1973). Only with time does client aggression decrease and the cleaner start to inspect their resident clients for food (Bshary 2002b). Thus, cleaners have to build up relationships with their resident clients, through initial investment rather than through increasing investment (Roberts & Sherratt 1998). Interactions with visiting clients, in contrast, are 'normal' right from the beginning, while predators seem to receive some extra tactile stimulation (see below) compared to what they normally receive (Bshary 2002b).

12.2.5.3   Use of tactile stimulation by cleaners to manipulate client decisions and reconcile after conflicts

As mentioned above, cleaners of the genus *Labroides* apply tactile stimulation to the clients' back (Potts 1973). Cleaner wrasse use this massage in particular during interactions with predatory clients; to make moving clients stop so that the cleaner can inspect them; and functionally to reconcile (de Waal & van Roosmalen 1979) with resident clients in follow up interactions after client punishment (Bshary & Würth 2001). The specific behaviour of cleaners in follow-up interactions makes it plausible that they really can recognize their clients and remember their last interaction with each of them. Tactile stimulation during interactions with predators can be seen as preconflict management (Aureli & de Waal 2000) that reduces the probability of predators cheating (see also Grutter 2004). Finally, the application of tactile stimulation to swimming clients is a means to manipulate the client's behaviour in favour of the cleaner as it gains access to a food source.

12.2.5.4   Indirect reciprocity based on image-scoring and tactical deception

Field observations revealed that clients arriving at a cleaning station invite inspections if the cleaner's current interaction ends without apparent conflict, but avoid an interaction if the current client flees or chases the cleaner (Bshary 2002a). Thus, the clients seem to observe ongoing interactions and extract information on the behaviour of the cleaner fish, which they use to make their decision on whether to invite inspection. In other words, cleaners are given an image-score or a social prestige (Alexander 1987; Zahavi 1995; Nowak & Sigmund 1998; Lotem *et al.* 2003) from bystanders or eavesdroppers (McGregor 1993) that is positive if they cooperate and negative if they cheat. As a consequence of image-scoring clients, cleaners seem to be more cooperative to their current client if eavesdropping clients are around (Bshary & D'Souza 2005). The benefits of this more cooperative behaviour are not accrued directly by the interacting client but indirectly, via access to the observing clients (hence 'indirect reciprocity'). There is now experimental evidence supporting the conclusions drawn from the field observations (Bshary & Grutter 2006). The main reason why clients image-score might be that some cleaners bite rather than clean (Bshary 2002a), and these biting cleaners can be avoided by image-scoring without having to make a personal negative experience first. Biting cleaners thus loose access to some potential victims but have found a way partly to cope with the image-scoring clients; they may raise their image by seeking interactions with small residents and giving them tactile stimulation (Bshary 2002a). These interactions are purely costly for the cleaners but benefits might be accrued through deceiving larger clients into visiting so that they can be exploited. Giving tactile stimulation to small residents thus functions as tactical deception (Hauser 1998); it is a signal out of context, directed at observing clients (rather than at the small resident that receives the stimulation) to make them choose an option that is costly to them but beneficial to the cleaner.

12.2.5.5   *What cognitive abilities might cleaners need to deal with their clients?*

It seems likely that cleaners learn most of their behavioural tactics through associative learning. A cleaner wrasse has about 2000 interactions per day (Grutter 1995) and, therefore, constantly receives feedback about its actions. Laboratory experiments confirmed that cleaners can easily learn to behave in a way that increases their foraging success (Bshary & Grutter 2002b, 2005). In these experiments, cleaners interacted with Plexiglas plates attached to levers that allowed them to make the plates 'behave' in certain ways. In one set of experiments, cleaners learned to choose the less preferred food items in order to stop the plates moving away or chasing them (Bshary & Grutter 2005). This strongly suggests that cleaners have some self control (as opposed to impulsiveness), defined as the ability to inhibit a natural tendency to reach for the greater or more attractive of two food items (Anderson 2001). This ability has been linked to self awareness in primates (Genty *et al.* 2004). In another experiment, cleaners learned to inspect a 'visiting' plate before a 'resident' plate because the former would not wait for inspection while the latter would (Bshary & Grutter 2002b). There were equal amounts of food on both plates and cleaners could invariably feed on the plate they visited first. So they learned to prefer the visitor plate even though they were rewarded for any action they took.

Field observations of one particular cleaner female revealed that a cleaner's behaviour can be completely independent of its internal state. This cleaner fish female was biting clients and, therefore, was not tolerated by the larger male at the cleaning station. The female, therefore, installed her station 3 m away but regularly visited the male's area until chased away. It turned out that she was a 'normal' cleaner while at her station and a biting cleaner whenever she visited the male's area (Bshary & D'Souza 2005). This switch in behaviour of the female might have been induced by the location, although it is not very clear how that should work. Thus, it seems more likely that cleaners might be able to adjust their behaviour to a specific external context ('interacting with the male's client or with own client') rather than merely behaving according to an internal state ('if I am hungry I bite').

Finally, cleaners seem to have some knowledge of the interspecific relationships between their clients. They exploit the presence of a predatory client to prevent punishing clients from chasing them further. They swim to the predator and provide tactile stimulation while the aggressive client apparently does not dare to approach the predator that close and terminates its pursuit (Bshary *et al.* 2002a). The critical question is whether this behaviour is the result of associative learning ('if swimming to species A, chasing stops while swimming to species B, C, and D does not end being chased') or whether cleaners know why their tactic is successful. In the latter case, cleaners might know that their chasing client is a potential prey of the predator and it therefore stays away.

12.3   Conclusion

The list of phenomena that are of interest with respect to the Machiavellian intelligence hypothesis is both long and poorly studied from a cognitive perspective. Most research on fishes has addressed functional questions and consequently neglected the underlying mechanisms. Therefore, we still know little about the kind and quantity of information that fishes use to make these decisions. Nevertheless, the little we do know indicates that a wealth of interesting results awaits description by scientists. This paper has focused on two broad systems that have been studied in detail from a functional perspective, namely cooperative breeding and cleaning mutualisms. But it is evident that the preconditions for Machiavellian-type intelligence, that is, individual recognition, living in groups/repeated interactions and knowledge about relations between other group members/third parties, is widespread in fishes. It is probable therefore that an array of interesting cognitive social abilities awaits to be found and studied in more detail in fishes. The research on the cleaner fish mutualism shows that if we use phenomenon-based definitions, we find reconciliation, punishment, indirect reciprocity based on image-scoring and tactical deception. The laboratory experiments show that cleaners can easily learn to respond adaptively to various behavioural responses of Plexiglas plates (thus excluding purely genetic strategies). They also show some self control (feed against their preference) and are flexible in their decisions rather than just dependent on their internal state. The various studies on cooperative breeding in fishes suggest that cichlids are also capable of fine-tuned adjustment of behaviour to ecological and social factors, and that social skills may enhance survival and reproductive success.

12.3.1   *Future avenues I: how Machiavellian is fish behaviour?*

Cognitive research on Machiavellian aspects of social behaviour could address detailed questions about the precise mechanism that is used for learning and decision making, as well as producing quantitative information on:

1   the number of individuals that can be known and their behaviours tracked;
2   how fast individual fish can learn social tasks;
3   how accurately they can learn to behave;
4   how fast they can integrate new additional or contradicting information.

These questions could be embedded in a comparative approach that compares many different species that differ with respect to the complexity of their social life. A comparison between two species might yield significant differences that are in line with predictions based on social complexity. However, alternative explanations for the differences cannot be excluded. Only if enough species are tested on the same question to allow statistical analysis on the number of species investigated, can we be able to find potential links between social complexity and cognitive capacities.

A highly relevant study on corvids with respect to learning mechanisms and speed of knowledge acquisition was produced by Bond *et al.* (2003). They compared semi-territorial western scrub jays (*Aphelocoma california*, Corvidae), with the highly social pinyon jays (*Gymnorhinus cyanocephalus*, Corvidae). The birds had to learn an arbitrary dominance hierarchy between symbols to obtain food. The more social pinyon jays learned some relations between symbols through transitive inference (if A > B and B > C, then A > C), while western scrub jays seem to have learned each pair through associative learning. The use of these different mechanisms may explain why the pinyon jays learned the tasks faster than the western scrub jays (Bond *et al.* 2003). Furthermore, it is argued that the capacity of transitive inference is more important in large social groups in order to be able to track relationships between other group members. Data on more species are needed to test whether the difference in leaning mechanism is indeed linked to the differences in sociality.

Fishes seem to be particularly suitable for study of interspecific social cognition in detail, as they often occur in mixed-species shoals (Krause 1993; and see Chapter 9), forage in mixed-species parties (Dugatkin & Godin 1992b), hunt with interspecific partners (Bshary *et al.* 2002) and often have cleaning interactions (Feder 1966). Often, closely related species differ markedly with respect to the importance of social (interspecific) interactions, offering ideal opportunities for large-scale interspecific comparisons that link a species' social environment and factors like habitat structure, diet or antipredator behaviour to specific cognitive capacities. Taking cleaning mutualism as an example, cleaner fish are found in many different fish families and can differ markedly in the degree to which they depend on interactions with clients for their diet (Feder 1966). It would therefore be interesting to investigate whether the degree of cleaning 'professionalism' correlates with the cognitive abilities of species in social contexts. Experiments using Plexiglas plates (Bshary & Grutter 2002b, 2005) could be used to test a variety of cleaning and non-cleaning species.

Three key questions have to be addressed. First, seemingly cunning social behaviour might have nothing to do with learning/cognition but is produced with complex chains of key stimuli and automatic responses, or with behaviour modified through endocrine responses (Oliveira 2005). So we have to show that learning plays a role. Second, any differences between species with respect to performance in the same task might be of a quantitative nature (speed of learning precision of behaviour), based on speed of associative learning. Are there any species that use more complex cognitive mechanisms? For example, is any fish species capable of transitive inference? Third, the social tasks could be transferred into environmental tasks to investigate whether social cognitive abilities correlate with environmental cognitive abilities or whether these abilities exist independently of each other. In a classic study on food caching in three corvid bird species (Balda & Kamil 1989; review by

Balda & Kamil 1998), the amount of food caching correlated well with a species' spatial memory capacity but not with colour memory.

Regarding within-species variation in social skills, one could try to use the fact that the degree of cooperation expressed by cleaner fish depends on ecological conditions. Grutter (1997) found that cleaner wrasse fed almost exclusively on parasitic isopods around Lizard Island but ate much more mucus around Heron Island at the southern end of the Great Barrier Reef, where ectoparasites are less abundant. Whether cleaners around Heron Island are, therefore, more Machiavellian in their cognitive abilities or whether the differences are purely resulting from an ecological factor (parasite abundance) remains an open question. Similarly, shoaling tendencies and sociality are likely to vary within species according to ecological factors and are thus amenable to being linked to cognitive skills.

12.3.2 *Future avenues II: relating Machiavellian-type behaviour to brain size evolution*

Currently, we do not know whether complex social behaviour in fishes is correlated with the size of relevant brain areas, as has been documented in several mammalian taxa (Barton & Dunbar 1997). Apparently, the telencephalon is an important area for higher decision-making processes and thus a prime candidate for future investigations (Roth & Wullimann 2001; and see Chapter 13). With experiments on three-spined sticklebacks, Schönherr (1954) and Seegar (1956) could show that after the lesion of the telencephalon, all behaviour that is necessary to build a nest is still performed but the proper coordination necessary to build the nest is lacking. Also the mesencephalon seems to be, at least in part, involved in learning and decision making (Healey 1957).

The link between the ecology of fish species and brain size has already been investigated (reviewed by Kotrschal *et al.* 1998). Cichlids of the East African Lakes vary greatly in the size of their telencephalon (van Staaden *et al.* 1995) but this variation has not yet been linked to the complexity of a species' social life. Cichlids are in general a very promising fish family for study of the relation between social organization and brain anatomy, because of the large interspecific variance in the amount of brood care and corresponding social organization. In addition, they also live in a great variety of habitats, and there is evidence that spatial complexity of habitat correlates with telencephalon size (van Staaden *et al.* 1995; Huber *et al.* 1997). What is still missing is the link between such information on fish ecology and brain structure, and cognitive skills, and including social complexity as a parameter.

With respect to intraspecific variation in social cognitive skills and the link to brain anatomy, fishes seem to be very suitable subjects. In particular, effects of ontogeny should become evident in brain structure as fishes can generate new brain cells throughout their entire life (reviewed in Kotrschal *et al.* 1998).

12.3.3 *Extending the Machiavellian intelligence hypothesis to general social intelligence*

The Machiavellian intelligence hypothesis puts the emphasis on the Prisoner's Dilemma type cooperation (where deception yields higher short-term benefits), deception and manipulation. However, group living may also promote the cognitive abilities for social learning that are generally of a cooperative nature. Aspects of social learning are discussed in several chapters of this volume (Chapters 3, 4, 9 and 10). These chapters provide plenty of evidence that fishes learn socially, although true imitation of tutor behaviour has not yet been documented. Social learning is potentially a practical tool in fish farms where hatchery fishes are trained in anti-predator responses with tutors (Suboski & Templeton 1989; Brown & Laland 2001). Naive fishes that are given the opportunity to watch a conspecific fleeing from a predator often show an escape response themselves (Magurran & Higham 1988). This process also occurs in mixed-species shoals and can result in the transmission of information between species (Krause 1993). Under field conditions, Fricke (1973b) described mobbing behaviour of the three-spotted damselfish (*Dascyllus trimaculatus*, Pomacentridae) against a variety of predators. Mobbing behaviour seems to be widespread in fishes (Dugatkin & Godin 1992b), and hence inexperienced fishes could learn about predators socially.

In addition, the spawning migrations found in many coral reef fish have been a model system for investigating the cultural transmission of behaviour. Coral reef fish repeatedly use specific spawning sites where individuals aggregate far away from their territories or home ranges on the reef. Prominent cases are the migrations of surgeonfish (*Acanthurus nigrofuscus*, Acanthuridae) in the Red Sea (Fishelson *et al.* 1986; Myrberg *et al.* 1988), French grunts (*Haemulon flavolineatum*, Haemulidae; Helfman & Schulz 1984) and blueheaded wrasses (*Thalassoma bifasciatum*, Labridae; Warner 1988, 1990). Translocation experiments confirmed that new immigrants pick up the traditional sites and keep them once the resident fishes are removed (Helfman & Schulz 1984), and that the exchange of entire populations leads to the start of new site traditions (Warner 1988, 1990). Traditions concerning routes to feeding sites were also readily established in controlled laboratory experiments on guppies (Laland & Williams 1997, 1998).

In conclusion, social life in fishes may select as much for the evolution of cognitive capacities, which allow the acquisition of freely available social information, as it does for Machiavellian intelligence.

12.4   Acknowledgements

I thank Culum Brown, Kevin Laland and Jens Krause for inviting me to write this chapter, and Ralph Bergmüller and Wolfgang Wickler for contributing relevant literature. Comments by Nicky Clayton and an anonymous reviewer greatly improved the manuscript. My research is currently supported by the Swiss Science Foundation and by a NERC grant.

12.5   References

Alexander, R.D. (1987) *The Biology of Moral Systems*. Aldine de Gruiter, New York.
Anderson, J.R. (2001) Self- and other-control in squirrel monkeys. In: T. Matsuzawa (ed.), *Primate Origins of Human Cognition and Behavior*, pp. 330–347. Springer-Verlag, Tokyo.

Aureli, F. & de Waal, F.B.M. (2000) *Natural Conflict Resolution*. Cambridge University Press, Cambridge.

Balda, R.P. & Kamil, A.C. (1989) A comparative study on cache recovery by three corvid species. *Animal Behaviour*, **38**, 486–495.

Balda, R.P. & Kamil, A.C. (1998) The ecology and evolution of spatial memory in corvids of the southwestern USA: the perplexing pinyon jay. In: R.P. Balda, P.A. Bednekoff & A.C. Kamil (eds), *Animal Cognition in Nature*, pp. 29–64. Academic Press, New York.

Balshine, S., Leach, B., Neat, F.C., Reid, H., Taborsky, M. & Werner, N. (2001) Correlates of group size in a cooperatively breeding cichlid fish. *Behavioural Ecology and Sociobiology*, **50**, 134–140.

Balshine-Earn, S. & Lotem, A. (1998) Individual recognition in a cooperatively breeding cichlid: evidence from video playback experiments. *Behaviour*, **135**, 369–386.

Balshine-Earn, S., Neat, F.C., Reid, H. & Taborsky, M. (1998) Paying to stay or paying to breed? Field evidence for direct benefits of helping behaviour in a cooperatively breeding fish. *Behavioral Ecology*, **9**, 432–438.

Barton, R.A. & Dunbar R.I.M. (1997) Evolution of the social brain. In: A. Whiten & D.W. Byrne (eds), *Machiavellian Intelligence II*, pp. 240–263. Cambridge University Press, Cambridge.

Bergmüller, R. & Taborsky, M. (2005) Experimental manipulation of helping in a cooperative breeder: helpers 'pay-to-stay' by pre-emptive appeasement. *Animal Behaviour*, **69**, 19–28.

Bergmüller, R., Heg, D., Peer, K. & Taborsky, M. (2005a) Extended safe havens and between group dispersal of helpers in a cooperatively breeding cichlid. *Behaviour*, **142**, 1643–1667.

Bergmüller, R., Heg, D. & Taborsky, M. (2005b) Helpers in a cooperatively breeding cichlid fish stay and pay, but choose to leave when independent breeding options are available. *Proceedings of the Royal Society of London, Series B*, **272**, 325–331.

Boesch, C. & Boesch, H. (1989) Hunting behaviour of wild chimpanzees in the Taï National Park. *American Journal of Physical Anthropology*, **78**, 547–573.

Bond, A.B., Kamil, A.C. & Balda, R.P. (2003) Social complexity and transitive inference in corvids. *Animal Behaviour*, **65**, 479–487.

Brown, C. & Laland, K.N. (2001) Social learning and life skills training for hatchery reared fish. *Journal of Fish Biology*, **59**, 471–493.

Brown, J.L. (1983) Cooperation: a biologists dilemma. In: Rosenblatt, J.S. (ed.) *Advances in the Study of Behavior*, pp. 1–37. Academic Press, New York.

Bshary, R. (2001) The cleaner fish market. In: R. Noë, J.A.R.A.M. van Hooff & P. Hammerstein (eds), *Economics in Nature*, pp. 146–172. Cambridge University Press, Cambridge.

Bshary, R. (2002a) Biting cleaner fish use altruism to deceive image scoring clients. *Proceedings of the Royal Society of London, Series B*, **269**, 2087–2093.

Bshary, R. (2002b) Building up relationships in asymmetric cooperation games between the cleaner wrasse *Labroides dimidiatus* and client reef fish. *Behavioral Ecology and Sociobiology*, **52**, 365–371.

Bshary, R. (2003) The cleaner wrasse, *Labroides dimidiatus*, is a key organism for reef fish diversity at Ras Mohammed National Park, Egypt. *Journal of Animal Ecology*, **72**, 169–176.

Bshary, R. & D'Souza, A. (2005) Cooperation in communication networks: indirect reciprocity in interactions between cleaner fish and client reef fish. In: P.K. McGregor (ed.), *Communication Networks*, pp. 521–539. Cambridge University Press, Cambridge.

Bshary, R. & Grutter, A.S. (2002a) Asymmetric cheating opportunities and partner control in a cleaner fish mutualism. *Animal Behaviour*, **63**, 547–555.

Bshary, R. & Grutter, A.S. (2002b) Experimental evidence that partner choice is the driving force in the payoff distribution among cooperators or mutualists: the cleaner fish case. *Ecology Letters*, **5**, 130–136.

Bshary, R. & Grutter, A.S. (2005) Punishment and partner choice cause cooperation in a cleaning mutualism. *Biology Letters*, **1**, 396–399.

Bshary, R. & Grutter, A.S. (2006) Image scoring and cooperation in a cleaner fish Mutualism. *Nature*, **441**, 975–978.

Bshary, R. & Noë, R. (2003) Biological markets: the ubiquitous influence of partner choice on cooperation and mutualism. In: P. Hammerstein (ed.), *Genetic and Cultural Evolution of Cooperation*, pp. 167–184. MIT Press, Cambridge.

Bshary, R. & Schäffer, D. (2002) Choosy reef fish select cleaner fish that provide high service quality. *Animal Behaviour*, **63**, 557–564.

Bshary, R. & Würth, M. (2001) Cleaner fish *Labroides dimidiatus* manipulate client reef fish by providing tactile stimulation. *Proceedings of the Royal Society of London, Series B*, **268**, 1495–1501.

Bshary, R., Wickler, W. & Fricke, H. (2002a) Fish cognition: a primate's eye view. *Animal Cognition*, **5**, 1–13.

Burchard, J.E. (1965) Family structure in the dwarf cichlid *Apistogramma trifasciatum*. *Zeitschrift für Tierpsychologie*, **22**, 150–162.

Byrne, R.W. & Whiten, A. (1988) *Machiavellian Intelligence*. Clarendon Press, Oxford.

Cheney, K.L. & Côté, I.M. (2001) Are Caribbean cleaning symbioses mutualistic? Costs and benefits of visiting cleaning stations to longfin damselfish. *Animal Behaviour*, **62**, 927–933.

Clayton, N.S., Bussey, T.J., Emery, N.J. & Dickinson, A. (2003) Prometheus to Proust: the case for behavioural criteria for 'mental time travel'. *Trends in Cognitive Sciences*, **7**, 436–437.

Clutton-Brock, T.H. & Parker, G.A. (1995) Punishment in animal societies. *Nature*, **373**, 209–215.

Côté, I.M. (2000) Evolution and ecology of cleaning symbioses in the sea. *Oceanography and Marine Biology: An Annual Review*, **38**, 311–355.

Croft, D.P., James, R., Ward, A.J.W., Botham, M.S., Mawdsley, D. & Krause, J. (2005) Assortative interactions and social networks in fish. *Oecologia*, **143**, 211–219.

Dierkes, P., Taborsky, M. & Kohler, U. (1999) Reproductive parasitism of broodcare helpers in a cooperatively breeding fish. *Behavioral Ecology*, **10**, 510–515.

Dierkes, P., Heg, D., Taborsky, M., Skubic, E. & Achmann, R. (2005) Genetic relatedness in groups is sex-specific and declines with age of helpers in a cooperatively breeding cichlid. *Ecology Letters*, **8**, 968–975.

Dugatkin, L.A. (1988) Do guppies play Tit for Tat during predator inspection visits? *Behavioral Ecology Sociobiology*, **25**, 395–399.

Dugatkin, L.A. (1997) *Cooperation Among Animals*. Oxford University Press, New York.

Dugatkin, L.A. & Godin, J.G.J. (1992a) Reversal of female mate choice by copying in the guppy *Poecilia reticulata*. *Proceedings of the Royal Society of London, Series B*, **249**, 179–184.

Dugatkin, L.A. & Godin, J.G.J. (1992b) Prey approaching predators: a cost-benefit perspective. *Annales Zoologici Fennici*, **29**, 233–252.

Dunbar, R.I.M. (1992) Neocortex size as a constraint on group size in primates. *Journal of Human Evolution*, **20**, 287–296.

Dutreland, C., McGregor, P.K. & Oliveira, R.F. (2001) The effect of an audience on intrasexual communication in male Siamese fighting fish, *Betta splendens*. *Behavioral Ecology*, **12**, 283–286.

Earley, R.L. & Dugatkin, L.A. (2002) Eavesdropping on visual cues in swordtail (*Xiphophorus helleri*) fights – a case for networking. *Proceedings of the Royal Society of London, Series B*, **269**, 943–952.

Earley, R.L. & Dugatkin, L.A. (2005) Fighting, mating and networking: pillars of poeciliid sociality. In: P.K. McGregor (ed.), *Animal Communication Networks*, pp. 84–113. Cambridge University Press, Cambridge.

Emlen, S.T. (1997) Predicting family dynamics in social vertebrates. In: J.R. Krebs & N.B. Davies (eds), *Behavioural Ecology*, 4th edn, pp. 228–253. Blackwell, Oxford.

Feder, H.M. (1966) Cleaning symbiosis in the marine environment. *Symbiosis*, **1**, 327–380.

Fishelson, L., Montgomory, L.W. & Myrberg, A.A. (1986) Social biology and ecology of acanthurid fish near Eilat (Gulf of Aqaba, Red Sea). In: T. Uyeno, R. Arai, T. Taniuchi & K. Matsuura (eds), *Indo-Pacific Fish Biology, Proceedings of the Second International Conference on Indo-Pacific Fishes*. Ichthyological Society of Japan, Tokyo.

Fricke, H. (1973a) Individual partner recognition in fish: field studies on *Amphiprion bicinctus*. *Naturwissenschaften*, **60**, 204–205.

Fricke, H. (1973b) Ökologie und sozialverhalten des korallenbarsches *Dascyllus trimaculatus* (Pisces, Pomacantridae). *Zeitschrift für Tierpsychologie*, **32**, 225–256.

Fricke, H. (1974) Öko-Ethologie des monogamen Anemonenfisches *Amphiprion bicinctus*. *Zeitschrift für Tierpsychologie*, **36**, 429–512.

Fricke, H. (1975) Sozialstruktur und oekologische spezialisierung von verwandten fischen (Pomacentridae). *Zeitschrift für Tierpsychologie*, **39**, 492–520.

Gazda, S.K., Connor, R.C., Edgar, R.K. & Cox, F. (2005) A division of labour with role specialization in group-hunting bottlenose dolphins (*Tursiops truncatus*) off Cedar Key, Florida. *Proceedings of the Royal Society of London, Series B*, **272**, 135–140.

Genty, E., Palmier, C. & Roeder, J.J. (2004) Learning to suppress responses to the larger of two rewards in two species of lemurs, *Eulemur fulvus* and *E. macaco*. *Animal Behaviour*, **67**, 925–932.

Grantner, A. & Taborsky, M. (1998) The metabolic rates associated with resting, and with the performance of agonistic, submissive and digging behaviours in the cichlid fish *Neolamprologus pulcher*. *Journal of Comparative Psychology*, **168**, 427–433.

Grutter, A.S. (1995) Relationship between cleaning rates and ectoparasite loads in coral reef fishes. *Marine Ecology Progress Series*, **118**, 51–58.

Grutter, A.S. (1997) Spatio-temporal variation and feeding selectivity in the diet of the cleaner fish *Labroides dimidiatus*. *Copeia*, **1997**, 346–355.

Grutter, A.S. (1999) Cleaners really do clean. *Nature*, **398**, 672–673.

Grutter, A.S. (2004) Cleaner fish use tactile dancing behavior as a preconflict management strategy. *Current Biology*, **14**, 1080–1083.

Grutter, A.S. & Bshary, R. (2003) Cleaner wrasse prefer client mucus: support for partner control mechanisms in cleaning interactions. *Proceedings of the Royal Society of London, Series B*, **270**, 242–244.

Grutter, A.S. & Bshary, R. (2004) Cleaner fish *Labroides dimidiatus* diet preferences for different types of mucus and parasitic gnathiid isopods. *Animal Behaviour*, **68**, 583–588.

Grutter, A.S., Murphy, J.M. & Choat, J.H. (2003) Cleaner fish drives local fish diversity on coral reefs. *Current Biology*, **13**, 64–67.

Hauser, M.D. (1998) Minding the behaviour of deception. In: A. Whiten & D.W. Byrne (eds), *Machiavellian Intelligence II*, pp. 112–143. Cambridge University Press, Cambridge.

Healey, E.G. (1957) The nervous system. In: M.E. Brown (ed.), *The Physiology of Fishes*, volume 2, pp. 1–119. Academic Press, New York.

Heg, D., Bachar, Z., Brouwer, L. & Taborsky, M. (2004) Predation risk is an ecological constraint for helper dispersal in a cooperatively breeding cichlid. *Proceedings of the Royal Society of London, Series B*, **271**, 2367–2374.

Helfman, G.S. & Schultz, E.T. (1984) Social transmission of behavioural traditions in a coral reef fish. *Animal Behaviour*, **32**, 379–384.

Hert, E. (1985) Individual recognition of helpers by the breeders in the cichlid fish *Lamprologus brichardi* (Poll, 1974). *Zeitschrift für Tierpsychologie*, **68**, 313–325.

Heyes, C.M. (1998) Theory of mind in non human primates. *Behavioral and Brain Sciences*, **21**, 101–134.

Hiatt, R.W. & Brock, V.E. (1948) On the herding of prey and the schooling of the black skipjack, *Euthynnus yaito* Kishinouye. *Pacific Science*, **2**, 297–298.

Huber, R., van Staaden, M.J., Kaufman, L.S. & Liem, K.F. (1997) Microhabitat use, trophic patterns, and the evolution of brain structure in African cichlids. *Brain Behavior and Evolution*, **50**, 167–182.

Johnstone, R.A. (2001) Eavesdropping and animal conflict. *Proceedings of the National Academy of Sciences of the USA*, **98**, 9177–9180.

Keenleyside, M.H.A. (1991) *Cichlid Fishes*. Chapman & Hall, London.

Kohler, U. (1997) Zur struktur und evolution des sozialsystems von *Neolamprologus multifasciatus* (Cichlidae, Pisces), dem kleinsten schneckenbuntbarsch des tanganjikasees. PhD thesis, Ludwig-Maximilian-Universität, Munich.

Kotrschal, K., van Staaden, M.J. & Huber, R. (1998) Fish brains: evolution and environmental relationships. *Reviews in Fish Biology and Fisheries*, **8**, 373–408.

Krause, J. (1993) Transmission of fright reaction between different species of fish. *Behaviour*, **127**, 37–48.

Laland, K.N. & Williams, K. (1997) Shoaling generates social learning of foraging information in guppies. *Animal Behaviour*, **53**, 1161–1169.

Laland, K.N. & Williams, K. (1998) Social transmission of maladaptive information in the guppy. *Behavioral Ecology*, **9**, 493–499.

Losey, G.C., Grutter, A.S., Rosenquist, G., Mahon, J.L. & Zamzow, J.P. (1999) Cleaning symbiosis: a review. In: V.C. Almada, R.F. Oliveira & E.J. Goncalves (eds), *Behaviour and Conservation of Littoral Fish*, pp. 379–395. Instituto Superior de Psicologia Aplicada, Lisbon.

Lotem, A., Fishman, M.A. & Stone, L. (2003) From reciprocity to unconditional altruism through signalling benefits. Proceedings of the Royal Society Series B, **270**, 199–205.

LoVullo, T.J., Stauffer, J.R. & McKaye, K.R. (1992) Diet and growth of a brood of *Bagrus meridionalis* (Siluriformes: Bagridae) in Lake Malawi, Africa. *Copeia*, **1992**, 1084–1088.

Luce, R.D. & Raiffa, H. (1957) *Games and Decisions*. Wiley, New York.

Magurran, A.E. & Higham, A. (1988) Information transfer across fish shoals under predator threat. *Ethology*, **78**, 153–158.

McGregor, P.K. (1993) Signalling in territorial systems: a context for individual identification, ranging and eavesdropping. *Philosophical Transactions of the Royal Society of London, Series B*, **340**, 237–244.

Milinski, M. (1987) Tit for tat in sticklebacks and the evolution of cooperation. *Nature*, **325**, 433–435.

Milinski, M., Külling, D. & Kettler, R. (1990a) Tit for tat: sticklebacks *Gasterosteus aculeatus* 'trusting' a cooperative partner. *Behavioral Ecology*, **1**, 7–11.

Milinski, M., Pfluger, D., Külling, D. & Kettler, R. (1990b) Do sticklebacks cooperate repeatedly in reciprocal pairs? *Behavioral Ecology and Sociobiology*, **27**, 17–21.

Milinski, M., Griffiths, S., Wegner, K.M., Reusch, T.B.H., Haas-Assenbaum, A. & Boehm, T. (2005) Mate choice decisions of stickleback females predictably modified by MHC peptide ligands. *Proceedings of the National Academy of Science of the USA*, **102**, 4414–4418.

Mun, C.A. (1986) Birds that 'cry wolf'. *Nature*, **319**, 143–145.

Myrberg, A.A. & Riggio, R.J. (1985) Acoustically mediated individual recognition by a coral reef fish (*Pomacentrus portitus*). *Animal Behaviour*, **33**, 411–416.

Myrberg, A.A., Montgomery, W.L. & Fishelson, L. (1988) The reproductive behaviour of *Acanthurus nigrofuscus* (Forskal) and other surgeon fishes (Fam. Acanthuridae) of Eilat, Israel (Gulf of Aqaba, Red Sea). *Ethology*, **79**, 31–61.

Noakes, D.L. (1979) Parent-touching behaviour by young fishes: incident, function and causation. *Environmental Biology of Fishes*, **4**, 389–400.

Noble, G.K. & Curtis, B. (1939) The social behaviour of the jewel fish, *Hemichromis bimaculatus* Gill. *Bulletin of the American Museum of Natural History*, **76**, 1–46.

Nowak, M.A. & Sigmund, K. (1998) Evolution of indirect reciprocity by image scoring. *Nature*, **393**, 573–577.

Oliveira, R.F. (2005) Hormones, social context and animal communication. In: P.K. McGregor (ed.), *Communication Networks*, pp. 481–520. Cambridge University Press, Cambridge.

Oliveira, R.F., McGregor, P.K. & Latruffe, C. (1998) Know thine enemy: fighting fish gather information from observing conspecific interactions. *Proceedings of the Royal Society of London, Series B*, **265**, 1045–1049.

Olsen, K.H., Grahn, M., Lohm, J. & Langefors, A. (1998) MHC and kin recognition in juvenile arctic charr, *Salvelinus alpinus* (L.). *Animal Behaviour*, **56**, 319–327.

Ostrom, E. (1990) *Governing the Commons: The Evolution of Institutions for Collective Action*. Cambridge University Press, New York.

Pitcher, T.J., Green, D.A. & Magurran, A.E. (1986) Dicing with death: predator inspection behavior in minnow *Phoxinus phoxinus* shoals. *Journal of Fish Biology*, **28**, 439–448.

Potts, G.W. (1973) The ethology of *Labroides dimidiatus* (Cuv. and Val.) (Labridae, Pisces) on Aldabra. *Animal Behaviour*, **21**, 250–291.

Premack, D. & Woodruff, G. (1978) Does the chimpanzee have a theory of mind? *Behavior and Brain Science*, **4**, 515–526.

Roberts, G. & Sherratt, T.N. (1998) Development of cooperative relationships through increasing investment. *Nature*, **394**, 175–179.

Roth, G. & Wullimann, M.F. (eds) (2001) Brain Evolution and Cognition. John Wiley & Sons, Inc., New York-Heidelberg, and Spektrum Akademischer Verlag, Heidelberg.

Schlupp, I., Marler, C. & Ryan, M.J. (1994) Benefit to male sailfin mollies of mating with heterospecific females. *Science*, **263**, 373–374.

Schmitt, R.J. & Strand, S.W. (1982) Cooperative foraging by yellowtail *Seriola lalandei* (Carangidae) on two species of fish prey. *Copeia*, **1982**, 714–717.

Schönherr, J. (1954) Über die abhängigkeit der instinkthandlungen vom vorderhirn und zwischenhirn (Epiphyse) bei *Gasterosteus aculeatus* L. *Zoologische Jahrbücher. Abteilung für Allgemeine Zoologie und Physiologie der Tiere*, **65**, 357–386.

Schradin, C. & Lamprecht, J. (2000) Female-biased immigration and male peace-keeping in groups of the shell-dwelling cichlid fish *Neolamprologus multifasciatus*. *Behavioral Ecology and Sociobiology*, **48**, 236–242.

Seegar, J. (1956) Brain and instinct with *Gaterosteus aculeatus*. *Proceedings of the Koninklijke Nederlandse Akademie van Wetenschappen, Series C– Biological and Medical Sciences*, **59**, 738–749.

Sette, O.E. (1950) Biology of the Atlantic mackerel (*Scomber scombrus*) of North America II: migration and habits. *Fish Bulletin 49, U. S. Fish Wildlife Service*, **51**, 251–358.

Stiver, K.A., Dierkes, P., Taborsky, M. & Balshine, S. (2004) Dispersal patterns and status change in a co-operatively breeding fish *Neolamprologus pulcher*: evidence from microsatellite analyses and behavioural observations. *Journal of Fish Biology*, **65**, 91–105.

Suboski, M.D. & Templeton, J.J. (1989) Life skills training for hatchery fish: social learning and survival. *Fisheries Research*, **7**, 343–352.

Taborsky, M. (1984) Broodcare helpers in the cichlid fish, *Lamprologus brichardi*: their costs and benefits. *Animal Behaviour*, **32**, 1236–1252.

Taborsky, M. (1985) Breeder–helper conflict in a cichlid fish with broodcare helpers: an experimental analysis. *Behaviour*, **95**, 45–75.

Taborsky, M. (1994) Sneakers, satellites, and helpers: parasitic and cooperative behavior in fish reproduction. *Advances in the Study of Behavior*, **23**, 1–100.

Taborsky, M. & Grantner, A. (1998) Behavioural time-energy budgets of cooperatively breeding *Neolamprologus pulcher* (Pisces: Cichlidae). *Animal Behaviour*, **56**, 1375–1382.

Taborsky, M. & Limberger, D. (1981) Helpers in fish. *Behavioral Ecology and Sociobiology*, **8**, 143–145.

Tebbich, S., Bshary, R. & Grutter, A.S. (2002) Cleaner fish *Labroides dimidiatus* recognize familiar clients. *Animal Cognition*, **5**, 139–145.

Thorndike, E.L. (1911) *Animal Intelligence: Experimental Studies*. Macmillan, New York.

Trivers, R.L. (1971) The evolution of reciprocal altruism. *Quarterly Review of Biology*, **46**, 35–57.

Tulving, E. (1983) *Elements of Episodic Memory*. Oxford University Press, New York.

Tulving, E. & Markowitsch, H.J. (1998) Episodic and declarative memory: role of the hippocampus. *Hippocampus*, **8**, 198–204.

van Staaden, M.J., Huber, R., Kaufman, L. & Liem, K. (1995) Brain evolution in cichlids of the African Great Lakes: brain and body size, general patterns and evolutionary trends. *Zoology*, **98**, 165–178.

de Waal, F.B.M. & van Roosmalen, A. (1979) Reconciliation and consolation among chimpanzees. *Behavioral Ecology and Sociobiology*, **5**, 55–66.

Warner, R.R. (1988) Traditionality of mating site preferences in a coral reef fish. *Nature*, **335**, 719–721.

Warner, R.R. (1990) Resource assessment versus traditionality in mating site determination. *American Naturalist*, **135**, 205–217.

Werner, N.Y., Balshine-Earn, S., Leach, B. & Lotem, A. (2004) Helping opportunities and space segregation among helpers in cooperatively breeding cichlids. *Behavioral Ecology*, **14**, 749–756.

Witte, K. & Ryan, M.J. (2002) Mate choice copying in the sailfin molly, *Poecilia latipinna*, in the wild. *Animal Behaviour*, **63**, 943–949.

Witte, K. & Noltemeier, B. (2002) The role of information in mate-choice copying in female sailfin mollies (*Poecilia latipinna*). *Behavioral Ecology and Sociobiology*, **52**, 194–202.

Zahavi, A. (1995) Altruism as a handicap – the limitations of kin selection and reciprocity. *Journal of Avian Biology*, **26**, 1–3.

Chapter 13
Neural Mechanisms of Learning in Teleost Fish

Fernando Rodríguez, Cristina Broglio, Emilio Durán,
Antonia Gómez, and Cosme Salas

13.1 Introduction

One of the most prevailing ideas about learning and memory is that these capabilities are attributes that distinguish the 'more recent' or 'more evolved' vertebrates, such as mammals and birds. According to this customary view, fishes represent the 'most primitive' and 'least evolved' vertebrate group and are supposed to have developed only relatively simple neural circuits, sustaining elemental forms of behaviour. In contrast, other vertebrates such as mammals would be characterized by higher cognitive capacity and behavioural flexibility, mainly associated with the expansion of the telencephalon and the emergence of the six-layered neocortex (Papez 1929; Romer 1962; Jerison 1973; MacLean 1990). Thus, the behaviour of fishes is thought to be mainly 'mechanical' or 'reflex' and lacking a significant degree of learning and memory capabilities. This pervasive idea, which is a consequence of the hybridizing of some Darwinian concepts with the traditional Aristotelian idea of scala naturae, has dominated the landscape of comparative psychology and neuroscience during the past century (Hodos & Campbell 1969, 1990; Deacon 1990). The earlier established theories about vertebrate brain evolution (Papez 1929; Ariëns-Kappers *et al.* 1936; Crosby & Schnitzlein 1983; MacLean 1990) considered the forebrain of actinopterygian fishes to consist mainly of a subpallium ('paleostriatum') and a very small pallium ('paleocortex'), both entirely dominated by olfactory inputs and consequently lacking a significant role, if any, in complex cognitive capabilities. According to these theories, the fish telencephalon should lack an 'archistriatum', a 'neostriatum', and an 'archicortex' (the proposed antecedents of the mammalian pallial amygdala, caudate and putamen, and hippocampus, respectively), and of course it should also lack a 'neocortex'. All these structures would have evolved later, supporting cognition and intelligent behaviour in more recent and complex vertebrate groups.

However, in light of recent developmental, neuroanatomical and functional data, it is necessary to revise this anagenetic, anthropocentric view concerning the evolution of brain, behaviour and cognition in vertebrates. Broad areas equivalent to those of land vertebrates can be recognized in the central nervous system of fishes (Fig. 13.1). Indeed, regardless of conspicuous morphological and cytoarchitectural differences, the telencephalon of every vertebrate radiation is characterized by equivalent pallial and subpallial zones. Moreover, phylogenetic analysis indicates that the main subdivisions of the pallium in the actinopterygian telencephalon are probably homologous to the hippocampus, the amygdala and the neocortex of tetrapods

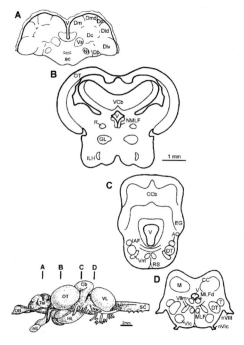

Fig. 13.1 The brain of a teleost fish showing a coronal section at the level of each main subdivision of the goldfish brain: (A) telencephalon; (B) mesencephalon; (C) cerebellum; (D) hindbrain. The insert shows a lateral view of the goldfish brain. The vertical lines indicate the level of each plate. Abbreviations: ac, anterior commissure; AO, anterior octaval nucleus; CC, crista cerebellaris; CCb, corpus cerebelli; Cb, cerebellum; Dc, area dorsalis telencephali pars centralis; Dd, area dorsalis telencephali pars dorsalis; Dld, area dorsalis telencephali pars lateralis; Dlv, area dorsalis telencephali pars lateralis ventralis; Dm, area dorsalis telencephali pars medialis; Dmd, area dorsalis telencephali pars medialis dorsalis; Dp, area dorsalis telencephali pars posterior; DT, descending trigeminal tract; EG, eminentia granularis; GL, nucleus glomerulosus; IAF, internal arcuate fibres; ILH, inferior lobe of the hypothalamus; HL, hypothalamic lobe; MLF, medial longitudinal fasciculus; MLFd, dorsal bundle of the medial longitudinal fasciculus; NMLF, nucleus of the medial longitudinal fasciculus; Nt, nucleus taeniae; nII, optic nerve; nVIc, caudal root of the abducens nerve; OB, olfactory bulb; OT, optic tectum; R, red nucleus; RS, superior reticular nucleus; SC, spinal cord; T, tangential nucleus; Tel, telencephalon; V, ventricle; VCb, valvula cerebelli; VL, vagal lobe; Vm, trigeminal motor nucleus; VIc, caudal subdivision of the abducens nucleus; VIIm, facial motor nucleus.

(Braford 1995; Northcutt 1995; Wullimann & Rink 2002; Butler & Hodos 2005; and see Fig. 13.2). We know now that the olfactory areas represent only a limited portion of the fish pallium, and that, as in land vertebrates, the hippocampal pallium of actinopterygian fishes is implicated in spatial memory and temporal attribute processing, whereas the amygdalar pallium is involved in emotional memory (Salas *et al.* 2003; Broglio *et al.* 2005). In the present review we summarize recent evidence on the behavioural and cognitive capabilities and their neural basis in teleost fishes, which challenges some traditional notions of brain and cognition evolution.

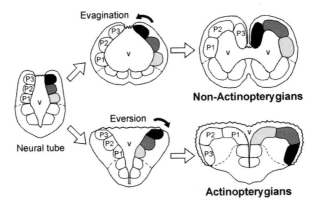

Fig. 13.2   Eversion versus evagination during the embryological development of the telencephalon of actinopterygian and non-actinopterygian vertebrates. The vertebrate forebrain shows an impressive range of morphological variation. One major variation is found in the telencephalon of actinopterygian fish (for instance, teleost fish), which undergoes a process of eversion during embryological development, relative to the telencephalon of other vertebrates (for instance, amniotes), which develops by a process of evagination. This schema represents the process of evagination that occurs in the telencephalon of non-actinopterygian vertebrates during embryological development, compared with the process of eversion or bending outward that occurs in actinopterygians. These different developmental processes produce notable morphological divergence, mainly paired telencephalic hemispheres with internal ventricles in non-actinopterygians, which contrast with the solid telencephalic hemispheres flanking a single ventricular cavity in the actinopterygian radiation. However, despite these conspicuous morphological and cytoarchitectural differences, the telencephala of actinopterygian and non-actinopterygian vertebrates present equivalent pallial and subpallial zones, and the pallium of the actinopterygian telencephalon probably contains subdivisions homologous to the hippocampus, amygdala and neocortex of land vertebrates (Braford 1995; Nieuwenhuys *et al.* 1998; Vargas *et al.* 2000; Butler & Hodos 2005). P1, P2, and P3 correspond to the three main subdivisions of the pallium; v, ventricle.

13.2   Pioneering studies

The pioneering research on the neural basis of learning and memory in teleost fishes has focused almost exclusively on the role of the telencephalon (for reviews see Savage 1980; Overmier & Hollis 1983, 1990). The first studies reported that ablation of the whole telencephalon in teleost fish produced few (if any) immediate effects on fish behaviour (Nolte 1932; Janzen 1933; Savage 1969b). Thus, gross sensory or motor deficits were not apparent after telencephalon ablation and the motivation of the ablated animals appeared to remain at normal levels (Polimanti 1913; Nolte 1932; Janzen 1933; Hosch 1936; Hale 1956). However, more careful analyses revealed the presence of significant alterations in emotional, social and reproductive behaviour after telencephalon ablation, as well as profound learning and memory deficits (Aronson 1970; Hollis & Overmier 1978; de Bruin 1980; Davis & Kassel 1983; Overmier & Hollis 1983). From these studies it became evident that telencephalon ablation in fishes produces severe impairments in different learning tasks, for example, in avoidance learning (Hainsworth *et al.* 1967; Overmier & Flood 1969; Savage 1968, 1969a), and in instrumental learning when there is a delay between the response and the reward (Savage & Swingland 1969; Overmier & Patten 1982), although it does not impair simple instrumental learning and classical condition-ing (Overmier & Curnow 1969; Flood & Overmier 1971; Frank *et al.* 1972; Overmier & Savage 1974; Farr & Savage 1978; Hollis & Overmier 1982). To account for these results, it was proposed that the fish telencephalon is implicated in specific psycho-logical processes (Flood *et al.* 1976; Hollis & Overmier 1978; Savage 1980), such as functions of arousal or attention (Aronson 1970, 1981), inhibition of dominant responses (Kholodov 1960; Savage 1968), working memory (Savage 1968; Beritoff 1971) and secondary or conditioned reinforcer processing (Flood *et al.* 1976; Hollis & Overmier 1978). One limitation of the pioneering studies is that they regarded the telencephalon as a single unitary entity, more than as the summation of different circuits and mechanisms. In consequence, they tended to interpret the effects of the lesions mainly as a deficit in a particular psychological function. As we will discuss below, the picture is probably more complex, and the deficits observed after ablation of the whole telencephalon in teleost fish is a combination of deficits in many differ-ent cognitive functions, subserved by multiple and separate cerebral circuits.

In addition, some early studies also identified the cerebellum as an important neural centre for learning and memory in teleost fishes, besides its well known involvement in motor coordination and in the regulation and plasticity of locomotor activity (Roberts *et al.* 1992, 2002). It has been reported that a lesion of the corpus and the valvula of the cerebellum disrupts reflex conditioning (Karamian 1956), instru-mental learning (Aronson & Herberman 1960), and avoidance conditioning (Kaplan & Aronson 1969). In the next paragraphs we will discuss recent evidence indicating that the cerebellum of teleost fish is essential for the conditioning of discrete motor responses and is involved also in emotional learning and spatial cognition.

13.3   Classical conditioning

Fish show reliable classical conditioning in a variety of reflexes and response systems, and in a wide range of conditions. Similarly to mammals, they show sensitivity to the predictive relationship between the conditioned and the unconditioned stimuli, and exhibit overshadowing, blocking, autoshaping, and higher-order conditioning (Davey 1989; Overmier & Hollis 1990). In addition, recent evidence suggests that at least some of the neural mechanisms underlying these learning phenomena in teleost fishes are shared with other vertebrates.

13.3.1   *Classical conditioning and teleost fish cerebellum*

In a typical classical conditioning paradigm (for instance, eye-blink classical condi-tioning), animals learn to express a conditioned response (CR; an eye-blink or eye retraction movement), to a predictive or conditioned stimulus (CS; light or sound) that is paired with a significant unconditioned stimulus (US; eye air-puff, or mild electric shock). In mammals, the essential circuit for acquisition and performance

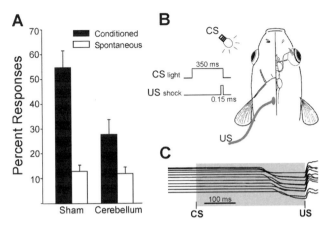

Fig. 13.3 Cerebellum lesions in goldfish impair the classical conditioning of a simple eye-retraction response equivalent to the eye-blink conditioning described in mammals. (A) Cerebellum ablation produces a severe impairment in delay eye-blink classical conditioning relative to the sham-operated animals. In contrast, no significant changes were observed in the percentage of spontaneous eye-movement responses. (B) Fishes were trained in a delay classical conditioning procedure analogous to the eye-blink classical conditioning model. In this procedure, the conditioned stimulus (CS) was a light (350 ms in duration), and the unconditioned stimulus (US) consisted of a mild shock (0.15 ms in duration). In this paradigm, the CS onset precedes the US, but both stimuli overlap in time and coterminate. The US evoked an unconditioned eye-retraction reflex or 'eye blink'. (C) The traces show some illustrative examples of conditioned responses (eye retractions) during paired presentations of the conditioned and the unconditioned stimuli in the control animals. Note the eye movements before the onset of the US. (Modified from Rodríguez *et al.* 2005.)

of this simple, learned reflex resides in the cerebellum and related brainstem struc-tures (McCormick & Thompson 1984; Thompson & Krupa 1994). As is the case in mammals, the teleost fish cerebellum is involved in the classical conditioning of motor responses. Cerebellum lesions in fishes impair the classical conditioning of a simple eye-retraction reflex analogous to the eye-blink conditioning described in mammals (Fig. 13.3). In a recent study on the contribution of teleost fish cerebellum in classical conditioning, cerebellum-ablated and sham-operated goldfish (*Carassius auratus*, Cyprinidae) were trained in a delay classical conditioning procedure ana-logous to the eye-blink classical conditioning model commonly used in mammals (Gómez 2003). A light was used as CS and a mild perifacial shock as US. A delay paradigm was used, that is, the CS onset preceded the US, but both stimuli over-lapped in time and coterminated. The US evoked an unconditioned eye-retraction or eye blink (unconditioned response, UR). Over the course of training the unoperated and control goldfish showed a progressive and significant increase in the percentage of CRs to the CS presentation, which were accurately timed to the onset of the US. In the control goldfish, as in mammals, the percentage of CRs increased with paired CS-US presentations and decreased with CS alone (extinction) or unpaired CS-US presentations. The sensitivity of the goldfish performance to these variations in train-ing conditions (i.e. in the CS-US relationships) indicates that learning was governed by associative rules, enabling them to discard the possibility of pseudoconditioning biases or other non-associative mechanisms. The results of this study showed that, as in mammals, cerebellar lesions almost completely and permanently abolish eye-blink conditioning in goldfish. The cerebellum-lesioned goldfish did not show significant changes in the percentage of CRs, independent of training conditions. In fact, the results indicate a severe learning impairment in as much as the number of CRs did not increase after 300-paired CS-US training trials. Furthermore, the deficit observed in the cerebellum-lesioned group was selective to the CRs, as the percentage of URs, and spontaneous eye-movement responses was similar in both groups.

A recent experiment, in which possible learning-related changes in the metabolic activity of the cerebellum of goldfish were studied by means of cytochrome oxidase (COX) histochemistry, provided additional evidence of the involvement of this structure in eye-blink classical conditioning (Álvarez *et al.* 2002). Goldfish in the eye-blink classical conditioning group (300 CS-US paired trials during a 4-hour session) were compared with two control groups, one subjected to unpaired presentations of the same stimuli and the other untrained. The animals were perfused immediately and their brains processed for COX histochemistry. Optical densitometry analysis showed an increase in the level of COX activity in the molecular and granular layers of the cerebellum, which was selective to the goldfish trained in the paired condition. That is, the COX activity did not increase in the cerebellum of the fishes in the unpaired CS-US condition. Note that the only training difference between both groups was restricted to the mode in which the CS and the US were presented. Consequently, the remarkable metabolic increment observed in the cerebellum of the animals in the paired CS-US procedure could not be caused by unspecific sensory or emotional factors. In contrast, these data reveal specific learning-related changes in the cerebellum of goldfish.

Although the cerebellum shows notable macroanatomical variability across vertebrate species, the pattern of cytoarchitectural organization, the basic circuitry and extrinsic connectivity, as well as the neurophysiological mechanisms, are highly conserved (Kotchabhakdi 1976; Moyer *et al.* 1990; Meek & Nieuwenhuys 1998; Butler & Hodos 2005). In addition, recent evidence from experimental and neuropsychological studies indicates that the cerebellum, traditionally associated with motor control, is implicated in a variety of cognitive and emotional functions in humans and other mammals (Lalonde & Botez 1990; Thompson & Krupa 1994; Petrosini *et al.* 1998). Similarly, the cerebellum of teleost fishes is not only involved in the classical conditioning of motor responses, but also, as we will review later in this chapter, is involved in emotional conditioning and in more complex, higher-order processes such as spatial cognition.

13.3.2   *Trace classical conditioning and teleost telencephalic pallium*

In mammals, the cerebellum and related brainstem circuits mediate eye-blink classical conditioning (see above); that is, brain structures above the level of the midbrain are not required for conditioning this simple motor response (Oakley & Russell 1977). However, this is valid for delay conditioning, but not for trace conditioning. Trace conditioning imposes additional task requirements, as in this procedure the CS and the US do not overlap; instead the end of the CS is separated from the onset of the US by a stimulus-free time gap (trace interval). In these conditions, some telencephalic structures, for example the hippocampus, become engaged in mammals (Moyer *et al.* 1990; Kim *et al.* 1995). As in mammals, the acquisition and maintenance of eye-retraction classical conditioning in goldfish is critically dependent on the cerebellum, irrespective of whether a delay or a trace procedure is used (Álvarez *et al.* 2003; Gómez 2003; Gómez *et al.* 2004). The cerebellum and associated brainstem circuitry seem to be sufficient for eye-retraction conditioning in the delay paradigm, as the complete ablation of the forebrain spares this form of learning in goldfish (Gómez *et al.* 2004). However, lateral pallium (LP) lesions in goldfish, like hippocampus lesions in mammals, selectively impair eye-retraction conditioning when a trace interval is introduced between the end of the CS and the onset of the US, without producing any significant deficit in delay conditioning (Gómez *et al.* 2004). In addition, medial pallium (MP) lesions do not produce any observable deficit in eye-retraction conditioning in goldfish, in either trace or delay conditioning procedures. Interestingly, the LP of teleost fishes has been proposed as the homologue of the medial cortex or hippocampus of amniotes on the basis of developmental, neuroanatomical and behavioural data (Northcutt & Braford 1980; Nieuwenhuys & Meek 1990; Northcutt 1995; Butler 2000; Rodríguez *et al.* 2002b; and see Fig. 13.2). Thus, these findings reveal that, in teleost fishes, as in mammals, the cerebellum plays an essential role in classical conditioning independently of the temporal requirements of the procedure, whereas the hippocampal pallium is specifically involved in trace conditioning but not in delay conditioning.

13.4   Emotional learning

It has long been known that the telencephalon of teleost fishes is involved in emotional, aggressive and reproductive behaviour (Segaar & Nieuwenhuys 1963; Overmier & Gross 1974; Shapiro *et al.* 1974; de Bruin 1980). In particular, a telencephalic structure that seems to play an important role in those aspects of behaviour in which the motivational and emotional factors must be taken into account is the MP. It has been reported that lesions to the MP disrupt or disorganize aggressive, reproductive and parental behaviour in teleost fish (Segaar & Nieuwenhuys 1963; de Bruin 1983). In addition, electrical stimulation in the MP in free-swimming fishes evokes arousal, defensive behaviour and escape responses (Savage 1971; Quick & Laming 1988). Interestingly, the medial telencephalic pallium of the actinopterygian fish has been proposed to be homologous to the amygdala of the land vertebrates on the basis of developmental and neuroanatomical comparative evidence (Marino-Nieto & Sabbatini 1983; Nieuwenhuys & Meek 1990; Braford 1995; Northcutt 1995; Butler 2000; and see Fig. 13.2). In mammals, the amygdala is involved in emotional behaviour and emotional learning (LeDoux 1995).

13.4.1   *Medial pallium and avoidance conditioning*

A recent series of experiments has specifically addressed the question of the possible involvement of the teleost telencephalic pallium in emotional learning and memory;

for example, one of these experiments showed that MP but not LP lesions impaired acquisition as well as retention of conditioned avoidance in goldfish (Portavella *et al.* 2004a). In the avoidance conditioning paradigm animals learn to prevent the presentation of an unpleasant unconditioned stimulus (i.e. a mild electric shock), by producing a particular response (such as jumping to a safe area) in response to the presentation of a conditioned stimulus (i.e. a light) that signals the presentation of the unconditioned stimulus. There is good evidence that avoidance learning is based on the acquisition of a mediational state of fear in goldfish; for example, prior Pavlovian pairings of the warning stimulus with a shock facilitate subsequent avoidance-conditioning acquisition (Gallon 1972; Overmier & Starkman 1974). Thus, it seems that in avoidance situations the Pavlovian contingency leads to the acquisition of an internal state of fear (conditioned fear) that, in turn, contributes to the development of the instrumental stimulus-response association because of a reduction of fear that follows the avoidance response (Mowrer 1947; Flood *et al.* 1976; Overmier & Hollis 1990; Zhuikov *et al.* 1994; Portavella *et al.* 2003). Interestingly, complete telencephalic ablation in teleost fishes produces devastating effects on the acquisition and maintenance of conditioned avoidance (Savage 1969a; Overmier & Papini 1985, 1986; Papini 1985; Overmier & Hollis 1990), suggesting that the teleost telencephalon is involved in the use of the emotional states as conditioned reinforcers to produce instrumental responses (Mowrer 1960; Flood *et al.* 1976). Thus, ablation of the telencephalon disrupts avoidance conditioning because it prevents engagement between these two learning processes (Pavlovian and instrumental). The study by Portavella *et al.* (2004a) has probably identified the critical areas of the teleost telencephalon involved in this function. This study showed that the damage to the MP, but not to other pallial areas, for example the LP, produces a deficit in the retention of conditioned avoidance as severe as that after ablation of the whole telencephalon (Fig. 13.4). Medial-pallial-lesioned animals are able to improve escape responses along the lines of postsurgical training but not to produce avoidance responses. The behavioural deficit caused by the MP lesion in goldfish could be caused by a deficit in the retrieval of the anticipatory fear response to the warning stimulus or by a deficit in the ability of an internal state of fear to induce the avoidance response (Portavella *et al.* 2004a). These results suggest that the MP could be the main telencephalic area involved in this kind of learning in teleost fishes, and indicate also that restricted telencephalic areas in teleost fishes are involved in different learning functions as parts of separate specialized memory systems, as has been proposed for land vertebrates (Nadel 1994; Schacter & Tulving 1994).

Although the results showing that lesion of the MP impairs avoidance learning in goldfish are analogous to those observed in mammals with amygdala lesions (O'Keefe & Nadel 1978; Aggleton 1992), the use of relatively massed training conditions (i.e. multiple trials per session) in the above-mentioned experiments may be, in itself, a source of experimental confounding, and may prevent an unambiguous interpretation of the results. In fact, in relatively massed training conditions, the possibility exists that the recent presentation of shocks and the performance of the shuttle response in the previous trials may become an additional source of control over behaviour (Bitterman 1975). These salient, recent events may enhance the retrieval and allow the information to carry over across trials and may facilitate the acquisition of fear in its own right, rendering ambiguous the claim that fear is induced

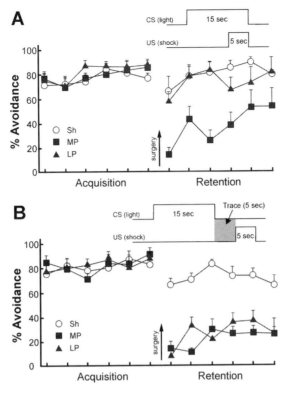

Fig. 13.4 Involvement of medial (MP) and lateral (LP) telencephalic pallium of teleost fishes in emotional and temporal learning. This experiment analysed the effects of MP and LP telencephalic lesions on the retention of an avoidance response previously acquired, in two different conditioning situations, one with stimuli overlapping (delay) and the other with an interstimuli gap (trace). A two-way active avoidance paradigm was used in a shuttle box adapted for goldfish conditioning. (A) In the first experiment, the discriminative stimulus (light) was turned on for a maximum of 15 sec in the compartment where the fish was located. If the fish did not respond (swimming across the barrier) within 10 sec of light onset, an electric shock was turned on for a maximum of 5 sec. Thus, the temporal separation between cue onset and shock onset was 10 sec. A response during the first 10 sec terminated the warning stimulus (light), and the shock was not delivered. A response during the 10–15 sec period cancelled both the warning stimulus and the shock. Once the animals learned the task, the specific pallial lesions were conducted. Results showed that damage to the goldfish MP produces a deficit in the retention of conditioned avoidance, as severe as that after ablation of the whole telencephalon (data not shown). In contrast LP lesion had no significant effects on the retention of avoidance in this non-trace procedure. (B) In the second experiment, a trace two-way active avoidance conditioning paradigm was used. The discriminative stimulus (light) was turned on for a maximum duration of 10 sec in the compartment where the fish was located, followed by a gap period of 5 sec after termination of the light. Thus, the temporal separation between cue onset and shock onset was 15 sec. If the fish did not respond within 15 sec, the electric shock was turned on for a maximum of 5 sec. A response during the first 15 sec terminated the warning stimulus (light), and the shock was not delivered. A response during the 15–20 sec period cancelled both the warning stimulus and the shock. Results showed that, unlike in the first experiment, the LP lesion impaired performance in the trace-conditioning procedure. Both figures show mean percentage of avoidance response in acquisition (six sessions after reaching the learning criterion before surgery) and retention (six sessions after surgery). These data support the presence of two different systems of memory in fishes, based on discrete telencephalic areas: the MP, involved in an emotional memory system; and the LP, involved in a spatial, relational, or temporal memory system. (Modified from Portavella *et al.* 2004a.)

by the warning stimulus. To eliminate such a source of experimental confounding, spaced-trial avoidance training was used in a follow-up series of experiments (Portavella *et al.* 2003, 2004b). In these experiments, the spacing of the trials (one trial per day) leaves little doubt that goldfish can retrieve the appropriate information to perform the shuttle response purely on the basis of the warning stimulus. The capacity of goldfish for avoidance learning under spaced-trial conditions implies that the acquired avoidance response can develop in the absence of stimulus carry-over effects from the shock, the discriminative stimulus, and the response-feedback stimuli occurring in previous trials (Bitterman 1975; Portavella *et al.* 2003). On the contrary, these data suggest that the avoidance behaviour depends on the ability of the discriminative stimulus associatively to reinstate a mediational state of fear, where the response-contingent termination maintains the conditioned response, as in the case of mammals (Mowrer 1960; Flood *et al.* 1976; Overmier & Papini 1986; Overmier & Hollis 1990; Zhuikov *et al.* 1994; Portavella *et al.* 2003). Moreover, these experiments show that MP lesions, but not LP lesions, disrupt spaced-trial avoidance learning in goldfish (Portavella *et al.* 2004b), indicating that MP lesions affect avoidance learning by impairing the retrieval of anticipatory fear response by the warning stimulus or by interfering with the ability of an internal state of fear to induce an avoidance response. Because a spaced-trial training procedure was used in this experiment, the effects of MP lesions cannot be explained in terms of disruption of sensory carry-over effects (Mowrer 1960; Overmier & Hollis 1990; Portavella *et al.* 2004b). The conditioned emotional response in mammals is acquired in few trials. Similarly, goldfish are able to learn to perform the avoidance response in few trials (Portavella *et al.* 2003). In conclusion, the absence of carry-over effects, the rapid acquisition of avoidance responses and the impairment of avoidance behaviour in the MP-lesioned animals suggest that the MP in teleost fishes is an essential component of an emotional system that is critical for fear conditioning. Thus, these results show that the deficit in avoidance learning produced by damage to the MP is strikingly similar to the deficits in avoidance learning caused by lesions of the amygdala in mammals (Sarter & Markowitsch 1985; Davis *et al.* 1992).

13.4.2   *Involvement of goldfish lateral pallium in trace avoidance conditioning*

In mammals, lesions of the hippocampal formation produce deficits in avoidance conditioning when contextual or temporal cues are significant for the conditioning situation, as in the case of trace avoidance learning and other trace emotional conditioning procedures (Woodruff & Kantor 1983; Moyer *et al.* 1990; Phillips & LeDoux 1992). Recent experimental evidence indicates that the LP of goldfish, proposed as homologous to the hippocampal formation of tetrapods on the basis of anatomical and developmental evidence, also shares a similar role in trace avoidance conditioning (Portavella *et al.* 2004a). When an inter-stimulus temporal gap (trace avoidance conditioning) is introduced in the two-way active avoidance procedure, LP lesions severely impair goldfish performance (Fig. 13.4). These results show that the LP of actinopterygian fishes plays a major role in the retention of conditioned avoidance in a trace procedure, indicating that the LP of teleost fishes, like the hippocampus of mammals, is involved in trace memories.

13.4.3   *Teleost cerebellum and emotional conditioning*

A growing number of studies suggest that the cerebellum of mammals participates not only in the conditioning of simple motor responses but also in emotional learning (Berntson & Torello 1982; Supple & Leaton 1990a,b; Supple & Kapp 1993; Gherladucci & Sebastiani 1996, 1997; Bobée *et al.* 2000; Sacchetti *et al.* 2002). Interestingly, recent data show that the cerebellum of goldfish, similar to mammals, is also involved in emotional learning. The involvement of the teleost fish cerebellum in emotional learning has been studied by means of analysing the effects of cerebellar lesions on fear heart rate conditioning in goldfish (Álvarez *et al.* 2003; Gómez 2003; Yoshida *et al.* 2004; Rodríguez *et al.* 2005; and see Fig. 13.5). In the control goldfish, paired CS-US presentations consistently produced a rapid increase in the percentage of conditioned bradycardia responses (a deceleration of the heart rate during CS-US interval relative to pre-CS baseline), which decreased quickly during extinction training. In contrast, goldfish with cerebellar lesions failed to acquire the conditioned bradycardia response. It is important to note that no deficit was observed either in the reflex response to the US or in the autonomic orientation response to the CS in cerebellum-ablated animals, indicating that the sensorial and motor neural

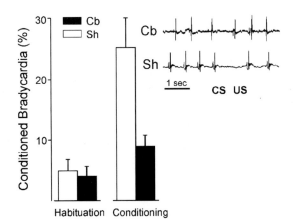

Fig. 13.5   Involvement of the goldfish cerebellum in emotional fear conditioning. The participation of the teleost fish cerebellum in emotional learning was studied by analysing the effects of cerebellar lesions on fear heart rate conditioning in goldfish. In control fish, the paired presentations of conditioned (light) and unconditioned stimuli (shock) consistently produce a conditioned bradycardia. The figure shows the percentage of deceleration of the heart rate during the CS-US (CS, conditioned stimulus; US, unconditioned stimulus) interval relative to pre-CS baseline during habituation and conditioning trials. However, goldfish with cerebellar lesions failed to acquire the conditioned bradycardia response. As in mammals, cerebellar lesions in goldfish impair the acquisition of the conditioned bradycardia without altering the heart rate baseline or the orientation response to the CS (habituation trials). The insert on the right shows electrocardiograms of representative cerebellar-lesioned (Cb) and sham (Sh) animals. Note that the cerebellar-lesioned animals do not exhibit the normal conditioned heart rate deceleration response to the CS. (Modified from Rodríguez *et al.* 2005.)

circuits underlying the expression of the unconditioned cardiac responses were spared in cerebellum-ablated goldfish. Thus, the effects of the cerebellar lesions on the cardiac activity of goldfish seem to be selective to the conditioned bradycardia response. Similarly, cerebellar lesions impair the acquisition of the conditioned bradycardia response in rats and rabbits, without altering the heart rate baseline or the orientation response to the CS (Supple & Kapp 1993; Gherladucci & Sebastiani 1996). Thus, the results from the cerebellar-lesioned fishes are remarkably similar to those obtained in mammals, suggesting that the cerebellum of teleost fishes, like the cerebellum of tetrapods, plays an essential role in the classical conditioning of emotional responses.

13.5   Spatial cognition

Early studies on the cerebral mechanisms of spatial learning and memory in teleost fishes were concentrated almost exclusively on the role of the telencephalon. Although the involvement of the telencephalon of teleost fishes in a variety of learning and memory processes was indicated by early lesion studies (see above), the reports concerning the effects of telencephalon ablation on the performance of fishes trained in alleys and different mazes were often contradictory; for example, following telencephalon ablation in fishes, studies investigating acquisition of spatial learning have reported results varying from impairments to no deficits and even some improvements (Zunini 1954; Hale 1956; Warren 1961; Ingle 1965; Frank *et al.* 1972; Farr & Savage 1978). Similarly, spatial reversal has been reported both as normal (Hosch 1936; Ingle 1965) and as impaired (Flood *et al.* 1976) in fishes with ablation of the telencephalon. The inconsistency of these results could be attributed to the fact that the tasks used in these pioneering ablation studies were not specifically aimed at analysing spatial cognition, and consequently lacked a precise definition of the spatial requirements of the task and specific control tests, essential for characterizing the impairments in spatial performance. Moreover, widely used terms such as 'spatial learning and memory' and 'spatial cognition' involve a variety of perceptive and cognitive processes, actually based in separate neural substrata.

In this regard, comparative evidence shows that separate brain systems contribute in different ways to spatial orientation and navigation. A variety of brain mechanisms are required for processing and integrating the sensory-motor information, and for translating it into a series of coordinate systems, from receptive surface, to head-centred, to body-centred coordinates, and finally to an allocentric, world-centred coordinated system; for example, in mammals the perception and action based on egocentric frames of spatial reference depend on brain circuits that extend from the superior colliculus and the cerebellum to the parietal, somatosensory cortex and the frontal motor cortical areas (Stein & Meredith 1993; Burgess *et al.* 1999). However, the use of allocentric frames of reference for navigation depends on other neural systems, mainly the hippocampal formation. Allocentric navigation is supported by complex map-like memory representations of the environment, based on the encoding of the reciprocal spatial relationships of the goal and multiple sensory features (O'Keefe & Nadel 1978; Burgess *et al.* 1999).

13.5.1   *Cognitive mapping in teleost fishes*

Substantial behavioural and neurobiological research shows that mammals, birds and other land vertebrates use a variety of egocentrically referenced mechanisms for orienting, and in addition, use allocentric memory representations (i.e. cognitive maps) for navigation (O'Keefe & Nadel 1978; Nadel 1991; Bingman 1992; López *et al.* 2000c, 2001; and see Chapter 7). A cognitive map is defined as a map-like, 'world-centred' representation of the objective space that provides a stable framework, allowing the subject to reach the goal independently of its own actual position. It is interesting to note that these map-like representations, which are true relational memories, can be considered the clearest animal equivalent of human declarative or episodic memory (Clayton & Dickinson 1998; Eichenbaum 2000). Although the traditional view is that cognitive mapping capabilities are an exclusive attribute of vertebrate groups that supposedly have evolved more complex associational structures (i.e. mammals and birds), a number of recent, thorough behavioural studies provide strong evidence indicating that teleost fishes are also able to use cognitive mapping strategies.

In fact data provided by some pioneering naturalistic and laboratory studies had already suggested the possibility that fish possessed such cognitive capabilities. The ability of fishes to travel efficiently over a wide range of geographic scales, from foraging excursions within their habitual living areas to intercontinental migrations, implies sophisticated spatial capabilities for navigating, orienting, piloting and recognizing their environment. This spatial behaviour is clearly a most flexible process, probably involving learning and memory mechanisms and cognitive phenomena (Helfman *et al.* 1982; Hallacher 1984; Helfman & Schultz 1984; Dodson 1988; Markevich 1988; Reese 1989; Odling-Smee & Braithwaite 2003). Also, some early laboratory studies showed that fishes, similar to mammals, use systematic exploration to extract spatial information in unfamiliar environments; for example, fishes use an organized pattern of exploration when introduced into a novel environment, avoiding previously visited locations (Kleerekoper *et al.* 1974), and react with increased exploratory activity to environmental modifications (Welker & Welker 1958; Russell 1967). The blind cave fish (*Anoptichthys jordani*, Characidae) swim around boundary features and increase swimming velocity when released into unfamiliar surroundings, or following changes in a familiar environment, possibly to optimize the lateral line organ stimulation (Teyke 1989). It has been suggested that these animals build a map-like memory representation of the environment, which allows them to swim at velocities below the optimal range for lateral-line stimulation, without observable deficits in obstacle avoidance (Campenhausen *et al.* 1981; Teyke 1989). Goldfish learn to swim in a constant direction relative to visual cues, even when approaching the goal from opposite directions (Ingle & Sahagian 1973), remember the spatial position of food patches in a tank (Pitcher & Magurran 1983) and use landmarks as indirect spatial reference points (Warburton 1990). All of these data point to the possibility that teleost fishes have the capacity for discriminating and encoding spatial relationships in the environment, allowing navigation independent of a body-centred reference system (see Chapter 7 for further discussion).

Spatial learning and memory capabilities of teleost fishes have been thoroughly examined more recently in controlled laboratory experiments designed to provide optimal conditions to reveal the spatial strategies, the cues and the mechanisms used by these vertebrates for orientation and navigation. Moreover, standard spatial tasks closely matching those used to analyse spatial cognition in mammals and birds, analogous laboratory settings and behavioural procedures, and thorough probe

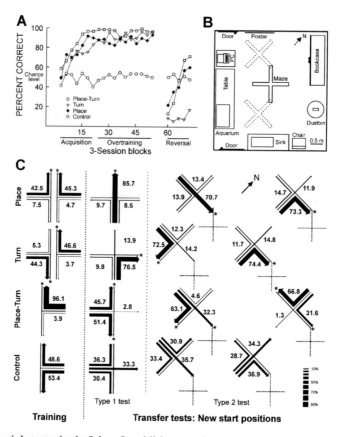

Fig. 13.6 Spatial strategies in fishes. In addition to using egocentric strategies, goldfish are able to build complex spatial representations of their environment and to solve spatial tasks on the basis of allocentric frames of reference. (A) Mean percentage of correct choices of goldfish trained in four different experimental conditions, and (B) in a plus maze located in a room with conspicuous spatial cues. In the place group, two different start positions were used and animals were trained to find the goal situated always in the same place in the room. Thus, animals in this group could only use the extra-maze cues to solve the task. No fixed turn direction was relevant to task solution, because, depending on the start arm, subjects were required to make a left-hand or right-hand turn. In the turn group, two different start positions were used, but in every trial the goal arm was determined by a fixed turn response (e.g. always left). Thus, in this task extra-maze cues were irrelevant to the task solution. The fishes in the place-turn group started always from the same place of the room and the goal arm was always in another constant place. Thus, this task allowed the selection of the correct arm on the basis of a specific turn direction and/or extra-maze cues. Finally, in the control group the location of the goal varied randomly across trials. Note that all groups except controls were able to solve the different tasks with high accuracy after 1 week of training. Once the animals had

and transfer trials, were used in order to increment the validity of the data. The first clear-cut and convincing data concerning the capacity of teleost fishes to use cognitive mapping strategies was provided by Rodríguez *et al.* (1994). In this study goldfish were trained to locate a baited feeder in a four-arm maze surrounded by an array of widely distributed distal visual cues. Goldfish trained in an allocentric procedure navigated directly to the rewarded place, adopting spontaneously the most direct routes to the goal from previously unvisited start locations, although the new routes involved navigating in different or even opposite directions and using shortcuts or detours, or navigating in the absence of any particular subset of visual cues (Rodríguez *et al.* 1994). The use of appropriate trajectories without a history of previous training, even when the new trajectories imply new (never experienced before) egocentric relations to landmarks provides distinct evidence for the capacity of these animals to represent spatial relationships in the environment, independent of a body-centred reference system (Fig. 13.6). Another distinctive piece of evidence concerning cognitive mapping in goldfish was provided by their accuracy in locating the goal in the absence of conspicuous environmental cues (Rodríguez *et al.* 1994). As cognitive maps store redundant environmental information when a subset of spatial cues becomes unavailable, accurate navigation is still possible on the basis of those cues that remain (O'Keefe & Nadel 1978). In addition, the goldfish could use orientation (egocentric) strategies, as indicated by their ability to reach the goal by using a fixed body turn (i.e. turn right or left), disregarding environmental information, or could implement simultaneously both body-centred and allocentric strategies and use one or the other according to experimental conditions (Fig. 13.6). Recently, using experimental procedures that closely matched those used in the study of Rodriguez *et al.* (1994) in goldfish, Schluessel & Bleckmann (2005) have obtained results that suggest that elasmobranchs can also use allocentric strategies for navigation. Like goldfish, ocellate river stingrays (*Potamotrygon motoro*, Potamotrygonidae) are able to reach the goal using novel routes starting from unfamiliar locations in addition to using egocentric strategies.

---

learned their tasks, different transfer tests (C) were run to elucidate whether the animals of the different groups solved their respective tasks on the basis of turn or place strategies. In the type 1 transfer tests the maze remained in its usual position but the animals were released from a novel start position. In the type 2 tests, the maze was displaced in the room (see dashed lines in B) in such a way that the end of one arm was located in the same place in the room where the fish was rewarded during training trials, but the start positions were different to those used during training. The figure shows the trajectories chosen by the animals in the different groups during training and transfer trials. The numbers and the relative thickness of the arrows denote the percentage of times that a particular choice was made. The dashed lines denote the neutral position of the maze before it was displaced for type 2 tests and the asterisks mark the goal location. Note that during the transfer trials the animals in the place group consistently chose the arm where the extreme was at the same place of the room that the fish was rewarded during training trials. In contrast, the most chosen arm by the fish in the turn group was that coinciding with the learned turn, independently of the location of the start arm. These results suggest that goldfish are able to make place responses, to establish new pathways towards a goal from unfamiliar start points and to reorganize their spatial strategies in response to an environmental change. (Modified from Rodríguez *et al.* 1994.)

Taking into account that an important requirement of the laboratory studies aimed at determining the structure and function of memory systems is to obtain convergent experimental evidence in a variety of experimental conditions, the spatial cognition abilities of fishes were analysed under a variety of training conditions and spatial tasks; for example, goldfish trained in a spatial constancy or a cue task to locate a goal in a small enclosure, where only proximal visual cues were available, also implemented orientation or relational strategies according to task demands (Salas *et al.* 1996b; López *et al.* 1999). Thus, whereas goldfish in the spatial

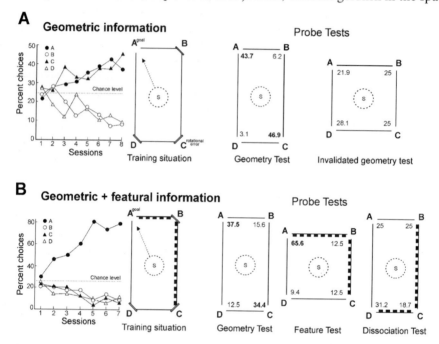

Fig. 13.7   Goldfish can encode geometric and featural information to navigate in the environment. (A) This experiment tested whether goldfish are able to encode and use the geometric information of the environment in a place-finding task. Fish were trained to find the exit door placed in a corner of a rectangular arena that had three identical blocked openings in the other three corners. The overall shape of the arena provided the spatial information. In this task, locating the goal (exit) required fish to determine the spatial relationships between the geometrical properties of the apparatus and the goal. Because of the geometric properties of the apparatus, the correct corner was indistinguishable from the diagonally opposite corner (rotational error). The curves show the percentage of choices of the four corners during the training sessions. The diagram on the right shows the starting position (S), and the location of the glass barriers (corners B, C and D) and the goal (corner A) during training trials. The figure shows that the animals progressively learned to choose the correct door (A) and the geometrically equivalent door (C). The two diagrams on the right show the percentage of selections of each corner (numbers) during the probe tests. In the geometry test, the glass barrier was removed, and the animals chose more frequently the geometrically correct doors. In the invalidated geometry test, the geometric cues were removed by using a new (square) apparatus, which modified the geometric properties of the experimental enclosure used during training. In this test, no significant differences were observed in the percentage of choices between the four doors. (B) In a second

constancy task navigated accurately to the goal from different start locations, regardless of route direction and response requirements (Salas *et al.* 1996b) and despite partial deletion of the visual cues (López *et al.* 1999), the performance of the fishes in the cue task was dramatically impaired when the cue associated directly with the goal was removed (López *et al.* 1999). Interestingly, spatial constancy, but not cued goldfish, was impaired when a modification was introduced into the experimental apparatus that altered its global shape and topography, but left unchanged the visual features of the areas corresponding to each of the doors (López *et al.* 1999). These findings suggested that these animals use geometric information to solve the spatial constancy task. In fact, a follow-up series of experiments demonstrated that fishes, as well as humans and other mammals trained in similar tasks, are able to use the geometry of the surroundings for spatial orientation and conjoin geometric and nongeometric information to reorient themselves (Broglio *et al.* 2000; Sovrano *et al.* 2002; Vargas *et al.* 2004; and see Fig. 13.7). These results are important because the use of the geometrical features of the environment for allocentric orientation indicates the knowledge of the situation as a whole, that is, the metric and geometrical relationships between the constituent elements (Cheng & Gallistel 1984; Cheng 1986; Gallistel 1990).

13.5.2 *Teleost fish telencephalon and spatial cognition*

The similarity of the data mentioned above to that obtained in mammals and birds (Tolman 1948; O'Keefe & Conway 1978; O'Keefe & Nadel 1978; Suzuki *et al.* 1980; Mazmanian & Roberts 1983; Fenton & Bures 1994; Thinus-Blanc 1996), indicates the presence of multiple spatial memory systems in fishes and outlines the central question of whether these cognitive abilities are supported by neural centres and circuits equivalent to those that underlie spatial cognition in land vertebrates. Recent lesion studies aimed at examining this question have provided strong evidence of the

---

experiment goldfish were tested on a different spatial configuration of cues to determine their capacity to codify jointly the geometric and non-geometric properties of the environment in a map-like representation. In this task, fish were trained in the same rectangular box but in which featural information was provided by two removable panels with alternate dark grey and white vertical stripes. These panels were placed on two adjacent walls of the arena. The goal was situated at one of the doors between a striped wall and an un-striped wall. The curves show the percentage of choices for the four corners during the training sessions. The diagram on the right shows the start position (**S**), and the location of the glass barriers, the feature cues (striped walls), and the goal, during training. In this task the animals readily learned to choose the correct door (**A**). The three diagrams on the right show the percentage of selections of each corner (numbers) during the probe tests. In the geometry test the striped panels were removed and the animals chose more frequently the two geometrically correct doors (**A** and **C**). In the feature test, where the geometry information was removed by testing the animal in a square arena, the fish chose the correct door significantly more (**A**). Finally, in the dissociation test, the information provided by the geometry and features were set in conflict by rotating the striped panels 90°. In this test, the fish did not show a preference for any particular door; neither the percentage of geometrical choices nor the feature choices were significantly different from that expected by chance. (Modified from Vargas *et al.* 2004.)

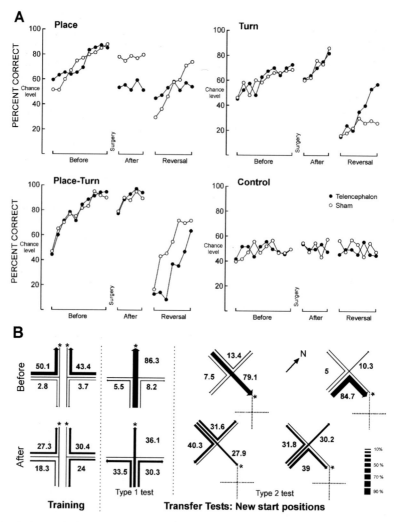

Fig. 13.8   Telencephalic ablation in goldfish produces spatial learning and memory deficits. Goldfish were trained in the same procedures described in Fig. 13.6. Following acquisition, complete ablation of both telencephalic hemispheres was carried out. Subsequently, the animals were retrained in the same task for 10 more sessions and then trained in the reversal of the task during 14 more sessions. (A) Percentage of correct choices of telencephalon-ablated and sham-operated animals for each group during the experiment. Note that ablation impaired performance exclusively in animals using place strategies; in these, accuracy fell to chance level during both post-surgery retraining and reversal periods. In the other groups, ablation of the telencephalon did not induce any significant deficit. (B) Trajectories chosen by animals in the place group during training and transfer tests where new start positions were used (for more explanations see Fig. 13.6). Note that after telencephalic ablation the fishes in the place task failed to navigate accurately to the goal from known locations (training trials) and from novel ones (transfer trials), whereas before ablation these animals accurately chose the arm where they were rewarded. These results suggest that the fish telencephalon plays a crucial role in complex place learning. (Modified from Salas *et al.* 1996b.)

importance of the telencephalon for spatial cognition in teleost fishes. Telencephalon ablation dramatically and irreversibly impaired the ability of the fishes trained in a place procedure to reach goal location (Salas *et al.* 1996a; and see Fig. 13.8). Conversely, telencephalon ablation did not alter the performance of the animals using egocentric (body-centred) orientation strategies. The ablation did not disrupt the overt performance of the fish trained in a mixed place-turn procedure in the same experiment. However, test trials revealed a notable deficit in these animals: whereas before ablation they used either place (allocentric) or turn (egocentric) strategies in a flexible and cooperative way according to the requirements of the tasks, after surgery they only used turn responses (Salas *et al.* 1996a). Consistent results were obtained in a subsequent experiment in which intact and telencephalon-ablated goldfish were trained in a mixed place-cue procedure (López *et al.* 2000a; and see Fig. 13.9). Interestingly, telencephalon-ablated goldfish showed better performance during training relative to sham animals; nonetheless, they were unable to reach the goal when shortcuts or flexible responses were required (control tests), indicating that their performance was based exclusively on egocentric strategies. These results reveal a profound learning deficit in the telencephalon-ablated fishes. In addition, telencephalon ablation disrupted the post-surgery performance (Salas *et al.* 1996b) and reversal learning (López *et al.* 2000b) of goldfish trained in a spatial constancy task, but did not produce any observable deficit in a cue procedure. In summary, the place memory impairments observed in these experiments, showing a selective but severe disruption in spatial cognition after telencephalon ablation, provide significant evidence concerning the presence of a telencephalon-dependent spatial memory system in teleost fish.

13.5.3   *Spatial learning and telencephalic pallium in actinopterygian fishes*

Considerable experimental evidence shows that in mammals, birds and reptiles the hippocampus, among other telencephalic structures, is critical for encoding the environmental information in map-like or relational memory representations for spatial navigation (O'Keefe & Nadel 1978; Sherry & Duff 1996; Bingman *et al.* 1998; Burgess *et al.* 1999; Rodríguez *et al.* 2002a,b; López *et al.* 2003a,b). As mentioned above (see Fig. 13.2) the telencephalon of ray-finned fishes (e.g. teleosts) presents major morphological differences relative to every other vertebrate group (Braford 1995; Northcutt 1995; Nieuwenhuys *et al.* 1998; Butler 2000), derived from a notable variation during embryological development (eversion versus evagination). The eversion process, which underlies the development of the actinopterygian telencephalon determines the reversal of the pallial medial-to-lateral topography observed in the evaginated telencephalon. Thus, the LP of the actinopterygian fishes is the most likely homologue of the amniote medial pallium or hippocampus (Nieuwenhuys 1963; Northcutt & Braford 1980; Nieuwenhuys & Meek 1990; Braford 1995; Northcutt 1995; Butler 2000). Recent functional studies agree with this embryological and anatomical hypothesis; for example, a study aimed at evaluating possible spatial learning-related changes in the transcriptional activity (protein synthesis) in the neurons of different pallial areas in goldfish trained in spatial or cue learning tasks, by means of a silver stain with high affinity for the argyrophilic proteins associated

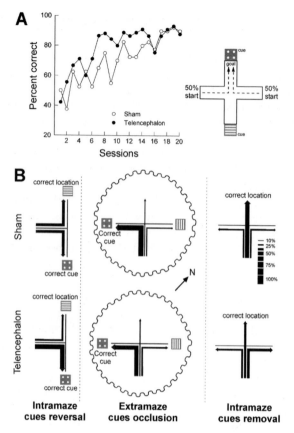

Fig. 13.9 Sham and telencephalon-ablated goldfish were trained in a test environment where they could simultaneously rely on the spatial information provided by extra-maze, distal landmarks on the periphery of the experimental room and on a single cue that directly indicated the location of food in a four-arm maze. (A) Percentage of correct responses of each group during training. Note that the telencephalon-ablated goldfish learned the task more quickly than controls. The insert on the right shows a schematic representation of the maze, the training procedure, and the position of the two intra-maze cues (dotted and striped panels). The maze was located in the centre of a room with abundant extra-maze distal cues (not shown). (B) Probe trials were run to examine the relative importance of the two sources of information in intact and telencephalon-ablated fishes. The figure shows the trajectories chosen by sham and telencephalon-ablated animals in these tests. In the intra-maze cues reversal test, the two sources of information were set in conflict; thus place and cue responses were incompatible. Results showed that control fishes did not have a significant preference between cue and place response. In contrast, telencephalon-ablated fishes showed a preference for the arm containing the cue that was associated with the goal during training. In the extra-maze cues occlusion test, the maze was completely surrounded by a curtain eliminating the use of extra-maze visual cues. In this test, both telencephalon-ablated and control fishes showed a strong preference for the arm containing the cue associated with the goal during training. In the last probe test the intra-maze cues were removed and the fish was released from a new start position. In this test controls consistently chose the arm placed at the location in the room where they were rewarded during training, even though they started from a novel position. In contrast, telencephalon-ablated fishes chose equally among the three arms, and no significant preference was observed. These results show that although both groups learned the task, they used different navigational strategies to find the goal. (Modified from López *et al.* 2000a.)

with the nucleolar organizing region of the neurones, provided evidence for the selective involvement of the LP of teleost fishes in spatial memory (Vargas *et al.* 2000). A significant and selective increase in the nucleolar organizing region was observed in the LP neurones in the animals trained in the spatial learning task, relative to the animals trained in cue learning or control procedures. Moreover, no differences were observed in the neurones of other pallial areas.

The implication of the teleost LP in spatial cognition has been confirmed in a series of experiments in which the effects of selective pallial lesions on spatial cognition were analysed in a variety of spatial learning and memory tasks. Lateral pallium lesions produced a dramatic impairment in place learning and memory in goldfish trained in a plus-maze located in a room with an array of distributed extra-maze visual cues (Rodríguez *et al.* 2002b; and see Fig. 13.10). In fact, the place memory deficit observed after LP lesions in goldfish is as severe as that produced by the complete ablation of both telencephalic hemispheres (Salas *et al.* 1996a; López *et al.* 2000a; Rodríguez *et al.* 2002a,b). In contrast, medial or dorsal pallium lesions do not produce any observable impairment in place learning (Rodríguez *et al.* 2002b; and see Fig. 13.10). Interestingly, the involvement of the LP of goldfish in spatial cognition seems to be selective to place learning, as damage to this area does not impair cue learning or other egocentric strategies (Salas *et al.* 1996a,b; López *et al.* 2000a; Rodríguez *et al.* 2002b).

Recently, Saito & Watanabe (2004) have obtained some results that are apparently inconsistent with this view, as they reported spatial learning deficits after MP rather than after LP lesions in goldfish. However, several facts can account for this inconsistency. First, the LP lesions in the experiment by Saito & Watanabe affect mainly the dorsal part of the area dorsalis telencephali (Dd) and the most dorsal subdivision of the lateral part of the area dorsalis (Dld), but spare the ventral subdivision of Dl (Dlv). In contrast, in the experiments where the damage of LP produces severe spatial cognition deficits, the lesions include the Dlv subdivision. In fact, the Dld area damaged in the Saito & Watanabe experiment probably includes the main sensory visual area of the pallium, and it is most likely homologous to the dorsal cortex of land vertebrates (Prechtl *et al.* 1998; Saidel *et al.* 2001), whereas the Dlv area is the most likely homologue to the hippocampus of land vertebrates on the basis of developmental, topological and histochemical considerations (Butler 2000; Le Crom *et al.* 2003). The performance deficit observed in the animals with MP lesions in the Saito & Watanabe experiment (i.e. mainly an increase in the latency to reach the goal) may not be the result of a specific spatial impairment, but may be the result of a decrease in general activity and motivation (Davis & Kassel 1983; Riedel 1998). Finally, as we have discussed before, fish can solve the same spatial task by a variety of spatial strategies, only some of which are hippocampal dependent. The use of a task 'analogous' to another used in mammals does not imply 'automatically' that the fish are using cognitive mapping strategies. Transfer and probe tests provide the only reliable instrument to identify the strategies employed by the animals to solve any spatial task. Unfortunately, the experiment by Saito & Watanabe lacks an adequate set of test trials to identify the spatial strategies that the control and the lesioned animals are using, making impossible an unambiguous interpretation of the data. Another recent experiment (Durán 2004) addressed these questions directly by comparing the effects of LP and MP lesions in a similar procedure, using adequate

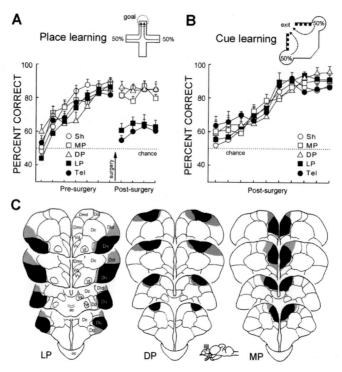

Fig. 13.10 Spatial memory deficits after lateral pallium lesion in goldfish. In this experiment, the effects of lesions to the lateral (LP), dorsal (DP), or medial (MP) pallium of goldfish were analysed in similar place and cue tasks. (A) Effects of selective pallial lesions on spatial memory in a place learning task. The insert shows a schematic representation of the training procedure (see Fig. 13.6 for additional comments). Lateral pallium lesion (LP) produced a dramatic impairment in goldfish trained in the place task, whereas MP and DP lesions did not decrease accuracy. Note that the LP lesion produced an impairment as severe as that observed in the animals with complete ablation of the telencephalon (Tel). (B) Effects of selective pallial lesions on goldfish spatial memory in a cued task. Following training in the place task, goldfish were trained in a cue-learning task in a different maze and room. In this task, fishes were trained to exit from a box maze with two start boxes and two exit doors. On each trial, one of the doors was blocked by a transparent glass barrier, leaving the other as the only exit (goal). The location of the goal varied pseudorandomly across trials, but it was always signalled by two striped panels (cue) that surrounded the goal, so that correct choices involved an alternation of turns. The insert shows a schematic representation of the training procedure. Results show that none of the pallial lesions produced deficits in cue learning relative to sham animals (Sh). (C) Schematic representation of the largest (grey shading) and smallest (black shading) extent of the LP, DP and MP lesions in goldfish, reconstructed in coronal sections. (Modified from Rodríguez *et al.* 2002b.)

probe and transfer tests. In this experiment goldfish with LP or MP lesions, complete telencephalon ablation or sham operation, were trained in a procedure analogous to the hole-board task employed with rats, to locate the baited feeder within a twenty-five-feeder matrix situated in a tank (Durán 2004). This matrix was surrounded by several conspicuous, widely distributed visual cues, which maintained stable spatial relationships relative to the goal. Although all the animals learned to reach the goal with accuracy, the control tests showed that, in contrast to the animals with MP lesions or who had undergone sham operation, the performance of the fish with lesions to the LP or with complete telencephalon ablation was based on a guidance strategy; these animals had learned to approach a particular visual cue from a particular direction. In fact, only LP and telencephalon ablated goldfish failed when one particular visual cue located in the proximity of the goal was excluded or displaced (probe trials). In addition, the data on the effects of MP lesions were consistent with previous results; these animals, like sham-operated fishes, navigated readily to the goal independently of environmental or procedural changes. Medial pallium and sham lesioned goldfish reached the goal when any of the particular visual cues was excluded, thus showing that they took advantage of whatever spatial information was actually available, regardless of direction of approach.

In summary, the data presented here provide strong support for the involvement of the LP of teleost fishes in spatial cognition. Furthermore, the lateral telencephalic pallium of teleost fishes, similar to the medial cortex or hippocampus of amniotes, underlies the ability of fishes to use allocentric representations of the environment (Tolman 1948; O'Keefe & Nadel 1978). These functional data, consistent with previous developmental and neuroanatomical evidence, provide additional support regarding the homology of the teleost LP with the hippocampal pallium in vertebrates with evaginated telencephala.

13.5.4 *Neural mechanisms for egocentric orientation*

Spatial memory and behaviour depend on both telencephalic and non-telencephalic brain structures and circuits. In teleost fishes, LP lesions do not impair the use of egocentric strategies for spatial orientation, such as body-turns, simple spatial discriminations, or approaching or avoiding a single cue, clearly indicating that other cerebral structures, such as the optic tectum and the cerebellum, are implicated in these processes.

The neuroanatomical and functional organization of the optic tectum (superior colliculus in mammals) is notably conserved in vertebrates; for example, marked similarities can be observed in the specialized cytoarchitecture and microcircuitry and the profuse connectivity with other motor and sensory centres (Vanegas 1984), as well as in the mechanisms for generating coordinated eye, head, and body movements, and for coding the metric and kinetic features of these movements (Du Lac & Knudsen 1990; Salas *et al.* 1997; Herrero *et al.* 1998; Sparks 2002; and see Fig. 13.11). In fact, this structure provides a common body-centred framework for multisensory integration and sensory-motor transformations (Stein & Meredith 1993; Sparks 2002) and is crucial for generating actions within an egocentric frame of spatial reference. Furthermore, it participates in the transformation of the information coded in spatial coordinates into a temporal signal in the brainstem motor generators (Isa & Sasaki 2002; Torres *et al.* 2002). As in other vertebrates, focal electrical stimulation elicits coordinated eye and body movements, postural adjustments and other motor patterns in teleost fishes (Demski 1983; Vanegas 1983; Al-Akel *et al.* 1986; Salas *et al.* 1995, 1997; Herrero *et al.* 1998). Similarly, there is a topographically ordered motor map in the deep tectal layers of teleosts, in correspondence with the retinotopic

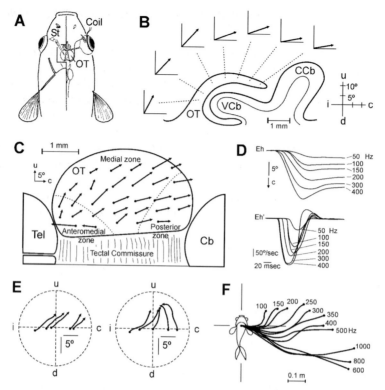

Fig. 13.11 The optic tectum of teleost fishes provides a common body-centred frame of reference for multisensory integration and for sensory-motor transformations and is a crucial centre for the generation of egocentrically referenced actions in space. (A) Focal electrical stimulation in the optic tectum of goldfish elicits coordinated eye and body movements, postural adjustments, and other motor patterns. (B) The amplitude and direction of eye movement vectors depend on the stimulation site within the tectum. Variation of the stimulation site in the rostro-caudal axis produced a systematic change in the amplitude of the horizontal component of the saccade, whereas variation of the stimulation site in the medial-lateral axis produced an increase in the vertical component of the eye movements (not shown). (C) Characteristic vectors of evoked saccades depending on the stimulation site in the right tectum of the goldfish. As in other vertebrates, goldfish orienting eye movement characteristics depend on the active tectal site, thus revealing a topographically ordered motor map within the optic tectum in alignment with the retinotopic visual map. (D) The direction and amplitude of the orienting responses depended not only on the tectal stimulation site but also on the stimulus parameters. The variation of the stimulation parameters (v.g. frequency) produced systematic changes in the metric and kinetic of the evoked orientation responses. (E) Stimulation of anatomically separated tectal areas evoked different types of eye movements. Left: Fixed vector movements, independent of the initial eye position, indicating that eye movements are coded retinotopically (Medial zone). Right: Goal directed movements, where the direction depends on the initial eye position, suggesting a craneotopic codification of the eye movement direction (Anteromedial zone). (F) The electrical microstimulation of the optic tectum in free-swimming fishes also produces body movements. Evoked movements consisted of complete orientation responses including coordinated movements of the axial musculature, fins and eyes, which closely resemble the natural responses. The direction and amplitude of the orienting responses depended on the tectal stimulation site and also on the stimulus parameters. Abbreviations: Cb, cerebellum; CCb, corpus cerebellum; Eh, horizontal component of eye position; Eh', eye velocity trace; OT, optic tectum; St, electrode for microstimulation; Tel, telencephalon; VCb, valvula cerebellum; d, u, i, c, downward, upward, ipsiversive and contraversive direction of evoked eye saccade, respectively. (Modified from Salas *et al.* 1997 and Herrero *et al.* 1998.)

visual map in the superficial layers, as revealed by the fact that the characteristics of the orienting eye movements depend on the active tectal site (Salas *et al.* 1997; Herrero *et al.* 1998; Sparks 2002). Moreover, electrical stimulation of the tectum in free-moving goldfish evokes different kinds of motor responses (orienting or escape, or a shift from one to the other) according to the location, intensity and frequency of the stimulus (Herrero *et al.* 1998). These data indicate that the tectal mechanisms of teleost fishes are strikingly similar to those of other vertebrates and that the tectum participates in the generation of orienting responses, providing egocentric frames of reference for perception and movement.

Increasing neuropsychological and experimental evidence indicate that the mammalian cerebellum, in addition to being a crucial centre for motor coordination, is also involved in spatial cognition and other cognitive processes (Lalonde & Botez 1990; Thompson & Krupa 1994; Petrosini *et al.* 1998). Interestingly, recent experiments show that the teleost cerebellum, similar to that of mammals, participates in spatial cognition, as indicated by lesion studies using a number of standard spatial tasks; for example, cerebellar-lesioned goldfish were trained to locate the only baited feeder within a 25-feeder matrix surrounded by an array of visual cues, which maintained stable spatial relationships relative to the goal (Durán *et al.* 2004). Although the cerebellar-lesioned animals improved slightly during training, they showed a stereotyped and inefficient search pattern. Moreover, these goldfish never reached the level of accuracy of the control and sham-operated animals. The poor performance of the cerebellum-ablated goldfish was probably related to an inability to generate or use a map-like representation of the environment, as indicated by the test trials. Thus, whereas the animals in the sham-operated group navigated readily to the goal regardless of the removal of any particular visual cue, the lesioned goldfish failed dramatically whenever a particular subset of visual cues was removed, revealing a profound spatial cognition deficit. Similar results have been observed in mammals with cerebellar lesions when they were trained to locate a goal in a variety of spatial tasks, such as the Morris water maze or the T-maze (Lalonde & Botez 1986; Goodlett *et al.* 1988; Petrosini *et al.* 1998; Rondhi-Reig *et al.* 2002). Moreover, when goldfish with cerebellar lesions, telencephalic lesions or sham operations, were trained in a spatial or a cue-learning task (Durán *et al.* 2004), cerebellar, but not telencephalic lesions, were equally disruptive, regardless of training conditions. As described above, in teleost fishes, telencephalic lesions, and in particular those damaging the hippocampal pallium, impair the performance in the spatial task, but spare cue learning (Salas *et al.* 1996a; López *et al.* 2000b; Rodríguez *et al.* 2002b; Salas *et al.* 2003; Broglio *et al.* 2005). In contrast, the post-surgery performance of the fishes with cerebellar lesions decayed to random levels in both the spatial and the cue task, indicating that the teleost cerebellum is also involved in the association of oriented motor responses with single landmarks and in other egocentric mechanisms. The most remarkable aspect of the effects of cerebellum cortex lesions in goldfish is that, whereas the spatial deficits are profound and widespread, impairing the use of both allocentric and egocentric strategies, they are restricted to spatial abilities; no sensory-motor-related differences were observed. In fact, posture and swimming ability, as well as distance travelled and obstacle avoidance were equally efficient in both the control and the cerebellar-lesioned animals (Durán *et al.* 2004; Rodríguez *et al.* 2005).

Interestingly, based on the comparative analysis of the relative development of different brain areas in a variety of fish groups showing particular adaptations to different habitats and ecological environments (Ridet & Bauchot 1990; Cambray 1994; Staaden *et al.* 1994), Demski & Beaver (2001) found that the development of the cerebellum, the tectum and the telencephalon are positively and strongly correlated. These findings suggest that the telencephalic–tectal–cerebellar neural associations form functional units that have been selected for specific behavioural capabilities and cognitive functions, especially for spatial cognition (for an interesting review on this topic, see Demski & Beaver 2001).

13.6   Conclusions

Historically, a dominant trend in comparative cognition and neuroscience, sustained by a pre-Darwinian notion of scala naturae of evolution and intelligence, has regarded fishes as the 'most primitive' or 'least evolved' vertebrate group, situated at the bottom of the so-called 'phylogenetic scale'. Consequently, fishes have been considered to lack most of the brain centres and neural circuits that support cognitive capabilities in the 'superior' vertebrate groups, that is, birds and, especially, mammals. We have reviewed here recent psychobiological and neurobiological evidence that challenges this traditional view. Taken as a whole, the results discussed above indicate that at least some learning and memory systems (including motor, emotional and spatial learning) of teleost fishes are strikingly similar to those of reptiles, birds and mammals. Such parallels in vertebrate groups that diverged millions of years ago suggest that the emergence of these memory systems could have occurred early in the phylogenetic history of vertebrates. In fact, these data suggest the possibility that extant fish, reptiles, birds and mammals, that share a common evolutionary ancestor (Lauder & Liem 1983; Carroll 1988; Northcutt 1995), could have inherited some behavioural and cognitive features from the common ancestor that would have been retained during phylogenesis.

13.7   Acknowledgements

We wish to thank Culum Brown, Ann B. Butler and Alice Powers for helpful critical reviews of this manuscript. Supported by the Spanish Ministerio de Educación (grant BFU2004-03219), and Junta de Andalucía (grant CVI-242).

13.8   References

Aggleton, J.P. (1992) *The Amygdala: Neurobiological Aspects of Emotion, Memory, and Mental Dysfunction.* Wiley-Liss, New York.
Al-Akel, A.S., Guthrie, D.M. & Banks, J.R. (1986) Motor responses to localized electrical stimulation of the tectum in the freshwater perch (*Perca fluviatilis*). *Neuroscience,* **19,** 1381–1391.

Álvarez, E., Gómez, A., Rodríguez, F., González, F., González-Pardo, J.A., Arias, J.L. & Salas, C. (2002) Effects of classical conditioning on cytochrome oxidase activity in the cerebellum of goldfish. *Abstract of the International Behavioral Neuroscience Society,* **11**, 49.

Álvarez, E., Gómez, A., Durán, E., Ocaña, F.M., Jiménez-Moya, F., Broglio, C., Rodríguez, F. & Salas, C. (2003) Brain substrates of 'eyeblink' classical conditioning in goldfish, *Acta Neurobiologiae Experimentalis,* **63** (Suppl.), 62.

Ariëns-Kappers, C.U., Huber, G.C. & Crosby, E.C. (1936) *The Comparative Anatomy of the Nervous System of Vertebrates, Including Man.* Macmillan, New York.

Aronson, L.R. (1970) Functional evolution of the forebrain in lower vertebrates. In: L.R. Aronson, E. Tobach, D.S. Lehrman & J. Rosenblatt, *Development and Evolution of Behavior,* pp. 75–107. W.H. Freeman, San Francisco.

Aronson, L.R. (1981) Evolution of telencephalic function in lower vertebrates. In: P.R. Laming (ed.), *Brain Mechanisms of Behavior in Lower Vertebrates,* pp. 33–58. Cambridge University Press, Cambridge.

Aronson, L.R. & Herberman (1960) Persistence of a conditioned response in the cichlid fish, *Tilapia macrocephala* after forebrain and cerebellar ablations. *Anatomical Record,* **138**, 332.

Beritoff, J.S. (1971) *Vertebrate Memory: Characteristics and Origin.* Plenum Press, New York.

Berntson, G.G. & Torello, M.W. (1982) The paleocerebellum and the integration of behavioral function. *Physiological Psychology,* **10**, 2–12.

Bingman, V.P. (1992) The importance of comparative studies and ecological validity for understanding hippocampal structure and cognitive function. *Hippocampus,* **2**, 213–220.

Bingman, V.P., Riters, L.V., Strasser, R. & Gagliardo, A. (1998) Neuroethology of avian navigation. In: R. Balda, I. Pepperberg & A. Kamil (eds), *Animal Cognition in Nature,* pp. 201–226. Academic Press, New York.

Bitterman, M.E. (1975) The comparative analysis of learning. *Science,* **188**, 699–709.

Bobée, S., Mariette, E., Tremblay-Leveau, H. & Caston, J. (2000) Effects of early midline cerebellar lesion on cognitive and emotional functions in the rat. *Behavioral Brain Research,* **112**, 107–117.

Braford, M.R. (1995) Comparative aspects of forebrain organization in the ray-finned fishes: touchstones or not? *Brain Behavior and Evolution,* **46**, 259–274.

Broglio, C., Gómez, Y., López, J.C., Rodríguez, F., Salas, C. & Vargas, J.P. (2000) Encoding of geometric and featural properties of a spatial environment in teleostean fish (*Carassius auratus*). *XXVII International Congress of Psychology.* Stockholm.

Broglio, C., Gómez, A., Durán, E., Ocaña, F.M., Jiménez-Moya, F., Rodríguez, F. & Salas, C. (2005) Hallmarks of a common forebrain vertebrate plan: specialized pallial areas for spatial, temporal and emotional memory in actinopterygian fish. *Brain Research Bulletin,* **66**, 277–281.

de Bruin, J.P.C. (1980) Telencephalon and behavior in teleost fish. A neuroethological approach. In: S.O.E. Ebbesson (ed.), *Comparative Neurology of the Telencephalon,* pp. 175–202. Plenum Press, New York.

de Bruin, J.P.C. (1983) Neural correlates of motivated behavior in fish. In: J.P. Ewert, R.R. Capranica & D.J. Ingle (eds), *Advances in Vertebrate Neuroethology,* pp. 969–995. Plenum Press, New York.

Burgess, N., Jeffery, K.J. & O'Keefe, J. (1999) *The Hippocampal and Parietal Foundations of Spatial Cognition.* Oxford University Press, London.

Butler, A.B. (2000) Topography and topology of the teleost telencephalon: a paradox resolved. *Neuroscience Letters,* **293**, 95–98.

Butler, A.B. & Hodos, H. (2005) *Comparative Vertebrate Neuroanatomy: Evolution and Adaptation,* 2nd edn. Wiley-Liss, New York.

Cambray, J.A. (1994) Effects of turbidity on the neural structure of two closely related redfin minnows, *Pseudobarbus afer* and *P. asper*, in the Gamtoos River system, South Africa. *South African Journal of Zoology*, **29**, 126–131.

Campenhausen, C.V., Riess, I. & Weissert, R. (1981) Detection of stationary objects by the blind cave fish *Anoptichtys jordani* (Characidae). *Journal of Comparative Physiology*, **143**, 369–374.

Carroll, R. (1988) *Vertebrate Paleontology and Evolution*. W.H. Freeman, New York.

Cheng, K. (1986) A purely geometric module in the rat's spatial representation. *Cognition*, **23**, 149–178.

Cheng, K. & Gallistel, C.R. (1984) Testing the geometric power of an animal's spatial representation. In: H.L. Roitblat, T.G. Bever & H.S. Terrace (eds), *Animal Cognition*, pp. 409–423. Hillsdale. Erlbaum, New Jersey.

Clayton, N.S. & Dickinson, A. (1998) Episodic-like memory during cache recovery by scrub jays. *Nature*, **395**, 272–274.

Crosby, E.C. & Schnitzlein, H.N. (1983) *Comparative Correlative Neuroanatomy of the Vertebrate Telencephalon*. Macmillan Publishing Co., New York.

Davey, G. (1989) Comparative aspects of conditioning: Pavlovian learning. In: G. Davey (ed.), *Ecological Learning Theory*, pp. 23–57. Routledge, London.

Davis, R.E. & Kassel, J. (1983) Behavioral functions of the teleost telencephalon. In: R.G. Northcutt & R.E. Davis (eds), *Fish Neurobiology*, pp. 237–263. The University of Michigan Press, Ann Arbor.

Davis, M., Hitchcock, J.M. & Rosen, J.B. (1992) A neural analysis of fear conditioning. In: E. Gormezano & E. Wasserman (eds), *Learning and Memory: the Behavioral and Biological Substrates*, pp. 153–181. Lawrence Erlbaum, Hillsdale.

Deacon, T.W. (1990) Rethinking mammalian brain evolution. *American Zoologist*, **30**, 629–705.

Demski, L.S. (1983) Behavioral effects of electrical stimulation of the brain. In: R.E. Davis & R.G. Northcutt (eds), *Fish Neurobiology*. Volume 2. *Higher Brain Areas and Functions*, pp. 317–359. The Michigan University Press, Ann Arbor.

Demski, L.S. & Beaver, J.A. (2001) Brain and cognitive function in teleost fishes. In: G. Roth & M.F. Wulliman (eds), *Brain Evolution and Cognition*, pp. 297–332. Wiley, New York.

Dodson, J.J. (1988) The nature and role of learning in the orientation and migratory behavior of fishes. *Environmental Biology of Fishes*, **23**, 161–182.

Du Lac, S. & Knudsen, E.I. (1990) Neural maps of head movement vector and speed in the optic tectum of the barn owl. *Journal of Neurophysiology*, **63**, 131–146.

Durán, E. (2004) Neural bases of spatial learning in goldfish. PhD thesis (Unpublished). University of Sevilla. September of 2004.

Durán, E., Gómez, A., Ocaña, F.M., Álvarez, E., Broglio, C., Jiménez-Moya, F., Rodríguez, F. & Salas, C. (2004) Cerebellum and spatial learning in teleost fish. *FENS Forum Abstracts*, p. A112.15.

Eichenbaum, H. (2000) A cortical-hippocampal system for declarative memory. *Nature Reviews: Neuroscience*, **1**, 41–50.

Farr, E.J. & Savage, G.E. (1978) First- and second-order conditioning in the goldfish and their relation to the telencephalon. *Behavioral Biology*, **22**, 50–59.

Fenton, A.A. & Bures, J. (1994) Interhippocampal transfer of place navigation monocularly acquired by rats during unilateral functional ablation of the dorsal hippocampus and visual cortex with lidocaine. *Neuroscience*, **58**, 481–491.

Flood, N.B. & Overmier, J.B. (1971) Effects of telencephalic and olfactory lesions on appetitive learning in goldfish. *Physiology & Behavior*, **6**, 35–40.

Flood, N.B., Overmier, J.B. & Savage, G.E. (1976) The teleost telencephalon and learning: an interpretative review of data and hypotheses. *Physiology & Behavior*, **16**, 783–798.

Frank, A.H., Flood, N.C. & Overmier, J.B. (1972) Reversal learning in forebrain ablated and olfactory tract sectioned teleost, *Carassius auratus*. *Psychonomic Science*, **26**, 149–151.

Gallistel, C.R. (1990) *The Organization of Learning*. MIT Press, Cambridge.

Gallon, R.L. (1972) Effects of pre-training with fear and escape conditioning on shuttle-box avoidance acquisition by goldfish. *Psychological Reports*, **31**, 919–924.

Gherladucci, B. & Sebastiani, L. (1996) Contribution of the cerebellar vermis to cardiovascular control. *Journal of Autonomic Nervous System*, **56**, 149–156.

Gherladucci, B. & Sebastiani, L. (1997) Classical heart rate conditioning and affective behavior: the role of the cerebellar vermis. *Archives Italiennes de Biologie*, **135**, 369–384.

Gómez, A. (2003) Neural bases of associative learning in goldfish. PhD thesis. University of Sevilla, December 2003.

Gómez, A., Álvarez, E., Durán, E., Ocaña, F.M., Broglio, C., Jiménez-Moya, F., Salas, C. & Rodríguez, F. (2004) Delay vs trace conditioning following pallium ablation in goldfish. *FENS Forum Abstracts*, p. A042.10.

Goodlett, C.R., Nonneman, A.J., Valentino, M.L. & West, J.R. (1988) Constraint on water maze spatial learning in rats: implications for behavioural studies of brain damage and recovery of function. *Behavioral Brain Research*, **28**, 275–286.

Hainsworth, F.R., Overmier, J.B. & Snowdon, C.T. (1967) Specific and permanent deficits in instrumental avoidance responding following forebrain ablation in the goldfish. *Journal of Comparative and Physiological Psychology*, **63**, 111–116.

Hale, E.B. (1956) Social facilitation and forebrain function in maze performance of green sunfish, *Lepomis cyanellus*. *Physiological Zoology*, **29**, 93–106.

Hallacher, L.E. (1984) Relocation of original territories by displaced black-and-yellow rockfish, *Sebastes chrysomelas*, from Carmel Bay, California. *Californian Fish Game*, **7**, 158–162.

Helfman, G.S. & Schultz, E.T. (1984) Social transmission of behavioral traditions in a coral reef fish. *Animal Behaviour*, **32**, 379–384.

Helfman, G.S., Meyer, J.L. & McFarland, W.N. (1982) The ontogeny of twilight migration patterns in grunts (pisces: Haemulidae). *Animal Behaviour*, **30**, 317–326.

Herrero, L., Rodríguez, F., Salas, C. & Torres, B. (1998) Tail and eye movements evoked by electrical microstimulation of the optic tectum in goldfish. *Experimental Brain Research*, **120**, 291–305.

Hodos, W. & Campbell, C.B.G. (1969) The scala naturae: Why there is no theory in comparative psychology. *Psychological Review*, **76**, 337–350.

Hodos, W. & Campbell, C.B.G. (1990) Evolutionary Scales and comparative studies of animal cognition. In: Kesner, R.P. & Olton, D.S. (eds) Neurobiology of Comparative Cognition, pp. 1–20. Hillsdale: Lawrence Erlbaum Associates.

Hollis, K.L. & Overmier, J.B. (1978) The function of the teleost telencephalon in behavior: a reinforcement mediator. In: D.I. Dostofsky (ed.), *The Behavior of Fishes and Other Aquatic Animals*, pp. 137–159. Academic Press, New York.

Hollis, K.L. & Overmier, J.B. (1982) Effect of telencephalon ablation on the reinforcing and eliciting properties of species specific events in *Betta splendens*. *Journal of Comparative and Physiological Psychology*, **96**, 574–590.

Hosch, L. (1936) Untersuchungen über Grosshirnfunktion der Elritze (*Phoxinus laevis*) und des grundlings (*Gobio fluviatilis*). *Zoologische Jarhbucher Abteilen Zoologie und Physiologie*, **57**, 57–70.

Ingle, D.J. (1965) Behavioral effects of forebrain lesions in goldfish. In: *Proceedings of the 73rd Annual Convention of the American Psychological Association*, pp. 143–144. Chicago: A.P.A.

Ingle, D.J. & Sahagian, D. (1973) Solution of a spatial constancy problem by goldfish. *Physiological Psychology*, **1**, 83–84.

Isa, T. & Sasaki, S. (2002) Brainstem control of head movements during orienting: organization of the premotor circuits. *Progress in Neurobiology*, **66**, 205–241.

Janzen, W. (1933) Untersuchungen über Grosshirnfunktionen des Goldfisches (*Carassius auratus*). *Zoologische Jahrbucher*, **52**, 591–628.

Jerison, H. (1973) *Evolution of the Brain and Intelligence*. Academic Press, New York.

Kaplan, H. & Aronson, L.R. (1969) Function of forebrain and cerebellum in learning in the teleost *Tilapia heudelotii macrocephala*. *Bulletin of The American Museum of Natural History*, **142**, 141–208.

Karamian, A.I. (1956) Evolution of the function of the cerebellum and cerebral hemispheres. *Fiziologicheskii Zhurnal SSSR*, **25**, 15–109.

Kholodov, Y.A. (1960) Simple and complex food obtaining conditioned reflexes in normal fish and in fish after removal of the forebrain. *Works of the Institute for Higher Nervous Activity Physiology Series*, **5**, 194–201.

Kim, J.J., Clark, R.E. & Thompson, R.F. (1995) Hippocampectomy impairs the memory of recently, but not remotely, acquired trace eyeblink conditioned responses. *Behavioral Neuroscience*, **109**, 195–203.

Kleerekoper, H., Matis, J., Gensler, P. & Maynard, P. (1974) Exploratory behaviour of goldfish *Carassius auratus*. *Animal Behaviour*, **22**, 124–132.

Kotchabhakdi, N. (1976) Functional circuitry of the goldfish cerebellum. *Journal of Comparative Physiology*, **112**, 47–73.

Lalonde, R. & Botez, M.I. (1986) Navigational deficits in weaver mutant mice. *Brain Research*, **398**, 175–177.

Lalonde, R. & Botez, M.I. (1990) The cerebellum and learning processes in animals. *Brain Research Review*, **15**, 325–332.

Lauder, G.V. & Liem, K.F. (1983) Patterns of diversity and evolution in ray-finned fishes. In: R.G. Northcutt & R.E. Davis (eds), *Fish Neurobiology*, pp. 1–24. The University of Michigan Press, Ann Arbor.

Le Crom, S., Kapsimali, M., Barome, P.O. & Vernier, P. (2003) Dopamine receptors for every species: gene duplications and functional diversification in Craniates. *Journal of Structural and Functional Genomics*, **3**, 161–176.

LeDoux, J.E. (1995) Emotions: clues from the brain. *Annual Review of Psychology*, **46**, 209–235.

López, J.C., Broglio, C., Rodríguez, F., Thinus-Blanc, C. & Salas, C. (1999) Multiple spatial learning strategies in goldfish (*Carassius auratus*). *Animal Cognition*, **2**, 109–120.

López, J.C., Bingman, V.P., Rodríguez, F., Gómez, Y. & Salas, C. (2000a) Dissociation of place and cue learning by telencephalic ablation in goldfish. *Behavioral Neuroscience*, **114**, 687–699.

López, J.C., Broglio, C., Rodríguez, F., Thinus-Blanc, C. & Salas, C. (2000b) Reversal learning deficit in a spatial task but not in a cued one after telencephalic ablation in goldfish. *Behavioural Brain Research*, **109**, 91–98.

López, J.C., Rodríguez, F., Gómez, Y., Vargas, J.P., Broglio, C. & Salas, C. (2000c) Place and cue learning in turtles. *Animal Learning and Behavior*, **28**, 360–372.

López, J.C., Gómez, Y., Rodríguez, F., Broglio, C., Vargas, J.P. & Salas, C. (2001) Spatial learning in turtles. *Animal Cognition*, **4**, 49–59.

López, J.C., Gómez, Y., Vargas, J.P. & Salas, C. (2003a) Spatial reversal learning deficit after medial cortex lesion in turtles. *Neuroscience Letters*, **341**, 197–200.

López, J.C., Vargas, J.P., Gómez, Y. & Salas, C. (2003b) Spatial and non-spatial learning in turtles: the role of medial cortex. *Behavioural Brain Research*, **143**, 109–120.

MacLean, P. (1990) *The Triune Brain in Evolution*. Plenum Press, New York.

Marino-Nieto, J. & Sabbatini, R.M. (1983) Discrete telencephalic lesions accelerate the habituation rate of behavioral arousal responses in Siamese fighting fish (*Betta splendens*). *Brazilian Journal of Medical and Biological Research*, **16**, 271–278.

Markevich, A.I. (1988) Nature of territories and homing in the eastern sea-perch *Sebastes taczanowski. Journal of Ichthyology,* **28**, 161–163.

Mazmanian, D.S. & Roberts, W.A. (1983) Spatial memory in rats under restricted viewing conditions. *Learning and Motivation,* **14**, 123–139.

McCormick, D.A. & Thompson, R.F. (1984) Cerebellum: essential involvement in the classically conditioned eyelid response. *Science,* **223**, 296–299.

Meek, J. & Nieuwenhuys, R. (1998) Holosteans and teleosts. In: R. Nieuwenhuys, H.J. ten Donkelaar & C. Nicholson (eds), *The Central Nervous System of Vertebrates,* pp. 759–937. Springer-Verlag, Berlin.

Mowrer, O.H. (1947) On the dual nature of learning a re-interpretation of 'conditioning' and 'problem solving'. *Harvard Educational Review,* **17**, 102–148.

Mowrer, O.H. (1960) *Learning Theory and Behavior.* Wiley, New York.

Moyer, J.R., Deyo, R.A. & Disterhoft, J.F. (1990) Hippocampectomy disrupts trace eyeblink conditioning in rabbits. *Behavioral Neuroscience,* **104**, 243–252.

Nadel, L. (1991) The hippocampus and space revisited. *Hippocampus,* **1**, 221–229.

Nadel, L. (1994) Multiple memory systems: What and why? An update. In: D.L. Schacter & E. Tulving (eds), *Memory Systems,* pp. 39–63. MIT Press, Cambridge.

Nieuwenhuys, R. (1963) The comparative anatomy of the actinopterygian forebrain. *Journal of Hirnforsch,* **6**, 171–192.

Nieuwenhuys, R. & Meek, J. (1990) The telencephalon of actinopterygian fishes. In: E.G. Jones & A. Peters (eds), *Comparative Structure and Evolution of the Cerebral Cortex,* pp. 31–73. Plenum, New York.

Nieuwenhuys, R., ten Donkelaar, H.J. & Nicholson, C. (1998) *The Central Nervous System of Vertebrates.* Springer-Verlag, Berlin.

Nolte, W. (1932) Experimentelle Untersuchungen zum Problem der Lokalisation des Assoziationsvermogens im Fischgehirn. *Zeitschrift für vergleichende Physiologie,* **18**, 255–279.

Northcutt, R.G. (1995) The forebrain of gnathostomes: In search of a morphotype. *Brain Behavior and Evolution,* **46**, 275–318.

Northcutt, R.G. & Braford, M.R. (1980) New observations on the organization and evolution of the telencephalon in actinopterygian fishes. In: S.O.E. Ebbesson (ed.), *Comparative Neurology Of The Telencephalon,* pp. 41–98. Plenum Press, New York.

Oakley, D.A. & Russell, I.S. (1977) Subcortical storage of Pavlovian conditioning in the rabbit. *Physiology & Behavior,* **18**, 931–937.

Odling-Smee, L. & Braithwaite, V.A. (2003) The role of learning in fish orientation. *Fish and Fisheries,* **4**, 235–246.

O'Keefe, J. & Conway, D.H. (1978) Hippocampal place units in the freely moving rat: why they fire where they fire. *Experimental Brain Research,* **31**, 573–590.

O'Keefe, J. & Nadel, L. (1978) *The Hippocampus as a Cognitive Map.* Clarendon Press, Oxford.

Overmier, J.B. & Curnow, P.F. (1969) Classical conditioning, pseudoconditioning, and sensitization in 'normal' and forebrainless goldfish. *Journal of Comparative and Physiological Psychology,* **68**, 193–198.

Overmier, J.B. & Flood, N.B. (1969) Passive avoidance in forebrain ablated teleost fish (*Carassius auratus*). *Physiology & Behavior,* **4**, 791–794.

Overmier, J.B. & Gross, D. (1974) Effects of telencephalic ablation upon nest-building and avoidance behaviors in East African mouth breeding fish, *Tilapia mossambica. Behavioral Biology,* **12**, 211–222.

Overmier, J.B. & Hollis, K.L. (1983) The teleostean telencephalon in learning. In: R.G. Northcut & R.E. Davis (eds), *Fish Neurobiology,* pp. 265–284. The University of Michigan Press, Ann Arbor.

Overmier, J.B. & Hollis, K.L. (1990) Fish in the think tank: learning, memory and integrated behavior. In: R.P. Kesner & D.S. Olton (eds), *Neurobiology of Comparative Cognition*, pp. 204–236. Lawrence Erlbaum Associates, Hillsdale.

Overmier, J.B. & Papini, M.R. (1985) Serial ablations of the telencephalon and avoidance learning by goldfish (*Carassius auratus*). *Behavioral Neuroscience*, **99**, 509–520.

Overmier, J.B. & Papini, M.R. (1986) Factors modulating the effects of teleost telencephalon ablation on retention, relearning, and extinction of instrumental avoidance behavior. *Behavioral Neuroscience*, **100**, 190–199.

Overmier, J.B. & Patten, R.L. (1982) Teleost telencephalon and memory for delayed reinforcers. *Physiological Psychology*, **10**, 74–78.

Overmier, J.B. & Savage, G.E. (1974) Effects of telencephalic ablation on trace classical conditioning of heart rate in goldfish. *Experimental Neurology*, **42**, 339–346.

Overmier, J.B. & Starkman, N. (1974) Transfer of control of avoidance behavior in normal and telencephalon ablated goldfish (*Carassius auratus*). *Physiology & Behavior*, **12**, 605–608.

Papez, J. (1929) *Comparative Neurology*. Crowell, New York.

Papini, M.R. (1985) Avoidance learning after simultaneous versus serial telencephalic ablations in the goldfish. *Bulletin of the Psychonomic Society*, **23**, 160–163.

Petrosini, L., Leggio, M.G. & Molinari, M. (1998) The cerebellum in spatial problem solving: a co-start or a guest start? *Progress in Neurobiology*, **56**, 191–210.

Phillips, R.G. & LeDoux, J.E. (1992) Differential contribution of amygdala and hippocampus to cued and contextual fear conditioning. *Behavioral Neuroscience*, **106**, 274–285.

Pitcher, T.J. & Magurran, A.E. (1983) Shoal size, patch profitability and information exchange in foraging goldfish. *Animal Behaviour*, **31**, 546–555.

Polimanti, O. (1913) Contributions a la physiologie du systeme nerveux central et du mouvement des poissons. *Archives Italiennes de Biologie*, **59**, 383–401.

Portavella, M., Vargas, J.P., Salas, C. & Papini, M. (2003) Involvement of the telencephalon in spaced-trial avoidance learning in the goldfish (*Carassius auratus*). *Physiology & Behavior*, **80**, 49–56.

Portavella, M., Torres, B. & Salas, C. (2004a) Avoidance response in goldfish: emotional and temporal involvement of medial and lateral telencephalic pallium. *Journal of Neuroscience*, **24**, 2335–2342.

Portavella, M., Torres, B., Salas, C. & Papini, M.R. (2004b) Lesions of the medial pallium, but not of the lateral pallium, disrupt spaced-trial avoidance learning in goldfish (*Carassius auratus*). *Neuroscience Letters*, **362**, 75–78.

Prechtl, J.C., von der Emde, G., Wolfart, J., Karamursel, S., Akoev, G.N., Andrianov, Y.N. & Bullock, T.H. (1998) Sensory processing in the pallium of a mormyrid fish. *Journal of Neuroscience*, **18**, 7381–7393.

Quick, I.A. & Laming, P.R. (1988) Cardiac, ventilatory and behavioural arousal responses evoked by electrical stimulation in the goldfish (*Carassius auratus*). *Physiology & Behavior*, **43**, 715–727.

Reese, E.S. (1989) Orientation behavior of butterflyfishes (family Chaetodontidae) on coral reefs: spatial learning of route specific landmarks and cognitive maps. *Environmental Biology of Fishes*, **25**, 79–86.

Ridet, J.M. & Bauchot, R. (1990) Analyse quantitative de l'encéphale des téléostéens: Caractères evolutifs et adaptatifs de l'encéphalisation. II. Les grandes subdivisions encé-phaliques. *Journal of Hirnforschung*, **31**, 433–458.

Riedel, G. (1998) Long-term habituation to spatial novelty in blind cave fish (*Astyanax hubbsi*): role of the telencephalon and its subregions. *Learning and Memory*, **4**, 451–461.

Roberts, B.L., van Rossem, A. & De Jager, S. (1992) The influence of cerebellar lesions on the swimming performance of the trout. *Journal of Experimental Biology*, **167**, 171–178.

Roberts, B.L., Dean, J.A. & Paul, D.H. (2002) Cerebellar regulation of sensorimotor activity in brown trout. *Brain Behavior and Evolution*, **60**, 241–248.

Rodríguez, F., Durán, E., Vargas, J.P., Torres, B. & Salas, C. (1994) Performance of goldfish trained in allocentric and egocentric maze procedures suggests the presence of a cognitive mapping system in fishes. *Animal Learning and Behavior*, **22**, 409–420.

Rodríguez, F., López, J.C., Vargas, J.P., Broglio, C., Gómez, Y. & Salas, C. (2002a) Spatial memory and hippocampal pallium through vertebrate evolution: insights from reptiles and teleost fish. *Brain Research Bulletin*, **57**, 499–503.

Rodríguez, F., López, J.C., Vargas, J.P., Gómez, Y., Broglio, C. & Salas, C. (2002b) Conservation of spatial memory function in the pallial forebrain of amniotes and ray-finned fishes. *Journal of Neuroscience*, **22**, 2894–2903.

Rodríguez, F., Durán, E., Gómez, A., Ocaña, F.M., Ávarez, E., Jiménez-Moya, F., Broglio, C. & Salas, C. (2005) Cognitive and emotional functions of the teleost fish cerebellum. *Brain Research Bulletin*, **66**, 365–370.

Romer, A.S. (1962) *The Vertebrate Body*. W.B. Saunders, Philadelphia.

Rondhi-Reig, L., Le Marec, N., Caston, J. & Mariani, J. (2002) The role of climbing and parallel fibers inputs to cerebellar cortex in navigation. *Behavioral Brain Research*, **132**, 11–18.

Russell, E.M. (1967) The effect of experience of surroundings on the response of *Lebistes reticulatus* to a strange object. *Animal Behaviour*, **15**, 586–594.

Sacchetti, B., Baldi, E., Lorenzini, C.A. & Bucherelli, C. (2002) Cerebellar role in fear-conditioning consolidation. *Proceedings of the National Academic of Sciences of the USA*, **99**, 8406–8411.

Saidel, W.M., Marquez-Houston, K. & Butler, A.B. (2001) Identification of visual pallial telencephalon in the goldfish, *Carassius auratus*: a combined cytochrome oxidase and electrophysiological study. *Brain Research*, **919**, 82–93.

Saito, K. & Watanabe, S. (2004) Spatial learning deficits after the development of dorsomedial telencephalon lesions in goldfish. *Neuroreport*, **15**, 2695–2699.

Salas, C., Herrero, L., Rodríguez, F. & Torres, B. (1995) On the role of goldfish optic tectum in the generation of eye movements. In: J.M. Delgado-García, E. Godaux & P.P. Vidal (eds), *Information Processing Underlying Gaze Control*, pp. 87–95. Pergamon Press, Oxford.

Salas, C., Broglio, C., Rodríguez, F., López, J.C., Portavella, M. & Torres, B. (1996a) Telencephalic ablation in goldfish impairs performance in a spatial constancy problem but not in a cued one. *Behavioural Brain Research*, **79**, 193–200.

Salas, C., Rodríguez, F., Vargas, J.P., Durán, E. & Torres, B. (1996b) Spatial learning and memory deficits after telencephalic ablation in goldfish trained in place and turn maze procedures. *Behavioral Neuroscience*, **110**, 965–980.

Salas, C., Herrero, L., Rodríguez, F. & Torres, B. (1997) Tectal codification of eye movements in goldfish studied by electrical microstimulation. *Neuroscience*, **78**, 271–288.

Salas, C., Broglio, C. & Rodríguez, F. (2003) Evolution of forebrain and spatial cognition in vertebrates: conservation across diversity. *Brain Behavior and Evolution*, **62**, 72–82.

Sarter, M. & Markowitsch, H.J. (1985) Involvement of the amygdala in learning and memory: a critical review, with emphasis on anatomical relations. *Behavioral Neuroscience*, **99**, 342–380.

Savage, G.E. (1968) Temporal factors in avoidance learning in normal and forebrainless goldfish (*Carassius auratus*). *Nature*, **218**, 1168–1169.

Savage, G.E. (1969a) Telencephalic lesions and avoidance behaviour in the goldfish (*Carassius auratus*). *Animal Behaviour*, **17**, 362–373.

Savage, G.E. (1969b) Some preliminary observations on the role of the telencephalon in food-reinforced behaviour in the goldfish, *Carassius auratus*. *Animal Behaviour*, **17**, 760–772.

Savage, G.E. (1971) Behavioural effects of electrical stimulation of the telencephalon of the goldfish, *Carassius auratus*. *Animal Behaviour*, **19**, 661–668.

Savage, G.E. (1980) The fish telencephalon and its relation to learning. In: S.O.E. Ebbesson (ed.), *Comparative Neurology of the Telencephalon*, pp. 129–174. Plenum, New York.

Savage, G.E. & Swingland, I.R. (1969) Positively reinforced behaviour and the forebrain in goldfish. *Nature*, **221**, 878–879.

Schacter, D.L. & Tulving, E. (1994) *Memory Systems*. MIT Press, Cambridge.

Schluessel, V. & Bleckmann, H. (2005) Spatial memory and orientation strategies in the elasmobranch *Potamotrygon motoro*. *Journal of Comparative Physiology A.*, **191**, 695–706.

Segaar, J. & Nieuwenhuys, R. (1963) New etho-physiological experiments with male *Gasterosteus aculeatus*, with anatomical comment. *Animal Behaviour*, **11**, 331–344.

Shapiro, S.M., Schuckman, H., Sussman, D. & Tucker, A.M. (1974) Effect of telencephalic lesions on the gill cover response of Siamese fighting fish. *Physiology & Behavior*, **13**, 749–755.

Sherry, D.F. & Duff, S.J. (1996) Behavioral and neural bases of orientation in food storing birds. *Journal of Experimental Biology*, **199**, 165–172.

Sovrano, V.A., Bisazza, A. & Vallortigara, G. (2002) Modularity and spatial reorientation in a simple mind: encoding of geometric and nongeometric properties of a spatial environment by fish. *Cognition*, **85**, B51–59.

Sparks, D.L. (2002) The brainstem control of saccadic eye movements. *Nature Reviews: Neuroscience*, **3**, 952–964.

Staaden, M.J., Huber, R., Kaufman, L.S. & Liem, K.F. (1994) Brain evolution in cichlids of the African Great Lakes: brain and body size, general patterns and evolutionary trends. *Zoology*, **98**, 165–178.

Stein, B.E. & Meredith, M.A. (1993) *The Merging of the Senses*. MIT Press, Cambridge.

Supple, W.F. Jr. & Kapp, B.S. (1993) The anterior cerebellar vermis: essential involvement in classically conditioned bradycardia in the rabbit. *Journal of Neuroscience*, **13**, 3705–3711.

Supple, W.F. Jr. & Leaton, R.N. (1990a) Cerebellar vermis: essential for classically conditioned bradycardia in the rat. *Brain Research*, **509**, 17–23.

Supple, W.F. Jr. & Leaton, R.N. (1990b) Lesions of the cerebellar vermis and cerebellar hemispheres: effects on heart rate conditioning in rats. *Behavioral Neuroscience*, **104**, 934–947.

Suzuki, S., Augerinos, G., & Black, A.H. (1980) Stimulus control of spatial behavior on the eight-arm maze in rats. *Learning and Motivation*, **11**, 1–18.

Teyke, T. (1989) Learning and remembering the environment in the blind cave fish *Anoptichthys jordani*. *Journal of Comparative Physiology A*, **164**, 655–662.

Thinus-Blanc, C. (1996) *Animal Spatial Cognition. Behavioral and Neural Approaches*. World Scientific, United Kingdom.

Thompson, R.F. & Krupa, D.J. (1994) Organization of memory traces in the mammalian brain. *Annual Review of Neuroscience*, **17**, 519–549.

Tolman, E.C. (1948) Cognitive maps in rats and men. *Psychological Review*, **55**, 189–208.

Torres, B., Pérez-Pérez, M.P., Herrero, L., Ligero, M. & Núñez-Abades, P.A. (2002) Neural substrate underlying tectal eye movement codification in goldfish. *Brain Research Bulletin*, **57**, 345–348.

Vanegas, H. (1983) Organization and Physiology of the teleostean optic tectum. In: R.E. Davis & R.G. Northcutt (eds), *Fish Neurobiology*. Volume 2. *Higher Brain Areas and Functions*, pp. 43–90. The University of Michigan Press, Ann Arbor.

Vanegas, H. (1984) *Comparative Neurology of the Optic Tectum*. Plenum Press, New York.

Vargas, J.P., Rodríguez, F., López, J.C., Arias, J.L. & Salas, C. (2000) Spatial learning-induced increase in the argyrophilic nucleolar organizer region of dorsolateral telencephalic neurons in goldfish. *Brain Research*, **865**, 77–84.

Vargas, J.P., Lopez, J.C., Salas, C. & Thinus-Blanc, C. (2004) Encoding of geometric and featural spatial information by Goldfish (*Carassius auratus*). *Journal of Comparative Psychology*, **118**, 206–216.

Warburton, K. (1990) The use of local landmarks by foraging goldfish. *Animal Behaviour*, **40**, 500–505.

Warren, J.M. (1961) The effects of telencephalic injuries on learning by Paradise fish, *Macropodus opercularis. Journal of Comparative and Physiological Psychology*, **54**, 130–132.

Welker, W.I. & Welker, J. (1958) Reaction of fish (*Eucinostomus gula*) to environmental changes. *Ecology*, **39**, 283–288.

Woodruff, M.L. & Kantor, H. (1983) Fornix lesions, plasma ACTH levels, and shuttle box avoidance in rats. *Behavioral Neuroscience*, **97**, 897–907.

Wullimann, M.F. & Rink, E. (2002) The teleostean forebrain: a comparative and developmental view based on early proliferation, Pax6 activity and catecholaminergic organization. *Brain Research Bulletin*, **57**, 363–370.

Yoshida, M., Okamura, I. & Uematsu, K. (2004) Involvement of the cerebellum in classical fear conditioning in goldfish. *Behavioural Brain Research*, **153**, 143–148.

Zhuikov, A.Y., Couvillon, P.A. & Bitterman, M.E. (1994) Quantitative two-process analysis of avoidance conditioning in goldfish. *Journal of Experimental Psychology: Animal Behavioral Processes*, **20**, 32–43.

Zunini, G. (1954) Researches on fish's learning. *Archives Néerlanden de Zoologische*, **10**, 127–140.

Chapter 14

The Role of Fish Learning Skills in Fisheries and Aquaculture

Anders Fernö, Geir Huse, Per Johan Jakobsen and Tore S. Kristiansen

14.1 Introduction

Humans have harvested fish resources for thousands of years, but only during the past century has the development of new fishing technologies produced detrimental impacts on fishery resources. The relative importance of commercial fisheries in regulating populations and marine ecosystems has increased dramatically. Over the same time period a worldwide aquaculture industry has developed, and within a few decades the production of farmed fishes will presumably exceed landings of wild fishes. The rapidly changing technology in both fisheries and aquaculture challenges fishes exposed to human impact in novel ways that differ from those that threaten their survival in their ancestral environments.

Animals categorize the world that surrounds them into objects of relevance to their survival and reproduction and objects of less or no relevance. The evolved ability of animals to categorize depends upon their sensory apparatus and their capacity to learn about and manipulate objects. Bodies and brains largely determine the kinds of categories and their structure and thus influence the conceptual systems of individual species (Lakoff & Johnson 1999). In other words, animals are neural units shaped by the ghost of selection past, and the way in which animals categorize events and objects may be preadapted and tailored to their ecological niche (Rozin & Kalat 1972). For instance, by transplanting pieces of quail neural tissue into chicken embryos, Balaban (1997) was able to demonstrate two different, but specific quail behaviours in chickens, suggesting that the functions of at least some neural circuits are innate, species-specific, and may thus limit the capacity to categorize evolutionarily novel situations adequately. As for automated responses, learning skills may be preprogrammed and adapted to environmental factors. This may be based on modification of sensory systems in the organism, and mammals show signs of cortical magnification – a sensory specialization to the adequate stimuli in the environment – with relevant stimuli being magnified and less important ones suppressed (Catania & Kaas 1997, 2001). Indications of similar specializations to relevant stimuli are also found in fishes (Kotrschal *et al.* 1998; Marchetti & Nevitt 2003). Brown *et al.* (2004) found differences in lateralization of the brain between fishes from high predatory areas and those from low predatory areas, with the former generally using their left eye to view novel objects and their right to view a cichlid predator. Species differences in lateralization also exist and may be generated by differential emotive responses to various stimuli (Bisazza *et al.* 2000; DeSanti *et al.* 2000). The development of cognitive skills (spatial learning, problem solving) in fishes seems to be associated with visual orientation and well-structured habitats (Kotrschal *et al.* 1998; Brown & Braithwaite 2005).

Although classical conditioning is widely applicable, various constraints may make it more adaptive. Fishes can be more readily conditioned to moving than to stationary stimuli (Wisenden & Harter 2001), and certain key stimuli are crucial when fishes learn to recognize and avoid predators (Csànyi 1986; Altbäcker & Csànyi 1990; Brown & Warburton 1997). Rats are able to associate taste with sickness, but not with electrical shock (Garcia & Koelling 1966), as inner pain presumably results from something ingested and surface pain is inflicted by an external agent (Johnston 1999). Similar surprises can be found in fisheries. A fish may perceive a fishing net, but associate it with vegetation. Consequently, no traumatic feelings are provoked, and because fishes can pass through vegetation, they may also try to pass through the net.

Lack of stimulus relevance from manmade constructions such as fishing gear and aquaculture technology may thus have profound consequences on fish learning abilities. Perceptual concepts encountered in modern fisheries and fish farming may not resemble those that occur in the natural environment, with selected responses towards manmade facilities coinciding with innate responses in some cases and contrasting with them in others (Brown & Warburton 1999a and refs therein for responses to trawls). Even though many traits necessary for learning may be preprogrammed, there may be a degree of compatibility between the adapted-innate and to-be-learned behaviour. This may greatly influence the ease with which certain responses are learned and behaviourally expressed in what we perceive as novel environments (Bolles 1970).

As a result of rapid changes in the equipment used and the holistic approach to the study of fish behaviour in fishery research, it is often a challenge to disentangle preprogrammed innate responses, adaptive predispositions to associate stimuli and to-be-learned behaviour based on general principles of learning. However, there is no doubt that the learning abilities of fishes could create problems for humans if fishes learn to avoid vessels and gear. Learning could also lead to less predictable spatial distributions and migration routes, making it more difficult to localize fish aggregations, with consequent implications for fisheries and stock assessment. Decadal changes in climatic conditions may favour the learning of migration patterns rather than more fixed migration strategies. Learning is particularly valuable for long-lived species that may rely on previous experience when making decisions. Thus, an Atlantic horse makerel (*Trachurus trachurus*, Carangidae) that lives to be 20 years of age could benefit from previous knowledge when making decisions about summer migration paths, while a capelin (*Mallotus villosus*, Osmeridae), which spawns only once in its lifetime has to rely on genetically inherited strategies when making its migratory decisions. At the most extreme, the implications of fisheries might be that we could wipe out whole cultural units of long-lived fish populations, and consequently important aspects of their adult behaviour, because there are no longer enough experienced fishes to copy (see Chapter 10). Frequency-dependent social learning may therefore be taken into account all the way up to a holistic ecosystem management perspective. In aquaculture and sea-ranching programmes on the other hand, the ability of fishes to learn may be beneficial, because we wish the fishes to adapt to both a farming environment and free-range conditions. The human ambition is to train fishes to deal with environmental cues and limit their distribution within a restricted area.

The fish species studied in applied research represent a diverse group that has not been systematically studied with respect to learning skills. Some species have also been studied in particular detail, biasing comparisons between species. Most of the research on fish learning has tended to focus on small freshwater species [guppy (*Poecilia reticulata*, Poeciliidae), goldfish (*Carassius auratus*, Cyprinidae), pumpkinseed sunfish (*Lepomis gibbosus*, Centrarchidae) etc.] that are easily maintained in small laboratories. This chapter puts forward examples where learning may play a role in fisheries, aquaculture and stock enhancement. The main topics deal with the implications for the efficiency of equipment used and the influence of fisheries on learned migrations in commercially exploited species. We also discuss the role of learning in reared fishes in aquaculture facilities and following release in the wild.

14.2　Fisheries

The adapted life history traits and behaviour of fishes in exploited fish populations are contested by fishing activity, and this may lead to the selection of new traits and modification of behaviour by learning. From the human point of view, the aim is to exploit fish populations in a responsible way in order to avoid population collapses and ecosystem breakdowns. To that end, it is vital to maintain control of population size as well as to understand how populations interact with other components of the ecosystem at higher and lower trophic levels. The next step is to exploit fish stocks using selective and environmentally friendly fishing gear. If we are to achieve these goals, both natural behaviour and reactions to stimuli from fishing operations and surveys must be taken into account. Fish anti-predator and feeding behaviour is shaped by an interaction between innate and learned components (see Chapters 2 and 3). Learning thus influences the species interactions and dynamics of ecosystems, with potentially powerful effects on the variations in abundance of commercially exploited species. The effect of relevant ecological processes, including learning, should ideally be incorporated into ecosystem-based management, but this is a complex task that demands quantitative knowledge of modifications of species interactions at several different levels. However, the effect of fishes' learning skills on the actual interactions between target and non-target species in an ecosystem should be borne in mind when evaluating the realism of population dynamic models and management reference points (Hall & Mainprize 2004). The following sections consider particularly the role played by learning processes in spatial dynamics and fish capture.

14.2.1　*Spatial dynamics*

14.2.1.1　*Learning skills and movement*

The capacity to learn provides flexibility that is crucial for the ability to utilize a variable environment (Dodson 1988). Habitat utilization and migration routes in fishes are presumably often influenced by experience, but our knowledge of this field in commercially exploited species is limited. Learning can control movements within a stationary home range as well as long-distance migrations (see Chapter 7). Our understanding of fish migrations is further constrained by the close interplay between navigation and orientation mechanisms, both of which rely to some extent on unknown sensory organs. Much of our insight into migrations in commercial fishes thus remains at a fairly descriptive level, gathered mainly using acoustic methods and tags, with limited knowledge of the motivation of the fishes and the state of the environment.

Demersal fishes often stay within a certain home range (Matthews 1990) and may therefore be assumed to rely on learned spatial cues (Reese 1989). For instance, the gadoid ling (*Molva molva*, Lotidae) occupies a home range in fjords and remains for about 65% of its time in a core area within the home range (Løkkeborg *et al.* 2000). A balance between positive and negative reinforcement could determine the selection of home range, with rewards being food availability and shelter, and punishment including predator attacks, intraspecific aggression and territory defensibility.

The pelagic environment is more homogeneous than the demersal zone, and pelagic fishes are, therefore, often assumed to display less fidelity to a home range. Tuna occur over large areas in tropical and subtropical waters, and may also undertake excursions into boreal and sub-Arctic waters during the summer (Block *et al.* 2005). Studies using electronic tags show the ability of yellowfin tuna (*Thunnus albacares*, Scombridae) to undertake precise navigation between fish aggregating devices (Brill *et al.* 1999), and to remain in an area over long periods. This behavioural pattern suggests that yellowfin tuna are able to maintain core areas around fish aggregating devices through learning much in the same manner as demersal fishes, although they are probably relying on different cues to aid navigation. On a larger scale, pelagic fishes may locate favourable habitats by using a combination of predictive orientation mechanisms based upon genetic factors and learning, and reactive mechanisms, such as memory-based state-location comparisons and orientation to gradients in the sea (Fernö *et al.* 1998; Kvamme *et al.* 2003). Field experiments on Atlantic salmon (*Salmo salar*, Salmonidae) have further demonstrated that smolts are imprinted with olfactory cues during seawards migration and that they use these cues during their return to their home river for spawning (see Chapter 7). Learning is, therefore, important to the movements of a wide range of commercial species.

14.2.1.2 Social learning of migration patterns

Social learning involves the transfer of knowledge between individuals of a population. The migration patterns of many fishes are learned at an early age, often from older individuals in the stock, so-called 'guided learning' (Brown & Laland 2003), and are then maintained over time. Helfman & Schultz (1984) performed a classic demonstration of social learning in fishes by showing that daily migrations by French grunts (*Haemulon flavolineatum*, Haemulidae) between resting and feeding grounds were maintained by guided learning. This was demonstrated by transplanting individuals from their home ground to a new resting area that already contained resident fishes. The transplanted fishes soon adopted the migration route to the feeding area utilized by the resident fishes. In a control experiment in which fishes were transplanted to a resting area from which the resident fishes had been removed, the transplanted fishes continued to use paths appropriate to their original resting site. There is growing evidence to suggest that guided learning plays a role in migration of important resources such as herring and cod.

A few centuries ago, Norwegian fishermen thought that the massive herring schools were guided to the coast by the oarfish (*Regalecus glesne*, Regalecidae), which is called the herring king in Norwegian. Indeed the herring (*Clupea harengus*, Clupeidae) do seem to be guided, but by their own elders rather than by serpent-like creatures. In some herring populations there appears to be strong inter-cohort learning (McQuinn 1997; Fernö *et al.* 1998; Corten 1999, 2001). Normally, a recruiting year class learns the migration pattern of the stock by schooling with the older component of the stock (McQuinn 1997). Under non-harvesting conditions, this is likely to be a fairly robust population mechanism. But if the stock collapses and most of the old individuals are lost, the interaction between recruiting and old cohorts may be disrupted, thus breaking the vertical social transmission chain (see Chapter 10). This is parallel to the experiment by Helfman & Schultz (1984), in which transplanted fishes did not encounter resident fishes. More generally, the adoption of the adult migration pattern by the recruiting cohorts can be interrupted if the proportion of recruits to the adult population is high (Huse *et al.* 2002b) or when there is lack of spatial overlap between cohorts (Corten 2001). The process whereby large cohorts 'repel' socially transmitted information from older cohorts, and thus inhibit leadership by experienced individuals, has been termed 'numerical domination' (Huse *et al.* 2002b). Numerically dominant herring year classes might not follow the behaviour of less abundant, older year classes simply because there are insufficient old individuals compared to abundant first-time spawners to elicit responses in the naive fishes. Reebs (2000) demonstrated experimentally that a minority of informed golden shiners (*Notemigonus crysoleucas*, Cyprinidae) were able to determine foraging movement in a shoal, and in the case of guppies, 'observers' were more faithful to the route to a foraging path with an increasing number of 'demonstrators' (Laland & Williams 1997). The probability of transmission of a novel behaviour to naive observers is generally believed to be a function of the number of demonstrators performing the behaviour. Numerical domination, on the other hand, refers to the relative proportion of naive observers to demonstrators. The disagreement about absolute and relative numbers is partly a matter of scale; more demonstrators will be needed to transmit novel behaviour to naive fishes in a school of thousands of fishes than in a school of tens of fishes. This matter can be most readily investigated by means of model simulations. In a simulation study, Couzin *et al.* (2005) found that the proportion of demonstrators needed to exert effective leadership in a school decreased as group size rose, while in a similar study, Huse *et al.* (2002b) found the proportion to be independent of group size. More studies are needed to clarify the dependence of transmission on absolute and relative numbers of demonstrators, focusing on a range of representative group sizes and sensitivity to simulation specifications. Lack of leadership as a result of numerical domination is likely to be a key factor in changing migration patterns of herring stocks. The wide spread of overwintering grounds of Norwegian spring-spawning herring in the past 50 years (Fig. 14.1) illustrates the great variation in the spatial location of overwintering grounds, with the changes co-occurring with the recruitment of large year classes to the spawning stock.

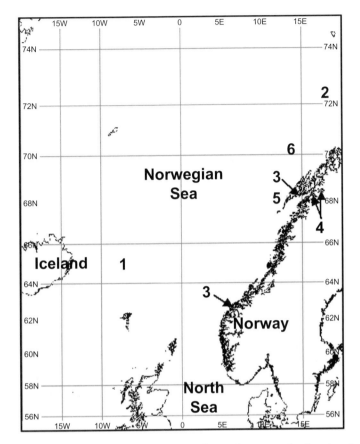

Fig. 14.1 Reported wintering locations of Norwegian spring-spawning herring during the past 50 years. The numbers indicate chronological development of the wintering locations. During the period 1950–2000 the herring has changed its wintering distribution five times, and at one time in the 1970s the population (3) was divided into two separate components (after Holst *et al.* 2002).

Maintenance of spawning areas may be also based on learning and tradition in cod (*Gadus morhua*, Gadidae). Harden Jones (1968) refers to evidence for 'dummy' spawning runs by juvenile cod, and spawning grounds seem to be learned by young Newfoundland cod by following the older spawning stock on a migration 'highway' (Rose 1993) in a similar manner to herring. Tagging studies have revealed that the cod repeatedly home to the same spawning grounds (Robichaud & Rose 2001). Like those of herring, cod traditions are therefore vulnerable to stock collapses, and there is reason to believe that local cod and tuna spawning grounds may be lost when populations are fished down (Cury & Anneville 1998). Bluefin tuna (*Thunnus thynnus*, Scombridae), migrating from the Mediterranean to the Norwegian Sea to feed, supported a Norwegian fishery for almost 30 years between the 1950s and the 1980s. The tuna consisted mainly of one very large cohort, and when this cohort had been fished out the migration was terminated and the fishery collapsed.

14.2.1.3 Implications of learning for fisheries management

Spatial management and the establishment of marine protected areas demand knowledge of spatial dynamics (Babcock *et al.* 2005), including the role of learning. Changes in migration patterns, as described earlier for herring and cod, appear to take place when the population is unstable or collapsing. While collapsed fish stocks may show signs of depensation (where a decrease in spawning stocks leads to reduced survival or reproduction), or Allee effects (Shelton & Healey 1999), the loss of culture associated with population crashes may well hamper the rebuilding of fish stocks. Thus, careful attention should be paid to the question of whether spatial patterns in fish stocks are governed by learning or through genetically mediated strategies, because the consequences of mismanagement might be greater if culture is lost. The collapsed stock of northern cod has yet to rebuild more than 10 years after the fishing moratorium was introduced in 1992 (Rose & O'Driscoll 2002). The collapse of this stock was accompanied by substantial changes in its distribution (Rose *et al.* 2000). Given the importance of collective movements in this stock (Rose 1993), the loss of culture associated with the demise of the spawning stock may play a role in the lack of rebuilding. In fact, there are many reasons for leaving big fishes in harvested populations, for the sake of increased fecundity, larval survival and guided learning (Birkeland & Dayton 2005). Cultural diversity in fish stocks can be seen as one aspect of biodiversity. While horizontal cultures can adapt rapidly within generations, oblique cultures that are transmitted across generations are more sensitive to anthropogenic effects (Whitehead *et al.* 2004). The question of culture, therefore, needs to be integrated into the management and conservation of fish populations.

Learning influences spatial dynamics and can thereby affect abundance estimates of fish stocks. Surveys should cover areas with high fish concentrations, and shifts in migration patterns present a challenge to survey design. One example of this is surveys that target Norwegian spring-spawning herring during the overwintering period. For many years, the whole population has overwintered in Norwegian fjords (numbers 4 and 5 in Fig. 14.1), but since 2002 abundant new year-classes have overwintered off the coast (number 6 in Fig. 14.1, Jens Christian Holst, Institute of Marine Research, Bergen, personal communication). Such pronounced changes in spatial pattern have consequences for surveys, which need to cover the entire distribution of the stock in order to obtain valid estimates of abundance. Furthermore, patterns of distribution can have consequences for stock ownership and thus access to exploitation.

14.2.2   *Fish capture*

Fishing can be regarded as an arms race between the fishes and the fishermen, but it is hardly a close race. Fishermen gain experience of the behaviour of the fishes, resulting over time in an accumulation of knowledge about the spatial and temporal dynamics of particular species and how this can be influenced to the advantage of the fishermen. With the advent of sonar and satellite navigation we often do not even need to have a historic knowledge base because we can find fishes so easily. Fishes have evolved a repertoire of behavioural patterns to meet the challenges of the natural environment, while fishermen utilize these adaptations by designing fishing gear that releases these normally adaptive reactions and 'cheats' the fishes into behaving maladaptively during the capture process (Fernö 1993). Given the rapid development of technology, humans are generally bound to be the winners of this race. Many fish stocks are currently under serious pressure (Christensen *et al.* 2003). Given the many experiences of failure to exploit fish stocks in a way that will sustain them (Mullon *et al.* 2005), we may be tempted to classify humans as slow learners, at least on a collective level.

Fishing is often the most important cause of mortality in exploited fish populations after recruitment to the fishery (Moav *et al.* 1978), and natural selection may, over time, result in evolutionary changes in genetically determined characteristics (Law & Grey 1989; Policansky 1993). However, fish stocks are often exploited by different fishing gears that change over time, creating conflicting and variable selection pressures. All the same, fishing has been shown to influence fish life-history parameters (Kristiansen & Svåsand 1998; Olsen *et al.* 2004). Fish behaviour may also have been influenced by selection, although this is difficult to establish. On a shorter time scale, however, learning can modify behaviour. A substantial number of fish survive contact with fishing gear, creating the potential for learning (Fréon & Misund 1998). Özbiljin & Glass (2004) have calculated that demersal fishes in the heavily fished North Sea may encounter towed fishing gears many times in the course of a year. If learning results in gear avoidance, the consequence could be decreased efficiency followed by increased fishing effort, with negative economic and environmental effects. Modification of the response will also result in biases when catch per unit effort is employed as an index of abundance (Freon & Misund 1998).

Fishing vessels and fishing gear emit a variety of stimuli. To elicit a response, a stimulus must first fall within the sensory capacity of a species. Second, both for the initial response and for its modification after experience, it is of critical importance what the perceived stimuli represents in the 'Umwelt' (the perceptible world of an animal: Uexküll 1921) (Fig. 14.2). Some stimuli may be novel in an evolutionary perspective and not release any initial response. Others may resemble stimuli from naturally occurring objects and release reactions adapted to these objects. The different phases in the reaction to baited fishing gear, for instance, are similar to the reactions to food items that emit chemical and visual stimuli (Løkkeborg 1994). Reactions to active fishing gear may be more complex and consist of a mixture of different responses. Avoidance and freezing released by trawls (Engås *et al.* 1998) may reflect anti-predator behaviour. Fishes that keep pace with the net wall of a trawl are displaying a visually triggered optomotor response (Kim & Wardle 2003) involved in the synchronization of individuals within schools.

The following sections give details of the five stages that have been identified at which learning can have an effect in fish capture.

14.2.2.1 *Natural variations in spatial distribution and behaviour*

A variable migration pattern influenced by learning (see above) could make it difficult to locate fishable concentrations. The level of activity and spatial use determine the probability of encountering static gear (Engås & Løkkeborg 1994), and at least the latter may be influenced by learning. A simulation study based on *in situ* movements of cod and ling in fjords showed that there was a low risk of encountering a fleet of gillnets set in a fixed area, and explained this finding by the limited search range of the fishes (Engås & Jørgensen 1997).

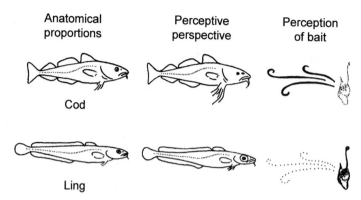

Anatomical proportions	Perceptive perspective	Perception of bait

Cod

Ling

Fig. 14.2   Schematic representation of how the gadoids cod and ling are believed to experience stimuli from baits. Both species react to both visual and chemical cues, but visual stimuli are regarded as being relatively more important in ling. This is based on a more pronounced diel activity rhythm, shorter attraction distance to baits and random swimming in relation to the current, in contrast to cod, which have a tendency to swim perpendicular to the current, thus increasing the probability of encountering odour plumes. Ling is also more stationary, hunting for mobile prey that emit visual stimuli (Skajaa 1997; Løkkeborg 1998; Løkkeborg & Fernö 1999; Løkkeborg *et al.* 2000; Vabø *et al.* 2004). Ling also have a more elongated body, permitting fast acceleration, and cod a greater inclination of the mouth, facilitating chemical food search (Mattson 1990). Finally, in a classification of gadoid species in a gradient from fish feeders relying on visual stimuli to invertebrate feeders relying on chemical stimuli, based on the anatomy of the brain, Kotrschal *et al.* (1998) have placed ling closer to piscivory.

14.2.2.2   *Avoidance and attraction before fishing*

Fishes seem to be able to learn about risky habitats from their own experience or from the reactions of others (Chivers & Smith 1995; Brown 2003), but at present we lack firm evidence that fishes avoid heavily fished areas. Chemosensory cues are generally more reliable than visual cues (see Chapter 4), and chemical stimuli from fishing gear could provide information about the local predation (fishing) risk. The release of alarm substances from injured fishes may reduce the number of fishes in an area and enhance the response to visual indicators of predation risk (Wisenden *et al.* 2004). Auditory stimuli and turbidity (Humborstad *et al.* 2006) generated by trawling activity may also provide fishes with reliable information. Interestingly, high concentrations of fishes have been observed in recently trawled areas, where fishing vessels have been observed to re-fish the same tow route, taking advantage of immigrating scavenging fishes (Kaiser & Spencer 1994). Fishes can also be encouraged to concentrate in a specific area before fishing by conditioning them to food (Zhuykov & Panyushkin 1991).

14.2.2.3   *Before physical contact with the gear*

The search image of the fishes and degree of novelty of the stimuli can be crucial. Bait attraction may be influenced by the olfactory and visual stimuli emitted by present prey organisms. A possible example is the low efficiency experienced in the longline fishery for cod feeding on capelin in the Barents Sea, when cod appear to react to visual stimuli from moving prey rather than to chemical stimuli from baits. Artificial baits are reported to be more effective for large than for small cod (Løkkeborg 1990). Small cod may experience the artificial bait as being more novel, as large cod have greater diet breadth (Pálsson 1994) and thus more experience of various prey organisms. The capture efficiency of pots can be influenced by habitat type. Pots that have been set close to structures and which therefore differ less from their surroundings than pots further away are more efficient (High & Beardsley 1970) and are presumably perceived by the fishes as less novel.

Socially transmitted learning may also lead to avoidance. Copying fishes can associate the response of others with the sound of trawler engines and thus acquire a modified response by observational conditioning (see Chapter 10). In the laboratory, fishes learn from experienced demonstrators the escape route from a model trawl apparatus (Brown & Laland 2002). Fishes in schools displayed learned avoidance of a model trawl, whereas pairs of fishes showed no evidence of learning (Brown & Warburton 1999b). Tropical sardines show conditioned avoidance reactions in the laboratory, with transmission of reactions from conditioned to naive fishes (Soria *et al.* 1993). Interestingly, fishermen claim that herring have became more difficult to catch since the start of purse seine fishing in the North Sea in the 1960s.

On the other hand, attacks on bait by other fishes and movements of hooked fishes seem to stimulate attacks by social facilitation in cod, haddock (*Melanogrammus aeglefinus*, Gadidae) and whiting (*Gadus merlangus*, Gadidae) (Fernö *et al.* 1986; Løkkeborg *et al.* 1989). Struggling fishes do not appear to trigger fear responses, and thus gadoid fishes do not seem to have any preprogrammed or learnt response to constrained fishes – a situation they presumably never encounter under natural conditions. Species reacting to chemical alarm cues released by injured fishes (Chivers & Smith 1998) may behave differently. Fishes approaching a pot could be socially reinforced by entrapped fishes. Munro *et al.* (1971) observed that when some individuals grunt (*Haemulon plumieri*, Haemulidae) were already in a pot, the daily ingress of fish rose sharply and suggested that the high between-pot variability in catches may be the result of conspecific attraction.

14.2.2.4   After physical contact with the gear

Response modification and learning after contact with gear may occur on various time-scales. Within a single encounter with gear fishes may modify their response based on events that took place seconds or minutes previously. The modification can be caused either by a change in motivational state, such as fright, or by learning. During classical (Pavlovian) conditioning (Lieberman 1990) fishes usually receive negative reinforcement from the gear. An example of classical fear conditioning may be fishes that learn to associate the sound from a trawl and later physical contact with the gear. Operant conditioning may take place when fishes attack baited hooks or when fishes find their way through the funnel of a pot and escape. Fishes seldom receive positive reinforcement, but haddock are known to 'steal' baits from long lines and thus be rewarded by bait attacks.

Decreased catchability by trawls during repeated hauls has been demonstrated in bream (*Abramis brama*, Cyprinidae) and has been explained by trawl avoidance observed in acoustically tagged fishes caught by trawl (Pyanov 1993). This indicates that fishes can learn to avoid trawls in the course of one-trial learning. Cod have been observed to swim in loops and return to baited nets from distances up to 400 m from different directions in relation to the current, indicating learning of spatial maps (Kallayil *et al.* 2003). This could reflect inspection behaviour after encountering a food item (bait bag) that is too large to swallow and is located in an unfavourable site (i.e. the net).

The behaviour resulting from learning can vary at the individual level. After biting a baited hook in the laboratory, cod response intensity decreases dramatically, irrespective of whether the fish was hooked or not (Fernö & Huse 1983; and see Fig. 14.3). However, a strong individual variation was observed. Some cod displayed only a few intensive attacks several days apart, whereas other individuals, after some initial strong responses, displayed several hundred low-intensity responses (approaches and tastings without biting). All the fishes had been starved, but previous experience or dominance rank may explain the variation. There are individual differences in the way in which fishes respond to novelty (Brown *et al.* 2005b), and wide individual variations in behaviour have also been observed in fishes learning to avoid toxic prey (Crossland 2001). The study on cod indicates that relatively minor individual differences can have important consequences in a conflict situation when a fish approaches an object that gives both positive (food) and negative (pain) reinforcement. The sharply decreased response intensity and 'neurotic' behaviour of some cod, which made a long series of approaches without bait contact, strongly

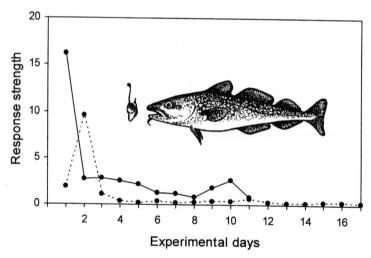

Fig. 14.3   The effect of experience on the behaviour of cod vis-à-vis a baited hook. The response strength is calculated as the ratio between the number of strong (complete bite, jerk, shake, pull, chew, rush) and weak (approach, taste, incomplete bite) responses on each day of the experiments (see Fernö & Huse 1983). Group 1 (solid line, 15 fishes) were caught by trawl and kept in the laboratory for 3 years and prior to the experiments were subjected to several bait bag preference tests, while group 2 (broken line, 20 fishes) were caught in traps and kept in the laboratory for 2 months prior to testing. The total number of responses per day varied from 22 to 178.

suggest that fishes experience a baited hook as aversive and painful; a possibility that has been much debated (Chandroo *et al.* 2004). Individual variations in cod responding to an approaching trawl (Engås *et al.* 1998) may also involve learning.

14.2.2.5   Behaviour after escaping the gear and long-term consequences

Fishes escaping from gear may suffer swimming impairment and behavioural deficits that subject them to elevated predation risk and reduced feeding behaviour (Ryer 2004). The degree to which learning influences later encounters depends on how fishes experience the reward or punishment and the number of encounters needed to establish an association. How long the modification persists may depend on the total reinforcement schedule in the variable situation in the sea, where fishes are exposed to a multitude of naturally occurring stimuli in addition to vessels and gear. In some cases the memory window can be long, but as the ability to forget in a dynamic environment may be as important as the ability to learn (Kraemer & Golding 1997; and see Chapter 2), infrequent reinforcement can lead to extinction of a learned response.

In fishes, learning about predators usually occurs after just one simultaneous presentation of the cue and the stimulus (Magurran 1989) and the response can be retained for several months (Chivers & Smith 1994). During fishing experiments with hook and lines and in catch-and-release fishing, lower catchability during the course of fishing has been observed in many species (Beukema & de Vos 1974; Hackney & Linkous 1978; O'Grady & Huges 1980; Yoneyama *et al.* 1996). One single hooking experience made carp (*Cyprinus carpio*, Cyprinidae) more difficult to catch for at least a year (Beukema 1970), and learned avoidance of a trawl apparatus in the laboratory persisted for at least 11 months (Brown 2001). However, Tsuboi & Morita (2004) found that white-spot char (*Salvelinus leucomaensis*, Salmonidae) that had been hooked and released were more likely to be caught than previously uncaught fishes, perhaps because they were initially the least risk-averse individuals. Small fishes can learn to penetrate meshes and in this way learn to escape through nets in trawls (Özbiljin & Glass 2004). In heavily fished areas with multiple gear contacts such modifications of behaviour can influence the selective properties of trawls.

When we are evaluating the role of learning in fish capture, we need to build on both laboratory and field studies. Laboratory studies can give valid qualitative information, but easily recognizable stimuli in a well-defined laboratory situation may facilitate learning. It is not a straightforward matter to transfer laboratory observations of learning skills to the variable environment in the sea. Field studies are generally needed to obtain reliable quantitative results (Løkkeborg *et al.* 1993), and one bottleneck in understanding the role of learning in fish capture is the lack of field observations of individual fishes over time.

14.3   Aquaculture

Aquaculture is a rapidly growing industry, and millions of tonnes of freshwater and marine fish species are produced annually. The farming environment ranges from semi-natural ponds to high-technology, intensive, recirculation farms. While pond-farmed fish experience near-natural environments, fishes reared in tanks and cages are kept in an environment very different from the natural habitats to which they are evolutionarily adapted. Intensively farmed fishes must be able to adapt to high fish densities, restricted space, and artificial and uniform food and, not least, to frequent disturbances and handling by man. The rearing environment has to be within the species' physiological range of tolerance, but the fishes must also cognitively process the sensory information presented by the farming systems. Fish species selected for intensive aquaculture need flexible behaviour to be able to adapt to the new environment. Considering only biological performance, the best aquaculture candidate species should, therefore, be social, generalist species with broad niches and large environmental tolerance intervals, like *Tilapia* spp. However, high market prices for more demanding species can also make it profitable to invest in expensive technology and tailor the rearing environment to the species' preferences (e.g. Atlantic halibut, *Hippoglossus hippoglossus*, Pleuronectidae).

Most aquaculture environments are structurally very simple, and the challenges for the fishes seem equally simple, with food easy to catch and an absence of predators. However, high stocking densities with frequent social interactions, suboptimal environmental conditions, very limited choice of habitat and food, and abundant, noisy and unpredictable sensory stimuli, may make the environment cognitively demanding with few learning opportunities and a high stress level. One way to reduce cognitive stress in these environments would be to hand over individual control and decision-making to the group. Information about rewards (food) and punishments (handling, social aggression) will often be received indirectly via the behaviour of other fishes through social learning, and the behaviour of individuals will trigger self-organizing group behaviour that may be adaptive or maladaptive.

14.3.1   *Ontogeny*

Most fishes are only a few millimetres long at hatching and their weight grows by several orders of magnitude before they reach maturity. They grow through several developmental stages in different environmental niches and habitats with different predators, prey and adaptive demands (Balon 1984; Fuiman 1994). The tiny larvae have few resources to allocate to brain development and learning, and must mostly rely on preprogrammed behaviour; for instance, halibut larvae that approach the start of feeding are positively phototactic, which in nature leads them towards the prey-rich surface layers of the sea (Naas & Mangor Jensen 1990). However, in tanks, such phototactic behaviour results in halibut larvae that are butting against light tank walls and the surface, as they are trapped in their preprogrammed behaviour without the capacity to learn to adapt this behaviour to the novel environment (Naas & Mangor Jensen 1990). However, learning ability may be better suited to associations that have a strong survival advantage in evolutionary terms; for example, most fish larvae rapidly develop clear prey preferences and improved prey-catching skills (Checkley 1982; Cox & Pankhurst 2000). Taste preferences are usually genetically controlled (Kasumyan & Døving 2003), but the association between the taste and visual characteristics of the prey and improved prey-catching skills ought to be a learned process with high survival value (see Chapter 2).

Farmed fishes accumulate experience during their lifespan, and it is crucial that behaviour promoting growth and welfare is established as early as possible, otherwise a suboptimal culture may result. Group behaviour often differs between identical rearing units, with fishes in one tank, for example, being consistently more fearful than those in other tanks. Mixing fishes between tanks or the removal of aggressive individuals may be able to turn a 'bad tank culture' into a 'good' one. However, a 'bad culture' may also occur when fishes from two 'good cultures' with different experience and habits are mixed. During the freshwater stage salmon school against the direction of the current in the tanks. When salmon smolts reared in tanks with either clockwise or anti-clockwise swimming directions were mixed during transfer to sea cages they were unable to school normally, resulting in an unordered group structure (see Fig. 14.4; Fernö *et al.* 1988; Juell 1995). This behaviour was explained by a conflict between different preferred swimming directions, which was supported by observations in one pen in which the fishes, after an initial period of unstructured swimming, swam clockwise in the upper part and anti-clockwise in the lower part of the pen. Fishes reared in larger tanks with a variable current direction schooled as normal soon after transfer to the sea (see Fig. 14.4) and had higher rates of food intake. Further studies on how early experience influences the development of behaviour in farmed fishes are needed.

14.3.2 *Habituation and conditioning*

Farmed fishes are exposed to many sudden stimuli that always elicit the same behavioural reactions (reflexes and fixed action patterns). These reactions can also trigger a physiological stress response (Conte 2004). An example is the startle reflex caused by sudden noise or approaching objects. If the stimulus is not associated with harmful events (e.g. noise from water pipes or people passing the tanks), repeated exposure results in response waning; a mechanism known as 'habituation' (Lieberman 1990). Rapid habituation should reduce stress and be a preferable trait in farmed fishes. Repeated responses to stressful events that are not associated with other fearful events (such as confinement), will also lead to reduced stress responses on later exposure to similar events. However, in most species certain stimuli, like silhouettes of bird or fish predators and approaching objects, release strong predisposed responses with slow habituation and should be avoided or associated with rewards (see below).

Classical (Pavlovian) conditioning occurs when two events overlap in time (and space) so that an originally neutral stimulus can be associated with an aversive (fear conditioning) or rewarding (reward conditioning) stimulus (Lieberman 1990). In aquaculture, reward conditioning often occurs automatically in connection with feeding. Fishes learn, for example, to associate the footsteps of the farmer or the sound of pellets in the feeding pipes with food and can show strong anticipatory behaviour before the food arrives. In planned conditioning procedures, light or sound signals are used to release anticipatory behaviour, for instance, leading fishes to a feeding area (Midling *et al.* 1997). The anticipatory behaviour functions as an arousal for appetitive responses and is a positive emotional event for the fishes that should increase feeding motivation and welfare (Lamb 2001; Spruijt *et al.* 2001).

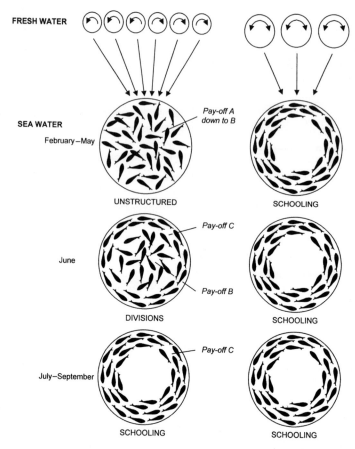

Fig. 14.4 A possible example of long-lasting effects of the early rearing environment on group structure in salmon in marine net pens (Fernö *et al.* 1988). Parr on the left were reared in small tanks with consistently different current directions and parr to the right in larger tanks with a variable current direction. The fish were transferred to marine net pens in May/June and the group structure recorded each weekday in February–September the following year. Unlike the parr reared in large tanks, those from small tanks were, for a long period, unstructured before eventually changing to school-like swimming. This was presumably caused by a conflict between different swimming directions. The shift in group structure can be explained by a change in the pay-off matrix, which was supported by data on food intake. The pay-off for unstructured fishes might have gradually decreased from A to B in connection with high and increasing perceived density as the fish grew, at the same time as the growth potential should increase when the temperature rises in the spring. Some fish then began to school along the net wall and could thereby attain a higher pay-off (C) than unstructured fishes (B) at the centre. Unstructured fishes may thus have chosen to join the structured division, and a frequency-dependent pay-off may have led to a rapid change until almost all fish swam in a polarized way, which co-occurred with increased food intake.

When a fish responds to a stimulus, both the perception of the stimulus and the memory of prior experiences are involved in the organization of the response (Barton 1997). Farmed (domesticated) fishes tend to display weaker stress responses to stressors than wild fishes reared under similar environmental conditions (Johnsson *et al.* 2001). Similarly, fishes from high-predation areas show weaker responses than fishes from low-predation areas (Brown *et al.* 2005a), and early or repeated exposure to stressful events can help fishes to manage their stress responses later on. Schreck (1981) suggested that the psychological component of a stressor was important in determining its severity, and proposed that the psychological wellbeing of a fish could be improved by positive conditioning. He demonstrated that chinook salmon (*Oncorhynchus tshawytscha*, Salmonidae) could be positively conditioned to the emptying of the tank until the water just covered the fishes, with the physiological stress response (cortisol) to subsequent stressors (transportation) significantly lowered compared to controls (Schreck *et al.* 1995). Other methods of stress control of farmed fishes that involve learning are reviewed by Lines & Frost (1999).

Fear is a primitive emotion resulting in physiological and behavioural changes and is caused by the perception of danger (Chandroo *et al.* 2004). Fear responses can easily be conditioned, and conditioned signals will lead to avoidance behaviour and stress responses (Yue *et al.* 2004). Unplanned fear conditioning occurs regularly in fish farms during handling; for example, the association between the sight of a dip net and confinement stress, resulting in strong startle responses as soon as a dip net is put into the tank. The uncertainty related to reward or punishment (handling) associated with husbandry routines may increase stress levels. Fish farmers should therefore try to avoid husbandry routines that lead to unplanned fear conditioning and to make their husbandry routines predictable and associated with positive events (see above). One possible way of doing so would be to clearly signal what is coming, for example, whether the fishes are about to be confined or not, perhaps by sounding a tone or turning on a light. This strategy was used by Bakken *et al.* (1993) in fox farms. By using differently coloured clothes during handling and during daily feeding, the stress level before feeding was reduced and anticipatory behaviour increased.

In classical conditioning there is an overlap in time and place between the conditioned stimulus (CS, the farmer) and the unconditioned stimulus (US, reward – food), and this is a type of learning also found in invertebrates. The offset of the CS and the onset of the US can also be separated by a space in time – a paradigm called trace conditioning (Lieberman 1990). Trace conditioning is memory-based learning and in mammals it is mediated by the hippocampal system. In fishes, trace conditioning has recently been found to be associated with the telencephalic pallium (Portavella *et al.* 2004). In trace conditioning experiments, farmed cod could associate a CS (light flash) with a US (food) separated by a trace period of up to 2 min and they could also remember the association for at least 3 months (Nilsson *et al.* 2005). This trace length is impressive even for mammals (Lieberman 1990), and indicates that fishes have better learning abilities than previously assumed. Trace conditioning can be crucial in a rearing unit as a means of enabling fishes to associate events separated in time (like self-feeding, see below).

Fishes in a tank or cage are often active in the absence of external stimulation. Exploratory behaviour is found in most animals, because it is crucial to obtain information about food resources and dangers in a constantly varying environment

(Inglis *et al.* 2001). On reward, the probability that the behaviour will be repeated gradually increases, resulting in operant conditioning – an association between a response (action) and a stimulus (reinforcement). In fish farming this way of learning is used in various types of self-feeding technology where the exploratory behaviour starts a feeding automat (Jobling *et al.* 2001). A trigger consisting of a rod to be pushed or a string to be pulled is usually involved, and the feeders may be purely mechanical or electronic (Alanärä 1996; Rubio *et al.* 2004). Many farmed fish species can learn to operate self-feeders (Divanach *et al.* 1986, 1993; Alanärä 1996), but most experiments have focused on feeding activity rather than learning ability, and the various aspects of learning rate and effects of reward value have not been described in any detail. The time that different species need to learn to operate the self-feeder varies from 10 to 45 days (Jobling *et al.* 2001). Because many fishes may be rewarded by the action of a single fish, and the fishes that operates the trigger may be not rewarded, the learning situation and correct adjustment of reward level are complex. A further complicating factor is that relatively few fishes learn to operate the triggers (Alanärä 1996; Gélineau *et al.* 1998). In some situations the operation of the trigger may also be rewarding by itself and release 'false triggering', leading to accidental feeding (Rubio *et al.* 2004).

Most fish farmers have observed that fish have an impressive ability to anticipate scheduled feeding times and that they must have some sort of an internal circadian clock (Reebs 2000). This is clearly demonstrated by sea bass (*Dicentrarchus labrax*, Moronidae) fed by time-restricted demand feeders with a clear increase in activation of the trigger when the time of feeding is approaching (Azzaydi *et al.* 1998; Sanchez-Vásques & Madrid 2001). Anticipation of multiple daily feeding times in individual fishes has been demonstrated in rainbow trout (Chen & Tabata 2002).

14.3.3   *Individual decisions and collective behaviour*

The behaviour of other fishes is a powerful stimulus for group-living fishes, and the copying behaviour has both inherent (preprogrammed) and learned components (Kieffer & Colgan 1992; and see Chapter 10). In fish tanks and cages, access to food is often unpredictable in time and space and visibility may be low as a result of turbidity and high fish density. The behaviour of other fishes is therefore an important source of information, and learning by a combination of classical and operant conditioning occurs when fishes learn to associate the behaviour of their shoalmates with rewards or aversive events. Learning to recognize group members may also modify inter-actions between farmed fishes, for instance by reducing aggression (Höjesjö *et al.* 1998) and thereby increase feeding activity (Griffiths *et al.* 2004). However, in the guppy the ability to recognize and associate with conspecifics decreases with group size (Griffiths & Magurran 1997), and the large number of fishes in a rearing unit may prohibit the development of individual recognition (but see Ward *et al.* 2005).

Social learning can lead to high local fish concentrations that may involve costs related to strong competition for food and low oxygen levels. This in turn may result in an ideal free distribution whereby animals exploiting patches distribute themselves over the area in proportion to patch profitability (Fretwell & Lukas 1970). The distribution of salmon in net cages (Juell *et al.* 1994; Fernö *et al.* 1995) may be regarded as the result of an ideal free distribution involving learning, with the fishes comparing the quality at different depths (see also Hakoyama & Iguchi 2001).

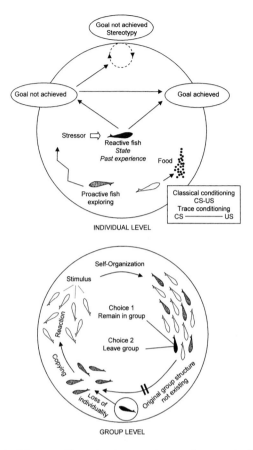

Fig. 14.5 Individual decisions and learning in a rearing environment of fishes with different coping styles (top) and consequences of copying and self-organization on collective behaviour (bottom). Maladaptive behaviour with loss of control and chronic stereotyped behaviours might develop on the individual level if the goal of the behaviour is not achieved. An individual joining a group of fishes that copy and mix with another group with a subsequent change in group structure by self-organization may be trapped in a maladaptive group structure.

Understanding the consequences of the behavioural decisions of an individual fish in a rearing unit is a challenge, particularly because social interactions will influence the end-result with regard to distribution and group behaviour. There is presumably a con.plicated intercalation of rigid and flexible components when fishes encounter a situation, respond to stimuli and subsequently enter a new situation that releases a new stimulus-behaviour complex (Fig. 14.5). If the outcome is not what the fishes 'expected', a shift from a high-level 'off-line' cognitive control to a low-level 'on-line' direct control (stimulus-response system) may occur, resulting in chronic stereotyped behaviours and behavioural pathology (Toates 2004). As in mammals, fishes have different personalities, such as bold and timid individuals (Magurran 1993; Sneddon 2003; Brown *et al.* 2005b) with proactive and reactive coping styles and probably different learning abilities and motivations (Koolhaas *et al.* 1999; Kristiansen & Fernö 2006).

Group decisions can reduce aggression and facilitate food localization, but can also have negative consequences. High densities can result in loss of individual cognitive control with the fishes switching from individual to school behaviour rules, leading to emergent school structures and self-organization (Camazine *et al.* 2001; Couzin *et al.* 2005; and see Chapter 9). In individual-based models, small changes in individual decision rules have a powerful impact on the emergent school behaviour (Nøttestad *et al.* 2004). In addition, in an environment where fishes are primarily influenced by social stimulation and feed, copying other fishes should strongly influence culturally mediated group behaviour, with various and unpredictable outcomes in species not adapted to a schooling lifestyle. Farmed fishes are in a way at the mercy of copying and self-organization (see Fig. 14.5). When a group structure has been established as a result of individual decisions, the original decisions may no longer be beneficial and fishes may find themselves trapped in a collective maladaptive pattern of behaviour, resulting in reduced growth rate. However, in some cases fishes can have a choice between different group structures. Two divisions of fishes have been observed to coexist in salmon net pens, with fishes at the centre swimming in different directions and fishes along the net walls in a polarized school-like way (Fernö *et al.* 1988; Juell 1995; and see Fig. 14.4). Over time the polarized group became larger and eventually took over the whole cage, indicating that individual fishes chose to join the structured division. School-like swimming should lead to fewer physical encounters, resulting in a lower perceived fish density and stress level and thus higher pay-off. In fact, schooling seems to be correlated with high food intake (Fernö *et al.* 1988), indicating that an adaptive collective behaviour developed in this case.

14.4   Stock enhancement and sea-ranching

During the past century, release of hatchery-reared fishes has been an important fisheries management tool in the restoration or creation of new fisheries and in increasing recruitment in areas in which the natural recruitment of juveniles is, or is believed to be, less than the body of water can sustain (Shelbourne 1964; Bowen 1970; Cowx 1994; White *et al.* 1995; Munro & Bell 1997). Over the course of the past few decades, rearing techniques for the mass production of many freshwater and marine fish species have been developed and a great amount of effort has been put into mass releases of more than 250 fish species (Howell *et al.* 1999). Currently releases of reared juveniles are taking place all over the world, ranging from small-scale releases to enhance recreational fisheries in rivers and lakes, to industrial-scale releases of billions of salmonids in the North Pacific (Howell *et al.* 1999; Brown & Laland 2001; Leber *et al.* 2004). In spite of some documented success stories (Aprahamian *et al.* 2003), such as the releases of chum salmon (*Oncorhynchus keta*, Salmonidae) in Japan (Kaeriyama 1999), the benefits of the releases have usually not been properly evaluated, and the use of reared fishes for stock enhancement is still a controversial issue (Cowx 1994; Blankenship & Leber 1995; White *et al.* 1995; Hilborn 1998; Welcomme 1998).

When hatchery-reared fishes are released into the wild, they must immediately cope with a novel and complex environment, identify and catch live prey, and avoid the risk of predator attacks. It is therefore not surprising that reared fishes run into problems after release, and a major problem with restocking is the high mortality of hatchery-reared fishes during the first period after release. Their poor performance can be explained by a combination of incomplete neural and sensory development resulting from prenatal stress and lack of appropriate environmental stimuli during critical periods in ontogeny, impaired prey-catching and predator-avoidance skills, and physiological and psychological stress responses after release, in addition to genetic selection (Blaxter 1976; Browman 1989; Svåsand *et al.* 1998; Huntingford 2004; Svåsand 2004).

Mortality rates of juvenile fishes are difficult to measure in nature, and direct comparisons between reared and wild fishes are few (Kristiansen 2001; Svåsand 2004). However, on the basis of comparisons of recapture rates of tagged wild fishes and released reared fishes and analysis of stomach contents of predators caught in the release area, it seems clear that released fishes are more vulnerable to predation than their wild counterparts (Miller 1954; Elson 1975; Blaxter 1976; Olla *et al.* 1998; Brown & Day 2002). Predator-avoidance skills improve rapidly with experience, and several laboratory studies have found better predator avoidance after relatively short experience of predators (Ginetz & Larkin 1976; Jakobsson & Järvi 1976; Patten 1977; Olla & Davis 1989; Suboski & Templeton 1989; Järvi & Uglem 1993; Hossain *et al.* 2002). Surprisingly few examples exist where predator-trained fishes have been released into the wild, so we need to be cautious in extrapolating findings from laboratory studies to natural conditions. In laboratory experiments, Nødtvedt *et al.* (1999) demonstrated differences in anti-predatory behaviour between predator-naive reared cod and wild cod. Reared cod went from a stage of excessively weak reactions to predators to a subsequent stage of excessively strong reactions, but presumably over time developed the 'balanced' response displayed by wild cod. However, in a release experiment with reared cod exposed just before release to large predators in their cages, no effects of predator training on later predation or recapture rates were found (Otterå *et al.* 1999). One explanation for this could be that the appropriate responses to predators are learned within a short period of time after release, and that the difference in mortality between trained and untrained fishes in that short period is too small to be detected in field studies. The fact that these were pond-reared fishes, raised in a semi-natural environment, may also help to explain the similar mortality of reared and wild cod and the negligible effect of training.

Foraging skills are acquired through learning based on innate predispositions (Kieffer & Colgan 1992; and see Chapter 2). After release, reared fishes develop the same prey preferences as their wild conspecifics relatively quickly, but may need many weeks or months to reach the same daily food energy intake as wild fishes (Sosiak *et al.* 1979; Johnsen & Ugedal 1986; Kristiansen & Svåsand 1992; Noreide & Fosså 1992). Poor prey-catching skills will also have indirect effects on predator avoidance, because more time and attention need to be devoted to prey search, and hunger may increase risk-taking behaviour (Hossain *et al.* 2002). Steingrund & Fernö (1997) showed that reared cod displayed much more active prey-catching behaviour than wild cod, which could lead to less predator attention and avoidance. Fishes trained to catch live prey in the hatchery before release have been shown to have better prey-catching abilities after release (Suboski & Templeton 1989).

Several studies of reared fishes have showed that the rearing process may lead to other morphological and behavioural deficiencies that increase mortality after release (see reviews by Blaxter 1976; Howell 1994; Ellis *et al.* 1997; Olla *et al.* 1998; Svåsand *et al.* 1998; Tsukamoto *et al.* 1999; Masuda 2004; Svåsand 2004; Huntingford 2004); for example, reared Japanese flounder (*Paralichthys olivaceus*, Paralichthyidae) swim longer distances and spend more time in the water column after take-off from the bottom and have low occurrence of burrowing behaviour (Yamashita & Yamada 1999). The larger the differences from natural food and environment in the rearing environment, the bigger will be the morphological and behavioural differences found, and such differences also increase with age at release (Tsukamoto *et al.* 1999; Masuda 2004).

In many cases reared fishes are in a poorer physiological state, thanks to inappropriate nutrition, less exercise and high stress levels in the tanks. Poor larval and juvenile nutrition might affect the development of the brain and nervous system, with lifelong effects on the sensory system and learning ability; for example yellowtails (*Seriola quinqueradiata*, Carangidae), fed *Artemia* enriched on a diet without the fatty acid docosahexaenoic acid did not school (Ishizaki *et al.* 2001). Rearing fishes in enriched or more natural environments has positive effects on prey-catching and anti-predator behaviour (Berejikian *et al.* 2001; Brown and Day 2002). Cod from enriched environments learn to switch earlier to new prey types, are more aggressive when feeding, use cover more often and recover sooner from a novel fright stimulus (Braithwaite & Salvanes 2005; Salvanes & Braithwaite 2005). It thus seems that simulated natural environments can help newly released fishes to categorize their environmental cues more properly and hence learn to adjust their behaviour vis-à-vis their new environmental concepts more readily. It is therefore reasonable to assume that some behavioural differences are caused by lack of appropriate stimuli during ontogeny (Svåsand *et al.* 1998).

In comparison with the enormous costs and efforts that have been allocated to restocking activities, surprisingly little has been done to develop full-scale methods for production of ecologically viable fry. As we have shown, improved juvenile quality with better survival skills can be attained by enriching the rearing environment and by various kinds of predator and live-feed training, and methods for large scale survival skill training using social (transmission chain) learning protocols should be developed and tested in full-scale experiments (Brown & Laland 2001; Brown & Day 2002).

14.5   Conclusions

Fishes may be more traumatized by stimuli relevant to their ancient survival traits than by novel stimuli. Hence, the ability to learn and speed of learning may be greatly influenced by the novelty of fishing gear, farming facilities and general breeding conditions. The ever-changing fishery technology and fishes culture environments have presumably brought about greater variance in individual fitness in commercially exploited populations. This may favour flexible traits (Caraco 1980; Real 1980), and among those, improved learning skills is one clear candidate. However, there is a discrepancy in the time-scale involved, and fishes are not necessarily able to change from preprogrammed behaviour that has evolved for a formerly more stable world, to behaviour appropriate to a new and rapidly changing environment. Another complication is that fisheries and fish stocking may modify competition between trophic levels and create indirect changes that are difficult to track and relate to consequences of manmade changes. Finally, as behavioural traits may be copied, selective fisheries may change whole cultures, for instance of herring populations, and consequently change migratory routes and national ownership of fishery resources and thereby have a great impact on fishery practice.

Our knowledge of the role of learning in the spatial dynamics of fishes and of where in the brain spatial learning is controlled has greatly expanded during the past decade (see Chapters 7 and 13; Broglio *et al.* 2003). This research, however, is predominantly based on laboratory experiments, and the steadily accumulating knowledge of the mesoscale and macroscale dynamics of commercially exploited species in the field is seldom considered. Likewise, field studies relate to only a limited extent to detailed laboratory findings. The interpretations of field data have been hampered by a primarily descriptive approach and by relatively crude data, resulting from limited accuracy during tracking, short observation periods and few replications. Recent technological advances, such as geographical information systems, multibeam sonars, passive integrated transponder tags, miniature acoustic tags and archive tags should enable us to acquire a better understanding of spatial dynamics and the role of learning. Individual-based modelling provides an ideal platform for linking individual fishes behaviour and learning to population dynamics (Huston *et al.* 1988; Huse *et al.* 2002a; Grimm & Railsback 2005). We are looking forward to a productive synthesis between the different scientific traditions in the laboratory, field and modelling.

In areas with strong fishing pressure where man challenges the fishes by repeatedly 'attacking' with an arsenal of fishing gear, it is even more vital for the fishes to react in an adaptive and balanced way than when it is under threat from natural predators alone (Lima & Bednekoff 1999). In order to avoid gear and at the same time continue fitness-related activities essential for growth and reproduction, it is crucial for the fishes to distinguish 'Danger' from 'No danger'; the benefits of learning skills should then be greater. It has been demonstrated that fishing has led to a rapid evolution of life-history traits (Olsen *et al.* 2004). Similarly, the idea cannot be excluded that an improved genetically controlled learning capacity can evolve in heavily exploited species. The general reaction to novel stimuli can also change if selection favours stronger neophobia. That selection has the potential to influence learning abilities is demonstrated by the finding that different populations within a fish species can be predisposed to learn specific associations appropriate to the specific predatory problems they may encounter (Kelly & Magurran 2003).

In fish farming there can also be genetically based changes in learning ability. The selection of fishes on the basis of rapid growth can lead to changes in a number of behavioural traits (Huntingford 2004), possibly also influencing innate learning skills. One might expect that a high capacity to learn should facilitate adaptation to the farming environment and thereby enhance growth and be selected for. However, the problems that farmed fishes face may still be simple compared to those of wild fishes. The brains of hatchery-reared fishes have in fact been observed to be relatively smaller in several critical measures than those of their wild counterparts

(Marchetti & Nevitt 2003), partly resulting from the influence of the early rearing environment (Kihslinger & Nevitt 2003). The scientific basis of many observations and interpretations of learning skills in fishes is still weak, and more effort should be put into studying both basic and applied aspects of learning.

14.6   Acknowledgements

We thank Svein Løkkeborg and Jonatan Nilsson for valuable comments on the manuscript. Ragnhild Jakobsen produced the figures skilfully.

14.7   References

Alanärä, A. (1996) The use of self-feeders in rainbow trout (*Oncorhynchus mykiss*) production. *Aquaculture*, **145**, 1–20.

Altbäcker, V. & Csànyi, V. (1990) The role of eye-spots in predator recognition and anti-predatory behaviour in the paradise fish (*Macropodus opercularis*). *Ethology*, **85**, 51–57.

Aprahamian, M.W., Martin-Smith, K., McGinnity, P., McKelvey, S. & Taylor, J. (2003) Restocking of salmonids – opportunities and limitation. *Fisheries Research*, **62**, 211–227.

Azzaydi, M., Madrid, J.A., Zamora, S., Sánchez-Vázquez, F.J. & Martínez, F.J. (1998) Effect of three feeding strategies (automatic, *ad libitum* demand-feeding and time-restricted demand-feeding) on feeding rhythms and growth in European sea bass (*Dicentrarchus labrax* L.). *Aquaculture*, **143**, 285–296.

Babcock, E.A., Pikitch, E.K., McAllister, M.K., Apostolaki, P. & Santora, C. (2005) A perspective on the use of spatialized indicators for ecosystem-based fishery management through spatial zoning. *ICES Journal of Marine Science*, **62**, 469–476.

Bakken, M., Moe, R. & Smith, A. (1993) Radio telemetry: a method of evaluating stress and learning ability in the silver fox (*Vulpes vulpes*). *Proceedings of the International Society of Applied Ethology, 3rd Joint Meeting*, **1993**, 591–594.

Balaban, E. (1997) Changes in multiple brain regions underlie species differences in a complex, congenital behaviour. *Proceedings of the National Academy of Sciences of the USA*, **94**, 2001–2006.

Balon, E.K. (1984) Reflections on some decisive events in the early life of fishes. *Transactions of the American Fisheries Society*, **113**, 178–185.

Barton, B.A. (1997) Stress in finfish: past, present and future – a historical perspective. Fish Stress and Health in Aquaculture. *Society of Experimental Biology Seminar Series*, **62**, 1–33.

Berejikian, B.A., Tezak, E.P., Risley, S.C. & LaRae, A. (2001) Competitive ability and social behaviour of juvenile steelhead reared in enriched and conventional hatchery tanks and a stream environment. *Journal of Fish Biology*, **59**, 1400–1413.

Beukema, J.J. (1970) Angling experiments with carp (*Cyprinus carpio* L.). II. Decreasing catchability through one-trial learning. *Netherlands Journal of Zoology*, **20**, 81–92.

Beukema, J.J. & de Vos, G.J. (1974) Experimental tests on a basic assumption of the capture-recapture method in pond populations of carp *Cyprinus carpio* L. *Journal of Fish Biology*, **6**, 317–329.

Birkeland, C. & Dayton, P.K. (2005) The importance in fishery management of leaving the big ones. *Trends in Ecology & Evolution*, **20**, 356.

Bisazza, A., Cantalupo, C., Capocchiano, M. & Vallortigara, G. (2000) Population lateralization and social behaviour: a study with 14 species of fish. *Laterality*, **5**, 269–284.

Blankenship, H.L. & Leber, K.M. (1995) A responsible approach to marine stock enhancement. *American Fisheries Society Symposium*, **15**, 147–175.

Blaxter, J.H.S. (1976) Reared and wild fish – how do they compare? In: G. Persone & E. Jaspers (eds), *Proceedings of the 10th European Symposium on Marine Biology*, volume 1, pp. 11–26. Universal Press, Wettern, Belgium.

Block, B.A., Teo, S.L.H., Walli, A., Boustany, A., Stokesbury, M.J.W., Farwell, C.J., Weng, K.C., Dewar, H. & Williams, T.D. (2005) Electronic tagging and population structure of Atlantic bluefin tuna. *Nature*, **434**, 1121–1127.

Bolles, R.C. (1970) Species-specific defense reactions and avoidance learning. *Psychological Review*, **77**, 32–48.

Bowen, J.T. (1970) A history of fish culture as related to the development of fishery programs. *American Fisheries Society Special Publication*, **7**, 71–93.

Braithwaite, V.A. & Salvanes, A.G.V. (2005) Environmental variability in early rearing environment generates behaviourally flexible cod: implications for rehabilitating wild populations. *Proceedings of the Royal Society of London, Series B – Biological Sciences*, **272**, 1107–1113.

Brill, R.W., Block, B.A., Boggs, C.H., Bigelow, K.A., Freund, E.V. & Marcinek, D.J. (1999) Horizontal movements and depth distribution of large adult yellowfin tuna (*Thunnus albacares*) near the Hawaiian Islands, recorded using ultrasonic telemetry: implications for the physiological ecology of pelagic fishes. *Marine Biology*, **133**, 395–408.

Broglio, C., Rodríguez, F. & Salas, C. (2003) Spatial cognition and its neural basis in teleost fishes. *Fish and Fisheries*, **4**, 247–255.

Browman, H.I. (1989) Embryology, ethology and ecology of ontogenetic critical periods in fish. *Brain, Behaviour and Evolution*, **34**, 5–12.

Brown, C. (2001) Familiarity with the test environment improves escape responses in the crimson spotted rainbowfish, *Melanotaenia duboulayi*. *Animal Cognition*, **4**, 109–113.

Brown, C. & Braithwaite, V.A. (2005) Effects of predation pressure on the cognitive ability of the poeciliid *Brachyraphis episcopi*. *Behavioural Ecology*, **14**, 482–497.

Brown, C. & Day, R. (2002) The future of stock enhancements: bridging the gap between hatchery practice and conservation biology. *Fish and Fisheries*, **3**, 79–94.

Brown, C. & Laland, K.N. (2001) Social learning and life skills training for hatchery reared fish. *Journal of Fish Biology*, **59**, 471–493.

Brown, C. & Laland, K.N. (2002) Social learning of a novel avoidance task in the guppy, *P. reticulata*: conformity and social release. *Animal Behaviour*, **64**, 41–47.

Brown, C. & Laland, K.N. (2003) Social learning in fishes: a review. *Fish and Fisheries*, **4**, 280–288.

Brown, C. & Warburton, K. (1997) Predator recognition and anti-predator responses in the rainbowfish *Melanotaenia eachamensis*. *Behavioral Ecology and Sociobiology*, **41**, 61–68.

Brown, C. & Warburton, K. (1999a) Differences in timidity and escape responses between predator-naive and predator-sympatric rainbowfish populations. *Ethology*, **105**, 491–502.

Brown, C. & Warburton, K. (1999b) Social mechanisms enhance escape responses in shoals of rainbowfish, *Melanotaenia duboulayi*. *Environmental Biology of Fishes*, **56**, 455–459.

Brown, C., Gardner, C. & Braithwaite, V.A. (2004) Population variation in lateralized eye use in the poeciliid *Brachyraphis episcopi*. *Proceedings of the Royal Society of London, Series B (Supplement)*, **271**, S455–S457.

Brown, C., Gardner, C. & Braithwaite, V.A. (2005a) Differential stress responses in fish from areas of high- and low-predation pressure. *Journal of Comparative Physiology*, **175**, 305–312.

Brown, C., Jones, F. & Braithwaite, V. (2005b) In situ examination of boldness-shyness traits in the tropical poeciliid, *Brachyraphis episcopi*. *Animal Behaviour*, **70**, 1003–1009.

Brown, G.E. (2003) Learning about danger: chemical alarm cues and local risk assessment in prey fishes. *Fish and Fisheries*, **4**, 227–234.

Camazine, S., Deneubourg, J.L., Franks, N.G., Sneyd, J., Theraulaz, G. & Bonebeau, E. (2001) *Self-organization in Biological Systems*. Princeton University Press, Princeton, New Jersey.

Caraco, T. (1980) On foraging time allocation in a stochastic environment. *Ecology*, **61**, 119–128.

Catania, K.C. & Kaas, J.H. (1997) Somatosensory fovea in the star-nosed mole: behavioral use of the star in relation to the innervation pattern and cortical representation. *Journal of Comparative Neurology*, **387**, 215–233.

Catania, K.C. & Kaas, J.H. (2001) Areal and callosal connections in the somatosensory cortex of the star-nosed mole. *Somatosensory & Motor Research*, **18**, 303–311.

Chandroo, K.P., Moccia, R.D. & Duncan, I.J.H. (2004) Can fish suffer? – Perspectives on sentience, pain, fear and stress. *Applied Animal Behaviour Science*, **86**, 225–250.

Checkley, D.M. Jr. (1982) Selective feeding by Atlantic herring (*Clupea harengus*) larvae on zooplankton in natural assemblages. *Marine Ecology Progress Series*, **9**, 245–253.

Chen, W.M. & Tabata, M. (2002) Individual rainbow trout can learn and anticipate multiple daily feeding times. *Journal of Fish Biology*, **61**, 1410–1422.

Chivers, D.P. & Smith, R.J.F. (1994) Fathead minnows, *Pimephales promelas*, acquire predator recognition when alarm substance is associated with the sight of unfamiliar fish. *Animal Behaviour*, **48**, 597–605.

Chivers, D.P. & Smith, J.F. (1995) Chemical recognition of risky habitats is culturally transmitted among fathead minnows, *Pimephales promelas* (Osteichthyes, Cyprinidae). *Ethology*, **99**, 286–296.

Chivers, D.P. & Smith, R.J.F. (1998) Chemical alarm signalling in aquatic predator-prey systems: a review and prospectus. *Écoscience*, **5**, 338–352.

Christensen, V., Gúanette, S., Heymans, J.J., Walters, C.J., Watson, R., Zeller, D. & Pauly, D. (2003) Hundred-year decline of North Atlantic predatory fishes. *Fish and Fisheries*, **4**, 1–24.

Conte, F.S. (2004) Stress and the welfare of cultured fish. *Applied Animal Behaviour Science*, **86**, 205–223.

Corten, A. (1999) The reappearance of spawning Atlantic herring (*Clupea harengus*) on Aberdeen Bank (North Sea) in 1983 and its relationship to environmental conditions. *Canadian Journal of Fisheries and Aquatic Sciences*, **56**, 2051–2061.

Corten, A. (2001) The role of 'conservatism' in herring migrations. *Reviews in Fish Biology and Fisheries*, **11**, 339–361.

Couzin, I.D., Krause, J., Franks, N.R. & Levin, S.A. (2005) Effective leadership and decision-making in animal groups on the move. *Nature*, **433**, 513–514.

Cowx, I.G. (1994) Stocking strategies. *Fisheries Management and Ecology*, **1**, 15–30.

Cox, E.S. & Pankhurst, P.M. (2000) Feeding behaviour of greenback flounder larvae, *Rhombosolea tapirina* (Gunther) with differing exposure histories to live prey. *Aquaculture*, **183**, 285–297.

Crossland, M.R. (2001) Ability of predatory native Australian fishes to learn to avoid toxic larvae of the introduced toad *Bufo marinus*. *Journal of Fish Biology*, **59**, 319–329.

Csànyi, V. (1986) Ethological analysis of predator avoidance by the Paradise fish (*Macropodus opercularis*). II. Key stimuli in avoidance learning. *Animal Learning and Behaviour*, **14**, 101–109.

Cury, P. & Anneville, O. (1998) Fisheries resources as diminishing assets: marine diversity threatened by anecdotes. In: M.H. Durand, P. Cury, R. Mendelssohn, C. Roy, A. Bakun & D. Pauly (eds), *Global Versus Local Changes in Upwelling Systems*, pp. 537–548. Orstom Editions, Paris.

DeSanti, A., Bisazza, A., Cappelletti, M. & Vallortigara, G. (2000) Prior exposure to a predator influences lateralization of cooperative predator inspection in the guppy, *Poecilia reticulata*. *Italian Journal of Zoology*, **67**, 175–178.

Divanach, P., Kentouri, M. & Dewavrin, G. (1986) The weaning and the development of biological performance of extensively reared sea bream, *Sparus auratus*, fry after replacing continuous feeders by self-feeding distributors. *Aquaculture*, **52**, 21–29.

Divanach, P., Kentouri, M., Charalambakis, G., Pouget, F. & Sterioti, A. (1993) Comparison of growth performance of six Mediterranean fish species reared under intensive farming conditions in Crete (Greece), in raceways with the use of self feeders. *Special Publication, European Aquaculture Society*.

Dodson, J.J. (1988) The nature and role of learning in the orientation and migratory behaviour of fishes. *Environmental Biology of Fishes*, **23**, 141–182.

Ellis, T., Howell, B.R. & Hayes, J. (1997) Morphological differences between wild and hatchery-reared turbot. *Journal of Fish Biology*, **50**, 1124–1128.

Elson, P.F. (1975) Atlantic salmon rivers, smolt production and optimal spawning: an overview of natural production. *Special Publication Series International Atlantic Salmon Foundation*, **6**, 96–119.

Engås, A. & Jørgensen, T. (1997) The probability of cod (*Gadus morhua*) and ling (*Molva molva*) to encounter gillnets based on *in situ* fish movements. *ICES C.M.* 1997/W:17.

Engås, A. & Løkkeborg, S. (1994) Abundance estimation using bottom gillnet and longline – the role of fish behaviour. In: A. Fernö & S. Olsen (eds), *Marine Fish Behaviour Related to Capture and Abundance Estimation*, pp. 134–145. Fishing News Books, London.

Engås, A., Kyrkjebø Haugland, E. & Øvredal, J.T. (1998) Reactions of cod (*Gadus morhua* L.) in the pre-vessel zone to an approaching trawler under different light conditions. Preliminary results. *Hydrobiologia*, **371/372**, 199–206.

Fernö, A. (1993) Advances in understanding of basic behaviour – consequences for fish capture. *ICES Marine Science Symposia*, **196**, 5–11.

Fernö, A. & Huse, I. (1983) The effect of experience on the behaviour of cod (*Gadus morhua* L.) towards a baited hook. *Fisheries Research*, **2**, 19–28.

Fernö, A., Solemdal, P. & Tilseth, S. (1986) Field studies on the behaviour of whiting (*Gadus merlangus* L.) towards baited hooks. *Fiskeridirektoratets Skrifter Serie Havundersøkelser*, **18**, 113–122.

Fernö, A., Furevik, D.M., Huse, I. & Bjordal, Å. (1988) A multiple approach to behaviour studies of salmon reared in marine net pens. *ICES C.M.* 1988/F: 15.

Fernö, A., Huse, I., Juell, J.E. & Bjordal, Å. (1995) Vertical distribution of Atlantic salmon in net pens: trade-off between surface light avoidance and food attraction. *Aquaculture*, **132**, 285–296.

Fernö, A., Pitcher, T.J., Melle, V., Nøttestad, L., Mackinson, S., Hollingworth, C. & Misund, O.A. (1998) The challenge of the herring in the Norwegian Sea: making optimal collective spatial decisions. *Sarsia*, **83**, 149–147.

Fréon, P. & Misund, O.A. (1998) *Dynamics of Pelagic Fish Distribution and Behaviour – Effects on Fisheries and Stock Assessment*. Blackwell Science, Oxford.

Fretwell, S.D. & Lucas, H.L. (1970) On territorial behaviour and other factors influencing habitat distribution in birds. *Acta Biotheoretica*, **19**, 14–36.

Fuiman, L.A. (1994) The interplay of ontogeny and scaling in the interactions of fish larvae and their predators. *Journal of Fish Biology*, **45**, 55–79.

Garcia, J. & Koelling, R.A. (1966) Relation of cue to consequence in avoidance learning. *Psychonomic Sciences*, **4**, 123–124.

Gélineau, A., Corraze, G. & Boujard, T. (1998) Effects of restricted ration, time-restricted access and reward level on voluntary food intake, growth and growth heterogeneity of rainbow trout (*Oncorhynchus mykiss*) fed on demand with self-feeders. *Aquaculture*, **147**, 247–258.

Ginetz, R.M. & Larkin, P.A. (1976) Factors affecting rainbow trout (*Salmo gairdneri*) predation on migrant fry of sockeye salmon (*Oncorhynchus nerka*). *Canadian Journal of Fisheries and Aquatic Sciences*, **33**, 19–24.

Griffiths, S.W. & Magurran, A.E. (1997) Schooling preferences for familiar fish vary with group size in a wild guppy population. *Proceedings of the Royal Society of London, Series B – Biological Sciences*, **264**, 547–551.

Griffiths, S.W., Brockmark, S., Höjesjö, J. & Johnsson, J.I. (2004) Coping with divided attention: the advantage with familiarity. *Proceedings of the Royal Society of London, Series B – Biological Sciences*, **271**, 695–699.

Grimm, V. & Railsback, S.F. (2005) *Individual-based modelling and ecology*. Princeton University Press, Princeton.

Hackney, P.A. & Linkous, P.E. (1978) Striking behaviour of the largemouth bass and use of the binomial distribution for its analysis. *Transactions of the American Fisheries Society*, **107**, 682–688.

Hakoyama, H. & Iguchi, K. (2001) Transition from a random to an ideal free to an ideal despotic distribution: the effect of individual difference in growth. *Journal of Ethology*, **19**, 129–137.

Hall, S.J. & Mainprize, B. (2004) Towards ecosystem-based fisheries management. *Fish and Fisheries*, **5**, 1–20.

Harden Jones, F.R. (1968) *Fish Migration*. Edward Arnold Ltd., London.

Helfman, G.S. & Schultz, E.T. (1984) Social tradition of behavioural traditions in a coral reef fish. *Animal Behaviour*, **32**, 379–384.

High, W.L. & Beardsley, A.J. (1970) Fish behaviour studies from an undersea habitat. *Commercial Fisheries Review*, **1970**, 31–37.

Hilborn, R. (1998) The economic performance of marine stock enhancement projects. *Bulletin of Marine Science*, **62**, 661–674.

Holst, J.C., Dragesund, O., Hamre, J., Misund, O.A. & Østvedt, O.J. (2002) Fifty years of herring migrations in the Norwegian Sea. *ICES Marine Science Symposia*, **215**, 352–360.

Hossain, M.A.R., Tanaka, M. & Masuda, R. (2002) Predator–prey interactions between hatchery reared Japanese flounder juvenile, *Paralichthys olivaceus*, and sandy shore crab, *Matuta lunaris*: daily rhythms, anti-predator conditioning and starvation. *Journal of Experimental Marine Biology and Ecology*, **267**, 1–14.

Howell, B.R. (1994) Fitness of hatchery reared fish for survival in the sea. *Aquaculture and Fisheries Management*, **25**, 3–17.

Howell, B.R., Moksness, E. & Svåsand, T. (eds) (1999) *Stock Enhancement and Sea Ranching*. Fishing News Books, Blackwell Science, Oxford.

Humborstad, O.B., Jørgensen, T. & Grøtmol, S. (2005) Exposure of cod (*Gadus morhua*) to resuspended sediment: an experimental study of the impact of bottom trawling. *Marine Ecology Progress Series*, **309**, 247–254.

Huntingford, F.A. (2004) Implications of domestication and rearing conditions for the behaviour of cultivated fishes. *Journal of Fish Biology*, **65** (Supplement A), 122–142.

Huse, G., Giske, J. & Salvanes, A.G.V. (2002a) Individual-based models. In: P.J.B. Hart & J. Reynolds (eds), *Handbook of Fish and Fisheries*, volume 2, pp. 228–248. Blackwell Science, Oxford.

Huse, G., Railsback, S.F. & Fernö, A. (2002b) Modelling changes in migration pattern of herring: collective behaviour and numerical domination. *Journal of Fish Biology*, **60**, 571–582.

Huston, M., DeAngelis, D. & Post, W. (1988) New computer models unify ecological theory. *BioScience*, **38**, 682–691.

Höjesjö, J., Johnsson, J., Petersson, E. & Järvi, T. (1998) The importance of being familiar: individual recognition and social behavior in sea trout (*Salmo trutta*). *Behavioural Ecology*, **9**, 445–451.

Inglis, I.R., Langton, S., Forkman, B. & Lazarus, J. (2001) An information primacy model of exploratory and foraging behaviour. *Animal Behaviour*, **62**, 543–557.

Ishizaki, Y., Masuda, R., Uematsu, K., Shimizu, K., Arimoto, M. & Takeuchi, T. (2001) The effect of dietary docosahexaenoic acid on schooling behaviour and brain development in larval yellowtail. *Journal of Fish Biology*, **58**, 1491–1703.

Jakobsson, S. & Järvi, T. (1976) Anti-predator behaviour of two-year-old hatchery-reared Atlantic salmon (*Salmo salar*) and a description of the predatory behaviour of burbot (*Lota lota*). *Zoologisk Revy*, **38**, 57–70.

Jobling, M., Covés, D., Damsgård, B., Kristiansen, H., Koskela, J., Petursdottir, T.E., Kadri, S. & Gudmunson, O. (2001) Techniques for measuring feed intake. In: D. Houlihan, T. Boujard & M. Jobling (eds), *Food Intake in Fish*, pp. 49–87. Blackwell Science, Oxford.

Johnsen, B.O. & Ugedal, O. (1986) Feeding by hatchery-reared and wild brown trout, *Salmo trutta* L., in a Norwegian stream. *Aquaculture and Fisheries Management*, **17**, 281–287.

Johnsson, J.I., Höjesjö, J. & Fleming, I.A. (2001) Behavioural and heart rate response to predation risk in wild and domesticated Atlantic salmon. *Canadian Journal of Fisheries and Aquatic Sciences*, **58**, 788–794.

Johnston, V.S. (1999) *Why we feel: The Science of Human Emotions*. Helix Books, Reading, USA.

Juell, J.-E. (1995) The behaviour of Atlantic salmon (*Salmo salar*) in relation to efficient cage rearing. *Reviews in Fish Biology and Fisheries*, **5**, 320–335.

Juell, J.-E., Fernö, A., Furevik, D.M. & Huse, I. (1994) Influence of hunger level and food availability on the spatial distribution of Atlantic salmon (*Salmo salar* L.) in sea cages. *Aquaculture and Fisheries Management*, **25**, 439–451.

Järvi, T. & Uglem, I. (1993) Predator training improves the anti-predator behaviour of hatchery reared Atlantic salmon (*Salmo salar*) smolts. *Nordic Journal of Freshwater Research*, **68**, 63–71.

Kaeriyama, M. (1999) Hatchery programmes and stock management of salmonid populations in Japan. In: B.R. Howell, E. Moksnes & T. Svåsand (eds), *Stock Enhancement and Sea Ranching*, pp. 153–147. Fishing News Books, Blackwell Science, Oxford.

Kaiser, M.J. & Spencer, B.E. (1994) Fish scavenging behaviour in recently trawled areas. *Marine Ecology Progress Series*, **112**, 41–49.

Kallayil, J.K., Jørgensen, T., Engås, Å. & Fernö, A. (2003) Behaviour of cod toward baited gill nets. *Fisheries Research*, **61**, 125–133.

Kasumyan, A.O. & Døving, K.B. (2003) Taste preferences in fish. *Fish and Fisheries*, **4**, 289–347.

Kelley, J.L. & Magurran, A.E. (2003) Learned predator recognition and antipredator responses in fishes. *Fish and Fisheries*, **4**, 214–226.

Kieffer, J.D. & Colgan, P.W. (1992) The role of learning in fish behaviour. *Reviews in Fish Biology and Fisheries*, **2**, 125–143.

Kihslinger, R.L. & Nevitt, G.A. (2003) The early rearing environment produces variation in the size of the brain subdivisions in steelhead trout (*Oncorhynchus mykiss*). *Integrative and Comparative Biology*, **43**, 944.

Kim, Y.-H. & Wardle, C.S. (2003) Optomotor response and erratic response: quantitative analysis of fish reaction to towed fishing gears. *Fisheries Research*, **60**, 455–470.

Koolhaas, J.M., Korte, S.M., De Boer, S.F., Van Der Vegt, B.J., Van Reenen, C.G., Hopster, H., De Jonga, I.C., Ruis, M.A.W. & Blokhuis, H.J. (1999) Coping styles in animals: current status in behaviour and stress-physiology. *Neuroscience & Biobehavioural Reviews*, **23**, 925–935.

Kotrschal, K., van Staaden, M.J. & Huber, R. (1998) Fish brains: evolution and environmental relationships. *Reviews in Fish Biology and Fisheries*, **8**, 373–408.

Kraemer, P.J. & Golding, J.M. (1997) Adaptive forgetting in animals. *Psychonomic Bulletin and Review*, **4**, 480–491.

Kristiansen, T.S. (2001) Enhancement studies of coastal cod in Norway. Stage and size dependent mortality of reared and wild cod (*Gadus morhua* L.). PhD thesis. University of Bergen, Norway.

Kristiansen, T.S. & Fernö, A. (2006) Individual behaviour and growth of halibut (*Hippoglossus hippoglossus* L.) fed sinking and floating feed: evidence of different coping styles. *Applied Animal Behaviour Science*. In press.

Kristiansen, T.S. & Svåsand, T. (1992) Comparative analysis of stomach contents of cultured and wild cod, *Gadus morhua* L. *Aquaculture and Fisheries Management*, **23**, 661–668.

Kristiansen, T.S. & Svåsand, T. (1998) Effect of size-selective mortality on growth of coastal cod illustrated by tagging data and an individual-based growth and mortality model. *Journal of Fish Biology*, **52**, 688–705.

Kvamme, C., Nøttestad, L., Fernö, A., Misund, O.A., Dommasnes, A., Axelsen, B.E., Dalpadado, P. & Melle, W. (2003) Age-specific migration patterns in Norwegian spring-spawning herring: why young fish swim away from the wintering area in late summer. *Marine Ecology Progress Series*, **247**, 197–210.

Lakoff, G. & Johnson, M. (1999) *Philosophy in the Flesh*. Basic Books, New York.

Laland, K.N. & Williams, K. (1997) Shoaling generates social learning of foraging information in guppies. *Animal Behaviour*, **53**, 1141–1149.

Lamb, C.F. (2001) Gustatation and feeding behaviour. In: D. Houlihan, T. Boujard & M. Jobling (eds), *Food Intake in Fish*, pp. 108–130. Blackwell Science, Oxford.

Law, R. & Grey, D.R. (1989) Evolution and yields from populations with age-specific cropping. *Evolutionary Ecology*, **3**, 343–359.

Leber, K.M., Kitada, S., Blankenship, H.L. & Svåsand, T. (eds) (2004) *Stock Enhancement and Sea Ranching. Developments, Pitfalls and Opportunities*. Blackwell Publishing, Oxford.

Lieberman, D.A. (1990) *Learning, Behaviour and Cognition*, 3rd edn. Wadsworth, Belmont, California.

Lima, S.L. & Bednekoff, P.A. (1999) Temporal variation in danger drives antipredator behavior: the predation risk allocation hypothesis. *American Naturalist*, **153**, 649–659.

Lines, J.A. & Frost, A.R. (1999) Review of opportunities for low stress and selective control of fish. *Aquacultural Engineering*, **20**, 211–230.

Løkkeborg, S. (1990) Reduced catch of under-sized cod (*Gadus morhua*) in longlining by using artificial bait. *Canadian Journal of Fisheries and Aquatic Sciences*, **47**, 1112–1115.

Løkkeborg, S. (1994) Fish behaviour and longlining. In: A. Fernö & S. Olsen (eds), *Marine Fish Behaviour Related to Capture and Abundance Estimation*, pp. 9–27. Fishing News Books, London.

Løkkeborg, S. (1998) Feeding behaviour of cod, *Gadus morhua*: activity rhythm and chemically mediated food search. *Animal Behaviour*, **56**, 371–378.

Løkkeborg, S. & Fernö, A. (1999) Diel activity pattern and food search behaviour in cod, *Gadus morhua*. *Environmental Biology of Fishes*, **54**, 345–353.

Løkkeborg, S., Bjordal, Å. & Fernö, A. (1989) Responses of cod (*Gadus morhua*) and haddock (*Melanogrammus aeglefinus*) to baited hooks in the natural environment. *Canadian Journal of Fisheries and Aquatic Sciences*, **46**, 1478–1483.

Løkkeborg, S., Bjordal, Å. & Fernö, A. (1993) The reliability and value of behaviour studies in longline gear research. *ICES Marine Science Symposia*, **196**, 41–46.

Løkkeborg, S., Skajaa, K. & Fernö, A. (2000) Food-search strategy in ling (*Molva molva* L.): crepuscular activity and use of space. *Journal of Experimental Marine Biology and Ecology*, **247**, 195–208.

Magurran, A.E. (1989) Acquired recognition of predator odour in the European minnow (*Phoxinus phoxinus*). *Ethology*, **82**, 214–233.

Magurran, A.E. (1993) Individual differences and alternative behaviours. In: T.J. Pitcher (ed.), *The Behaviour of Teleost Fishes*, pp. 441–477. Chapman and Hall, London.

Marchetti, M.P. & Nevitt, G.A. (2003) Effects of hatchery rearing on brain structures of rainbow trout, *Oncorhynchus mykiss*. *Environmental Biology of Fishes*, **66**, 9–14.

Masuda, R. (2004) Behavioural approaches to fish stock enhancement: a practical review. In: K.M. Leber, S. Kitada, H.L. Blankenship & T. Svåsand (eds), *Stock Enhancement and Sea Ranching. Developments, Pitfalls and Opportunities*, 2nd edn, pp. 83–90. Blackwell Publishing, Oxford.

Matthews, K.R. (1990) An experimental study of the habitat preferences and movement patterns of copper, quillback, and brown rockfishes (*Sebastes* spp.). *Environmental Biology of Fishes*, **29**, 141–178.

Mattson, S. (1990) Food and feeding habitats of fish species over a soft sublittoral bottom in the northeast Atlantic. *Sarsia*, **75**, 247–260.

McQuinn, I.H. (1997) Metapopulations and the Atlantic herring. *Reviews in Fish Biology and Fisheries*, **7**, 297–329.

Midling, K.Ø., Kristiansen T.S., Ona, E. & Øiestad, V. (1987) Fjord ranching with conditioned cod (*Gadus morhua* L.). *ICES CM* 1987/F:10.

Miller, R.B. (1954) Comparative survival of wild and hatchery-reared cutthroat trout in a stream. *Transactions of the American Fisheries Society*, **83**, 120–130.

Moav, R., Brody, T. & Hulatin, G. (1978) Genetic improvement of wild fish populations. *Science*, **201**, 1090–1094.

Mullon, C., Fréon, P. & Cury, P. (2005) The dynamics of collapse in world fisheries. *Fish and Fisheries*, **6**, 111–120.

Munro, J.L. & Bell, D. (1997) Enhancement of marine fisheries resources. *Reviews in Fisheries Sciences*, **5**, 185–222.

Munro, J.L., Reeson, P.H. & Gant, V.C. (1971) Dynamic factors affecting the performance of the Antillean fish trap. *Proceedings of Gulf Caribbean Fisheries Institute*, **23**, 184–194.

Naas, K. & Mangor Jensen, A. (1990) Positive phototaxis during late yolk sack stage of Atlantic halibut larvae *Hippoglossus hippoglossus*. *Sarsia*, **75**, 243–246.

Nilsson, J., Kristiansen, T.S. & Fosseidengen, J.E. (2005) Sett fisken på skolebenken! *Norsk Fiskeoppdrett*, **30**, 60–61.

Nødtvedt, M., Fernö, A., Gjøsæter, J. & Steingrund, P. (1999) Anti-predator behaviour of hatchery-reared and wild juvenile Atlantic cod (*Gadus morhua* L.) and the effect of predator training. In: B.R. Howell, E. Moksness & T. Svåsand (eds), *Stock Enhancement and Sea Ranching*, pp. 350–362. Fishing News Books, Blackwell Science, Oxford.

Noreide, J.T. & Fosså, J.H. (1992) Diet overlap between two subsequent year-classes of juvenile coastal cod (*Gadus morhua* L.) and wild and released cod. *Sarsia*, **77**, 111–117.

Nøttestad, L., Fernö, A., Misund, O.A. & Vabø, R. (2004) Understanding herring behaviour: Linking individual decisions, school patterns and population distribution. In: H.R. Skjoldal, R. Sætre, A. Fernö, O.A. Misund & I. Røttingen (eds), *The Norwegian Sea Ecosystem*, pp. 227–262. Tapir, Trondheim.

O'Grady, K.T. & Huges, P.C.R. (1980) Factorial analysis of an experimental comparison of three methods of fishing for rainbow trout, *Salmo gairdneri* Richardson, in still water. *Journal of Fish Biology*, **14**, 257–264.

Olla, B.L. & Davis, M.W. (1989) The role of learning and stress in predator avoidance of hatchery reared coho salmon (*Oncorhynchus kisutch*) juveniles. *Aquaculture*, **76**, 209–214.

Olla, B.L., Davies, M.W. & Ryer, C.H. (1998) Understanding how the hatchery environment represses and promotes the development of behavioral survival skills. *Bulletin of Marine Science*, **62**, 531–550.

Olsen, E.M., Heino, M., Lilly, G.R., Morgan, M.J., Brattey, J., Ernande, B. & Dieckmann, U. (2004) Maturation trends indicative of rapid evolution preceded the collapse of northern cod. *Nature*, **428**, 932–935.

Otterå, H., Kristiansen, T.S., Svåsand, T., Nødtvedt, M. & Borge, A. (1999) Sea-ranching of Atlantic cod (*Gadus morhua* L.), effects of release strategy on survival. In: B.R. Howell, E. Moksness & T. Svåsand (eds), *Stock Enhancement and Sea Ranching*, pp. 293–305. Fishing News Books, Blackwell Science, Oxford.

Özbiljin, H. & Glass, C.W. (2004) Role of learning in mesh penetration behaviour of haddock (*Melanogrammus aeglefinus*). *ICES Journal of Marine Science*, **61**, 1190–1194.

Pálsson, Ó.K. (1994) A review of trophic interactions of cod stocks in the North Atlantic. *ICES Marine Science Symposia*, **198**, 553–575.

Patten, B.G. (1977) Body size and learned avoidance as factors affecting predation on coho salmon, *Oncorhynchus kisutch*, fry by torrent sculpin, *Cottus rhotheus*. *Fishery Bulletin*, **75**, 457–459.

Policansky, D. (1993) Fishing as a cause of evolution in fishes. In: T.K.A. Stokes, R. Law & J. McGlade (eds), *The Exploitation of Evolving Resources*, pp. 2–13. Springer Verlag, Berlin.

Portavella, M., Torres, B. & Salas, C. (2004) Avoidance response in goldfish: emotional and temporal involvement of medial and lateral telencephalic pallium. *The Journal of Neuroscience*, **24**, 2335–2342.

Pyanov, A.I. (1993) Fish learning in response to trawl fishing. *ICES Marine Science Symposia*, **196**, 12–14.

Real, L.A. (1980) Fitness, uncertainty, and the role of diversification in evolution and behavior. *American Naturalist*, **115**, 623–638.

Reebs, S.G. (2000) Can a minority of informed leaders determine the foraging movements of a fish shoal? *Animal Behaviour*, **59**, 403–409.

Reese, E.S. (1989) Orientation behaviour of butterflyfishes (family Chaetodontidae) on coral reefs: spatial learning of route specific landmarks and cognitive maps. *Environmental Biology of Fishes*, **25**, 79–86.

Robichaud, D. & Rose, G.A. (2001) Multiyear homing of Atlantic cod to a spawning ground. *Canadian Journal of Fisheries and Aquatic Sciences*, **58**, 2325–2329.

Rose, G.A. (1993) Cod spawning on a migration highway in the North-West Atlantic. *Nature*, **366**, 458–461.

Rose, G.A. & O'Driscoll, R.L. (2002) Capelin are good for cod: can the northern stock rebuild without them? *ICES Journal of Marine Science*, **59**, 1018–1026.

Rose, G., deYoung, B., Kulka, D., Goddard, S. & Fletcher, G. (2000) Distribution shifts and overfishing the northern cod (*Gadus morhua*): a view from the ocean. *Canadian Journal of Fisheries and Aquatic Sciences*, **57**, 644–663.

Rozin, P. & Kalat, J. (1972) Learning as a situation-specific adaption. In: M.E.P. Seligman & J. Hager (eds), *Biological Boundaries of Learning*, pp. 66–97. Appleton, New York.

Rubio, C., Vivasa, M., Sanchez-Mut, A., Sanchez-Vazquez, F.J., Coves, D., Dutto, G. & Madrid, J.A. (2004) Self-feeding of European sea bass (*Dicentrarchus labrax*, L.) under laboratory and farming conditions using a string sensor. *Aquaculture*, **233**, 393–403.

Ryer, C.H. (2004) Laboratory evidence of behavioural impairments of fish escaping trawls. *ICES Journal of Marine Science*, **61**, 1144–1157.

Salvanes, A.G.V. & Braithwaite, V.A. (2005) Exposure to variable spatial information in the early rearing environment generates asymmetries in social interactions in cod (*Gadus morhua*). *Behavioural Ecology & Sociobiology*, **59**, 250–257.

Sanchez-Vásques, F.J. & Madrid, J.A. (2001) Feeding anticipatory activity. In: D. Houlihan, T. Boujard & M. Jobling (eds), *Food Intake in Fish*, pp. 217–232. Blackwell Science, Oxford.

Schreck, C.B. (1981) Stress and compensation in teleostean fishes: response to social and physical factors. In: A.D. Pickering (ed.), *Stress and Fish*, pp. 295–321. Academic Press, New York.

Schreck, C.B., Jonsson, L., Feist, G. & Reno, P. (1995) Conditioning improves performance of juvenile Chinook salmon, *Oncorhynchus tshawytscha*, to transportation stress. *Aquaculture*, **135**, 99–110.

Shelbourne, J.E. (1964) The artificial propagation of marine fish. *Advances in Marine Biology*, **2**, 1–83.

Shelton, P.A. & Healey, B.P. (1999) Should depensation be dismissed as a possible explanation for the lack of recovery of the northern cod (*Gadus morhua*) stock? *Canadian Journal of Fisheries and Aquatic Sciences*, **56**, 1521–1524.

Skajaa, K. (1997) Basic movement pattern and chemo-oriented search towards baited gears in demersal species; a field study on ling and edible crab. Masters thesis. University of Bergen.

Sneddon, L.U. (2003) The bold and the shy: individual differences in rainbow trout. *Journal of Fish Biology*, **62**, 971–975.

Soria, M., Gerlotto, F. & Fréon, P. (1993) Study of learning capabilities of tropical clupeoids using an artificial stimulus. *ICES Marine Science Symposia*, **196**, 17–20.

Sosiak, A.J., Randall, R.G. & McKenzie, J.A. (1979) Feeding by hatchery-reared and wild Atlantic salmon (*Salmo salar*) parr in streams. *Journal of the Fisheries Research Board of Canada*, **36**, 1408–1412.

Spruijt, B.M., Bos, R. van den & Pijlman, F.T.A. (2001) A concept of welfare based on reward evaluating mechanisms in the brain: anticipatory behaviour as an indicator for the state of reward systems. *Applied Animal Behaviour Science*, **72**, 145–171.

Steingrund, P. & Fernö, A. (1997) Feeding behaviour of reared and wild cod and the effect of learning: two strategies of feeding on two spotted goby. *Journal of Fish Biology*, **51**, 334–348.

Suboski, M.D. & Templeton, J.J. (1989) Life skills training for hatchery fish: social learning and survival. *Fisheries Research*, **7**, 343–352.

Svåsand, T. (2004) Why juvenile quality and release strategies are important factors for stock enhancement and sea ranching. In: K.M. Leber, Kitada, S., H.L. Blankenship, & T. Svåsand (eds), *Stock enhancement and sea ranching. Developments, pitfalls and opportunities*, 2nd edn, pp. 61–70. Blackwell Publishing, Oxford.

Svåsand, T., Skilbrei, O., van der Meeren, G.I. & Holm, M. (1998) Review of morphological and behavioral differences between reared and wild individuals: implications for searanching of Atlantic salmon, *Salmo salar* L., Atlantic cod, *Gadus morhua* L., and European lobster, *Homarus gammarus* L. *Fisheries Management and Ecology*, **5**, 1–18.

Toates, F. (2004) Cognition, motivation, emotion and action: a dynamic and vulnerable interdependence. *Applied Animal Behaviour Science*, **86**, 173–204.

Tsuboi, J. & Morita, K. (2004) Selectivity effects on wild white-spotted char (*Salvelinus leucomaensis*) during a catch and release fishery. *Fisheries Research*, **69**, 229–238.

Tsukamoto, K., Kuwada, H., Uchida, K., Masuda, R. & Sakakura, Y. (1999) Fish quality and stocking effectiveness: behavioural approach. In: B.R. Howell, E. Moksness & T. Svåsand (eds), *Stock Enhancement and Sea Ranching*, pp. 306–314. Fishing News Books, Blackwell Science, Oxford.

Uexküll, J.V. (1921) *Umwelt und Innenwelt der Tiere*, 2nd edn. J. Springer, Berlin.

Vabø, R., Huse, G., Fernö, A., Jørgensen, T., Løkkeborg, S. & Skaret, G. (2004) Simulating search behaviour of fish towards bait. *ICES Journal of Marine Science*, **61**, 1224–1232.

Ward, A.J.W., Holbrook, R.I., Krause, J. & Hart, P.J.B. (2005) Social recognition in sticklebacks: The role of direct experience and habitat cues. *Behavioral Ecology & Sociobiology*, **57**, 575–583.

Welcomme, R.L. (1998) Evaluation of stocking and introductions as management tools. In: I.G. Cowx (ed.), *Stocking and Introduction of Fish*, pp. 397–413. Fishing News Books, Blackwell Science, Oxford.

White, R.J., Karr, J.R. & Nehlsen, W. (1995) Better roles for fish stocking in aquatic resource management. In: H.L. Schramm & R.G. Piper (eds), *Uses and Effect of Cultured Fishes in Aquatic Ecosystems: American Fisheries Society Symposium*, pp. 527–547, volume 15. American Fisheries Society, Bethesda, USA.

Whitehead, H., Rendell, L., Osborne, R.W. & Wursig, B. (2004) Culture and conservation of non-humans with reference to whales and dolphins: review and new directions. *Biological Conservation*, **120**, 427–437.

Wisenden, B.D. & Harter, K.R. (2001) Motion, not shape, facilitates association of predation risk with novel objects by fathead minnows. *Ethology*, **107**, 357–364.

Wisenden, B.D., Vollbrecht, K.A. & Brown, J.L. (2004) Is there a fish alarm cue? Affirming evidence from a wild study. *Animal Behaviour*, **67**, 59–67.

Yamashita, Y. & Yamada, H. (1999) Release strategy for Japanese flounder fry in stock enhancement programmes. In: B.R. Howell, E. Moksness & T. Svåsand (eds), *Stock Enhancement and Sea Ranching*, pp. 191–204. Fishing News Books, Blackwell Science, Oxford.

Yoneyama, K., Matsuoka, T. & Kawamura, G. (1996) The effect of starvation on individual catchability and hook-avoidance learning of rainbow trout. *Nippon Suisan Gakkaishi*, **62**, 236–242.

Yue, S., Moccia, R.D. & Duncan, I.J.H. (2004) Investigating fear in domestic rainbow trout, *Oncorhynchus mykiss*, using an avoidance learning task. *Applied Animal Behaviour Science*, **87**, 343–354.

Zhuykov, A.Y. & Panyushkin, S.N. (1991) Use of conditioned-reflex concentration of fish for fishing artificial reefs. *Journal of Ichthyology*, **31**, 50–54.

List of Fishes

Common name	Species	Family
African mouth-breeding cichlids	*Pundamilia nyererei* *P. pundamilia*	Cichlidae
Alaskan pollock	*Theragra chalcogramma*	Gadidae
Amazon molly	*Poecilia formosa*	Poeciliidae
Anemonefish (*see* clownfish)		
Arctic charr	*Salvelinus alpinus*	Salmonidae
Atlantic cod		
Atlantic halibut	*Hippoglossus hippoglossus*	Pleuronectidae
Atlantic salmon	*Salmo salar*	Salmonidae
Azorean rockpool blennies	*Parablennius sanguinolentus*	Blenniidae
Banded killifish	*Fundulus diaphanus*	Cyprinodontidae
Banggai cardinal fish	*Pterapogon kauderni*	Apogonidae
Barramundi	*Lates calcarifer*	Centropomidae
Belted sandfish	*Serranus subligarius*	Serranidae
Black hamlet fish	*Hypoplectrus nigricans*	Serranidae
Blenny	*Aidablennius sphinx*	Blenniidae
Blind Mexican cave fish	*Astyanax fasciatus* *Anoptichthys jordani*	Characidae
Blue tang surgeonfish	*Acanthurus coeruleus*	Acanthuridae
Bluefin tuna	*Thunnus thynnus*	Scombridae
Bluegill sunfish	*Lepomis macrochirus*	Centrarchidae
Bluehead wrasse	*Thalassoma bifasciatum*	Labridae
Bluntnose minnows	*Pimephales notatus*	Cyprinidae
Bream	*Abramis brama*	Cyprinidae
Brook charr	*Salvelinus fontinalis*	Salmonidae
Brook stickleback	*Culaea inconstans*	Gasterosteidae
Brook trout	*Salvelinus fontinalis*	Salmonidae
Brown surgeonfish	*Acanthurus nigrofuscus*	Acanthuridae
Brown trout	*Salmo trutta* morph?	Salmonidae
Bullhead goby	*Cottus gobio*	Cottidae
Bully	*Gobiomorphus cotidianus*	Eleotridae
Butterflyfish		Chaetodontidae
Capelin	*Mallotus villosus*	Osmeridae
Carp	*Cyprinus carpio*	Cyprinidae
Chain pickerel	*Esox niger*	Esocidae

Common name	Species	Family
Chalk bass	*Serranus tortugarum*	Serranidae
Chinook salmon	*Oncorhynchus tshawytscha*	Salmonidae
Chromis	*Chromis caeruleus*	Pomacentridae
Chub	*Leuciscus cephalus*	Cyprinidae
Chum salmon	*Oncorhynchus keta*	Salmonidae
Cichlid	*Aequidens pulcher*	Cichlidae
Princess of Burundi	*Neolamprologus pulcher*	
	Lamprologus brichardi	
Cleaner wrasse	*Labroides dimidiatus*	Labridae
Clownfish/anemonefish	*Amphiprion melanopus*	Pomacentridae
Atlantic Cod	*Gadus morhua*	Gadidae
Coho salmon	*Oncorhynchus kisutch*	Salmonidae
Convict cichlids	*Archocentrus nigrofasciatus*	Cichlidae
Corkwing wrasse	*Crenilabrus melops*	Labridae
Crucian carp	*Carassius carassius*	Cyprinidae
Damselfish (Stout chromis)	*Chromis chrysurus*	Pomacentridae
Threespot damsel	*Stegastes planifrons*	Pomacentridae
Dusky damselfish	*Stegastes dorsopunicans*	Pomacentridae
Three-spotted damselfish	*Dascyllus trimaculatus*	Pomacentridae
Dwarf cichlid	*Apistogramma trifasciatum*	Cichlidae
Elephantnose fish	*Gnathonemus petersii*	Mormyridae
Eurasian perch	*Perca fluviatilis*	Percidae
European minnows	*Phoxinus phoxinus*	Cyprinidae
Fantail darter	*Etheostoma flabellare*	Percidae
Fathead minnows	*Pimephales promelas*	Cyprinidae
Fifteen-spined sticklebacks	*Spinachia spinachia*	Gasterosteidae
Finescale dace	*Phoxinus neogaeus*	Cyprinidae
French grunts	*Haemulon flavolineatum*	Haemulidae
Frogfish	*Antennarius marmaoratus*	Antennariidae
Giant moray eels	*Gymnothorax javanicus*	Muraenidae
Glowlight tetras	*Hemigrammus erythrozonus*	Characidae
Gobies (Frillfin goby)	*Bathygobius soporator*	Gobiidae
Starry goby	*Asterropteryx semipunctatus*	Gobiidae
Eyebar goby	*Gnatholepsis anjerensis*	Gobiidae
Two-spotted goby	*Gobiusculus flavescens*	Gobiidae
Common goby	*Pomatoschistus microps*	Gobiidae
Blackeye gobies	*Coryphopterus nicholsi*	Gobiidae
Sand goby	*Pomatoschistus minutus*	Gobiidae
Zebra goby	*Lythrypnus zebra*	Gobiidae
Golden shiners	*Notemigonus crysoleucas*	Cyprinidae
Goldfish	*Carassius auratus*	Cyprinidae
Gouramis (blue)	*Trichogaster trichopterus*	Osphronemidae
Grey moray eel	*Siderea grisea*	Muraenidae
Grouper	*Epinephelus striatus*	Serranidae
Grunts		Haemulidae

Common name	Species	Family
Guppies	*Poecilia reticulata*	Poeciliidae
Haddock	*Melanogrammus aeglefinus*	Gadidae
Herring	*Clupea harengus*	Clupeidae
Horse mackerel	*Sarda australis*	Scombridae
Humpback limia	*Limia nigrofasciata*	Poeciliidae
Inangas	*Galaxias maculates*	Galaxiidae
Iowa darters	*Etheostoma exile*	Percidae
Japanese flounder	*Paralichthys olivaceus*	Paralichthyidae
Japanese medaka	*Oryzias latipes*	Adrianichthyidae
Kokanee	*Oncorhynchus* spp.	Salmonidae
Largemouth bass	*Micropterus salmoides*	Centrarchidae
Leaf fish	*Monocirrhus polyacanthus*	Polycentridae
Leatherjacket	*Paraluteres prionurus*	Monacanthidae
Ling	*Molva molva*	Lotidae
Lizardfish	*Synodus intermedus*	Synodontidae
Longtail knifefish	*Sternopygus macrurus*	Sternopygidae
Lunartail groupers	*Variola louti*	Serranidae
Mackerel Atlantic	*Scomber scobrus*	Scombridae
Mangrove killifish	*Rivulus marmoratus*	Cyprinodontidae
Mollies	*Poecilia latipinna*	Poeciliidae
Mosquitofish	*Gambusia affinis*	Poeciliidae
Mouth almighty	*Glossamia aprion*	Apogonidae
Mozambique tilapia	*Oreochromis mossambicus*	Cichlidae
Nile tilapia	*Oreochromis niloticus*	Cichlidae
Oarfish/king of herring	*Regalecus glesne*	Regalecidae
Ocellate river stingrays	*Potamotrygon motoro*	Potamotrygonidae
Pacific herring	*Clupea pallasi*	Clupeidae
Panamanian bishop	*Brachyraphis episcopi*	Poeciliidae
Paradise fish	*Macropodus opercularis*	Anabantidae
Perugia's limia	*Limia perugiae*	Poeciliidae
Pike/northern pike	*Esox lucius*	Esocidae
Pikeperch	*Sander lucioperca*	Percidae
Pipefishes	*Siphostoma*	Syngnathinae
Platyfish southern	*Xiphophorus maculatus*	Poeciliidae
Variable platyfish	*Xiphophorus variatus*	
Pollock	*Pollachius pollachius*	Gadidae
Pufferfish	*Canthigaster valentine*	Tetraodontidae
Pumpkinseed sunfish	*Lepomis gibbosus*	Centrarchidae
Rainbow/steelhead trout	*Oncorhynchus mykiss*	Salmonidae
Rainbowfish	*Melanotaenia duboulayi*	Melanotaeniidae
Red sea coral groupers	*Plectropomus pessuliferus*	Serranidae
Redbelly dace	*Phoxinus eos*	Cyprinidae
Roach	*Rutilus rutilus*	Cyprinidae
Rock bass	*Ambloplites rupestris*	Centrarchidae
Sailfin molly	*Poecilia latipinna*	Poeciliidae
Sea bass	*Dicentrarchus labrax*	Moronidae

Common name	Species	Family
Sea trout	*Salmo trutta trutta*	Salmonidae
Seadragon	*Phyllopteryx eques*	Syngnathidae
Sergeant major damselfish	*Abudefduf troschelli*	Pomacentridae
Siamese fighting fish	*Betta splendens*	Osphronemidae
Silver perch	*Bidyanus bidyanus*	Terapontidae
Skipjack tuna	*Katsuwonus pelamis*	Scombridae
Sole	*Solea solea*	Soleidae
Sooty grunter	*Hephaestus fuliginosus*	Terapontidae
Striated surgeonfish	*Ctenochaetus striatus*	Acanthuridae
Swordtail/green swordtail	*Xiphophorus helleri*	Poeciliidae
(Three-spined) sticklebacks	*Gasterosteus aculeatus*	Gasterosteidae
Nine-spined sticklebacks	*Pungitius pungitius*	
Three-spot damselfish	*Stegastes planifrons*	Pomacentridae
Tilapia	*Sarotherodon melanotheron*	Cichlidae
Tobacco fish	*Serranus tabacarius*	Serranidae
Trahira	*Hoplias malabaricus*	Erythrinidae
Two-spotted gobies	*Gobiusculus flavescens*	Gobiidae
Variegated pupfish	*Cyprinodon variegates*	Cyprinodontidae
White-spot char	*Salvelinus leucomaensis*	Salmonidae
Whiting	*Sillago maculata*	Sillaginidae
Whiting	*Gadus merlangus*	Gadidae
Winter flounder	*Pseudopleuronectes americanus*	Pleuronectidae
Wrasse	*Thalassoma lucasanum*	Labridae
Yellow perch	*Perca flavescens*	Percidae
Yellowfin tuna	*Thunnus albacares*	Scombridae
Yellowtail amberjack	*Seriola lalandei*	Carangidae
Japanese amberjack	*Seriola quinqueradiata*	
Zebra danios	*Brachydanio rerio*	Cyprinidae

Index